Beam Dynamics

THE PHYSICS AND TECHNOLOGY OF PARTICLE AND PHOTON BEAMS (formerly ACCELERATORS AND STORAGE RINGS)

a series of monographs edited by Swapan Chattopadhyay, Lawrence Berkeley Laboratory, California, USA

VOLUME 1
The Microtron, S.P. Kapitza and V.N. Melekhin

VOLUME 2
Collective Methods of Acceleration, N. Rostoker and M. Reiser

VOLUME 3
Recirculating Electron Accelerators, Roy E. Rand

VOLUME 4
Particle Accelerators and Their Uses, Waldemar Scharf

VOLUME 5
Theory of Resonance Linear Accelerators, I.M. Kapchinskiy

VOLUME 6
The Optics of Charged Particle Beams, David C. Carey

VOLUME 7
Getter and Getter-Ion Vacuum Pumps, G.L. Saksaganskii

VOLUME 8
Beam Dynamics: A New Attitude and Framework
Étienne Forest

This book is part of a series. The publisher will accept continuation orders which may be cancelled at any time and which provide for automatic billing and shipping of each title in the series upon publication. Please write for details.

Beam Dynamics

A New Attitude and Framework

Étienne Forest

High Energy Accelerator Research Organization
Ibaraki, Japan

Graduate University for Advanced Studies
Kanagawa, Japan

and

Ernest Orlando Lawrence
Berkeley National Laboratory
California, USA

CRC Press
Taylor & Francis Group
Boca Raton London New York

CRC Press is an imprint of the
Taylor & Francis Group, an **informa** business

CRC Press
Taylor & Francis Group
6000 Broken Sound Parkway NW, Suite 300
Boca Raton, FL 33487-2742

Visit the Taylor & Francis Web site at
http://www.taylorandfrancis.com

and the CRC Press Web site at
http://www.crcpress.com

À ma famille

Masami, Guillaume et Natalie

CONTENTS

ABOUT THE SERIES

The series *The Physics and Technology of Particle and Photon Beams* is intended to promote the science and technology of the production, manipulation, storage and control of systems of charged particles and photons — often in the form of 'beams' with 'directed' energy — as applied to studies of (1) the fundamental structure and processes of the physical and living world, (2) alternative energy sources and (3) the health and environmental sciences. Its primary mission will be to expose developments in particle accelerators and storage rings, which are important both for high energy and nuclear physics and for the wide range of disciplines now turning to synchrotron radiation and free-electron lasers. The series will be dedicated to disseminating advanced and matured knowledge on the physics, instrumentation, methods, devices and technology that deal with acceleration of and radiation from charged particle and photon beams, useful for research in such areas as high energy and nuclear physics, materials and chemical sciences, atomic physics, combustion, life sciences, fusion energy, medical and surgical monitoring, diagnostics and treatment, and radiation monitoring and control. Volumes treating both fundamental physics and advanced engineering and technology topics on accelerators and lasers and plasmas will be commissioned for the series.

FIGURES AND TABLES

EXERCISES

Interactive versions of these exercises are available on the World Wide Web. Information can be found at the URL's

http://www-acc-theory.kek.jp/members/forest.html

http://www-afrd.lbl.gov/forestbook.htm

courtesy of the High Energy Accelerator Research Organization and the Ernest Orlando Lawrence Berkeley National Laboratory.

PREFACE
THE POWER OF METAPHORS

An anecdote from Don Edwards:

> ...By mid 1952, Bob Wilson and his colleagues had decided that the 300 MeV electron synchrotron should be replaced by a ring capable of three or four times higher energy. Just as we started work on the magnet design in the summer of that year, news of the invention of the alternating gradient principle reached us. We started looking at modifications to the magnet design to use strong focusing, and at the same time held some meetings to understand the advantages of this approach. Hans Bethe talked at one of these. He wrote down the Kerst-Serber equations, modified so that the field index was a periodic function of position. (I'm dropping back into the language of the time.) He then pointed out that the problem was formally identical to the time independent Schrödinger equation for the motion of an electron in a one degree of freedom periodic potential, and that the solutions could be expressed in the Bloch function form familiar from solid state physics. I don't recall him using the matrix form during his talk, but the matrix treatment was in use in a matter of days. The thing that so impressed me as a graduate student was the fact that *all* Bethe needed was the one or two sentences stating the principle; its application was *immediate* and *transparent* to him. By the time the Courant-Livingston-Snyder paper that discussed their invention appeared in Physical Review, we were well down the road to an alternating gradient synchrotron...
>
> I didn't attempt to learn the Hamiltonian language till about 20 years later, after I left Cornell to join Fermilab. By then, I had worked on four electron synchrotrons at Cornell and didn't really feel a need to change my ways.

- But because a number of the people at Fermilab (e.g., Teng, Ohnuma, Snowden) talked in that language I picked up enough to communicate with them.

The sad thing about this story is that Edwards and others did not try to understand fully why everything was so "immediate and transparent" to Bethe. About 30 years later, without any knowledge of the Bethe anecdote, Dragt gave a seminar at the University of Maryland[1] which generalized Bethe's remark to nonlinear perturbation theory in the full six dimensional phase space. A bit later and independently, Turchetti, a mathematician from Bologna, Italy, realized the same thing [2] at least for the transverse phase space.

The most important point of this book is to realize how Bethe's remark extends to virtually all of ring dynamics. The fundamental quest of ring theory can be expressed by the following equation:

$$\mathcal{A}_1\mathcal{M}_{12}\mathcal{A}_2^{-1} = \mathcal{R}_{12}. \tag{P.1}$$

In this equation, the entity $\mathcal{M}_{12}$ is a propagator from location "1" to location "2" within a ring. The map $\mathcal{R}_{12}$ is a Jordan normal form of this propagator or at least a simplified form. This equation encompasses

- the trivialities of the Courant-Snyder theory in one-degree of freedom,
- all of Hamiltonian perturbation theory,
- all of non-Hamiltonian perturbation theory,
- Sands integrals,
- moment dynamics with stochasticity,
- "One-resonance" calculations,
- etc.

The reader will see that I make a big fuss about "gauge invariance," about things that do not depend on the choice of the particular $\mathcal{A}_i$ used to produce the map $\mathcal{R}_{12}$ in Equation (P.1). I go as far as calling these objects "measurables" by analogy to quantum mechanics. Many readers will be annoyed since, after all, we ought to be able to measure any coefficient of any particular parametrization of the map. The reader will forgive me if I adopt this bias. It is a bias derived from the question of stability and our desire to answer it by certain procedures such as normal form. Because of our interest in long term behavior, we hope for or assume the existence of quantities which are invariant under certain changes of the initial conditions. This is even true when, due to chaos, we do not assume the existence of a normal form.

The major conceptual difficulty in generalizing Equation (P.1) beyond Courant-Snyder resides in figuring out the exact nature of the space in which the $\mathcal{M}_{12}$ lives.[0.1] Once this is clearly explained the utter simplicity of Bethe's assertion explodes in one's face. Unfortunately, standard accelerator theory is an effort at obscuring all conceptual links between its various components using "the need for intuitive formalism" as a false excuse.

This brings me to the second title for this preface: the power of metaphors. Too often in the pure sciences we look at our theories with a pure objectivist point of view. Our theory is a mathematical edifice of pure objectivity describing the objective reality using the language of mathematics which, in turn, is supposed to be devoid of any subjective content. In particular, the very metaphors which are used in our daily lives are supposed to have no place in science ("Time is money," etc..). However, one of the underlying themes of this book, the contrast between two *mathematically* equivalent theories, can only be understood if one tries to understand how human knowledge is structured in our brains, and by extension, in the indispensable silicon brain—the computer.

Lakoff and Johnson in a book[3] entitled "Metaphors We Live by" declare:

> The most important claim we have made so far is that metaphor is not just a matter of language, that is, of mere words. We shall argue that on the contrary, human *thought processes* are largely metaphorical. This is what we mean when we say that the human conceptual system is metaphorically structured and defined. Metaphors as linguistic expressions are possible precisely because there are metaphors in a person's conceptual system.

[0.1]Michelotti points out in his book that it is not appreciated that Floquet theory is valid in more than one-degree of freedom. He then applies Floquet theory to the full six by six matrices of phase space. However, we will see that nonlinear perturbation theory is an application of Floquet theory to a much larger space: the space of monomials in "2N" variables of degree "k" where N is the number of degrees of freedom and k is the truncation degree.

My experience over the years has convinced me that Lakoff's claim, which I am not competent to judge in toto, is of extreme relevance to the question of preference between standard accelerator theory and the theory presented in this book. The two theories differ fundamentally at the metaphorical level. Consider examples of everyday experimental language:

- Propagation through a beam line or magnet,
- Propagation of the envelope or beam ellipse,
- Watch the beam at the beam position monitor,
- Matching a beam line,
- Alignment monuments,
- Magnet misalignments,
- Interaction region,
- Survey the beam line,
- Fourier analysis of the signal at a beam position monitor.

Although we share a lot with celestial mechanicians on the fundamental theory, one would not find these expressions in celestial mechanics. Celestial mechanicians do not make the moon appear and disappear while keeping the other planets in the simulation. In rings, however, we can switch on a bypass and avoid a beam line for awhile using fast electronics. Hence, the metaphor "avoiding-the-beam-line" is far more in tune with one's understanding of the experimental process (i.e., someone or some computer turning on a switch), than imagining a global time-dependent Hamiltonian of extreme complexity describing all the beam lines and all the switches. Finally, the magnets themselves in an s-dependent Hamiltonian are equivalent to planets which appear, disappear and reappear at a given frequency: a situation undoubtedly bizarre if you are studying the solar system. These short lived changes of the Hamiltonian are just a manifestation of the localized object we call a magnet.

From a pure mathematical point of view, one could consider all the objects mentioned in the above list, including the magnets themselves, to be metaphors of the trivial kind—just a matter of language: only the global Hamiltonian for the ring exists and it suffices. This point of view is correct but highly theoretical. However, standard "advanced" accelerator theories of single particle dynamics do exactly that. One needs only to look at Guignard's work: one is hard pressed to find any magnet amongst the orgy of Greek indices describing all the Fourier modes entering in his perturbation theory.

The goal of the theory presented in this book is to structure the conceptual framework metaphorically in a way consistent with the metaphors of accelerator physicists. If some magnets are intuitively relevant to a calculation, then they should be present in the calculation. If a beam line is relevant, then it should be there also, not the Fourier modes which miraculously sum up to an extremely localized fluctuation of the Hamiltonian. As Lakoff[3] says in the same book:

> The intuitive appeal of a scientific theory has to do with how well its metaphors fit one's experience.

The introductory literature in accelerator dynamics is filled with "intuitive introductory" treatments which authors use because they are metaphorically consistent with one's experience and thus have a great illustrative appeal. For example, here is a revealing quotation[4] from Edwards and Syphers's book titled "An Introduction to the Physics of High Energy Accelerators":

> Only the most basic methods of dynamics have been used thus far, because we feel that the physics at work is most transparently illustrated in that way. But much of accelerator physics makes use of one form or another of higher[0.2] dynamics. The Hamiltonian approach is the method most frequently encountered in the literature.
>
> In this section, we review the Hamiltonian form of dynamics, and then recast much of the material of the earlier discussions in this language.[0.3] No new physics is introduced, [0.4] but the generality obtained may be helpful[0.5] to the reader who wishes to pursue this approach further.

Edwards and Syphers used, in their intuitive treatment, derivatives with respect to turn number in trying to study the stability of a ring—

[0.2]Metaphor, please read: hardly understandable and probably useless. The reader may argue that I used it also when I said "highly theoretical."

[0.3]The word "language" is itself a metaphor.

[0.4]Ah! Ah! Of course, the "language" will switch from plain English to Greek (literally), thus the authors are lowering the expectation to zero.

[0.5]Why "may be"? Do the authors themselves believe it to be useless? I bet they would not put their heads on the log to save standard Hamiltonian theory!

they observed the map at one point. Thus in perfect tune with the metaphors of experimental accelerator physics, they look at the ray at one beam position monitor. Our theory extends the "intuitive" treatments so that the discussion of more complex phenomena can still be in the same "language." This intuitive language forces us to concentrate on the objects consistent with the metaphors. As Lakoff says:

> The very systematicity that allows us to comprehend one aspect of a concept in terms of another will necessarily hide other aspects of the concept. In allowing us to focus on one aspect of a concept, a metaphorical concept can keep us from focusing on the other aspects of the concept that are inconsistent with the metaphor.

It is my belief that a theory structured metaphorically in a way compatible with the intuitive treatment of Edwards and Syphers is far more efficient than one borrowed from the language of celestial mechanics. The theory should focus on the primary metaphor—the magnet—and build mathematical objects out of these magnets—beam lines.

In the traditional advanced treatments, the "Hamiltonian of the Universe" is the only real object and, thus, the magnets are metaphorical constructs. However, we reject this view as counterintuitive and ill adapted to accelerator physics. In the metaphorical language of this book, the magnets are not localized fluctuations of the time-dependent grandiose Hamiltonian, but it is the grandiose Hamiltonian itself that is the limit of a ring containing an infinite number of tiny *related* magnets. This Hamiltonian limit is still accessible (the theories are mathematically equivalent) but, as in the real experimental world, it is not part of the "normal experience." Thus, one can regain Guignard's Fourier modes if needed. Real finite length magnets and beam lines are part of the "experiential gestalts" of accelerator theory and experiment, and thus, our theory, like all the so-called "intuitive" theories, is structured around them.

Acknowledgments

Anyone who writes about "his way of thinking" is either a fool or arrogant, and probably an arrogant fool. Actually, at a fundamental level it was Dragt's way of thinking first. While some have used Lie methods and geometrical concepts (Michelotti, for example) and others have tried to use maps, Dragt is, to my knowledge, the first person with a clear understanding of and unshakable belief in the virtue of merging maps and Lie methods in accelerators. This book owes more to him than to anyone else. Without him, I do not think I would be the author of this book.

Secondly, I had the chance to spend a couple of years at the SSC Central Design Group at the Lawrence Berkeley Laboratory under the direction of Maury Tigner. The theory group was headed then by Alex Chao. It was during those years that I learned to hate standard accelerator theory. In a perverse and strange way, everything I hated and loved about the field seemed to add coherently to produce the theoretical concept of this book. At the same time, John Irwin and Johan Bengtsson joined the SSC-CDG and, with Alex Chao's blessings, were the first scientists to take a serious look at the ideas of this book. Finally, during my sojourn at the SSC Alex Chao decided to let Martin Berz spend a year in our group. His introduction of automatic differentiation to the field (the "TPSA" package later relabeled as "DA" package) was fundamental in making the production of approximate maps out of "kick" codes a reality. I am grateful to Berz for this contribution to our field. For the record I should say that my first attempts to get maps out of kick codes made use of the manipulation routines of MARYLIE[5]. This code analyzes maps and thus has a Taylor series manipulation scheme optimized for its purposes. Those routines of MARYLIE were inserted into private versions of the integrator codes TEAPOT and RACETRACK[6, 7].

Then, I joined the Lawrence Berkeley Laboratory Exploratory Studies Group, now known as the Center for Beam Physics. In this group I found scientists who were willing to support me unconditionally. This put an end to sterile discussions. My time could be spent efficiently on day to day simulation tasks and on theory without having to argue the trivial and the self-evident. I think I owe most to my former supervisor Swapan Chattopadhyay, but also to Michael Zisman who headed some ALS studies and the initial design effort of the B-factory. It is during these last years that the entire theory blossomed into a coherent framework. Automatic differentiation and advanced programming techniques have propelled some of my ideas into the forefront of accelerator simulations. I owe a great deal to Johan Bengtsson, whose discipline in matters of physics and computers was unique and irreplaceable. Theories may be equivalent "in theory" but, in practice, when numbers must be extracted, one can pull them apart; they are metaphorically different. At a fundamental level it was always clear to me (and Dragt) that our methods were equivalent to standard accelerator methods. I believe that we obtained this clear understanding by viewing the theory from "an information theory" point of view. In contrast, most accelerator theorists need to see formulas. This kind of thinking made it very easy for me to collaborate over the years with Johan Bengtsson. We always shared a common understanding of computers and their relation to physics.

Ronald Ruth, Robert Warnock, Kohji Hirata and Katsunobu Oide have directly or indirectly affected the ideas of this book. For example, without the concept of symplectic integration introduced by Ruth, I do not know how this book could stand on its own. In passing, I am grateful to Peter Kahn and Alex Dragt as well as Diana Murray for reading carefully my draft. I am also grateful to David Whittum, Wolfram Fischer, Christoph Iselin and others who read part of the manuscript. Finally, I am grateful to David Sutter of the US Department of Energy, who always supported my work morally and financially.

I also acknowledge my first teachers, James Rouillon of the Collège Stanislas and Subal Das Gupta of McGill University, as well as my parents Bernard and Sylvie Forest. I probably would not be in physics in the first place if it had not been for their support.

CHAPTER 1
INTRODUCTION

In this book, I use single particle dynamics in rings to introduce the concepts of a map based theory. "Why single particle dynamics ?" is quite obvious: it is simple enough and the theory is therefore complete at least from a simulation point of view. This is true because the magnet metaphor is exact if fringe fields are not overlapping. (See the preface and, of course, Part III.) "Why rings and not linacs?" is a more subtle question, but a question whose answer is at the core of what this book is about. From a mathematical point of view rings are beautiful objects which are virtually self-contained. It suffices to postulate very few things about rings in order to derive the core of the theory necessary for the analysis of a ring. For example, if we start with the hypothesis of confinement in a one degree of freedom periodic symplectic system, then the Courant-Snyder ansatz

$$\begin{aligned} x(s) &= \sqrt{\beta(s)\varepsilon}\cos\left(\varphi(s)+\varphi_0\right) \\ x'(s) &= -\sqrt{\frac{\varepsilon}{\beta(s)}}\left(\sin\left(\varphi(s)+\varphi_0\right)+\alpha\cos\left(\varphi(s)+\varphi_0\right)\right) \end{aligned} \tag{1.1}$$

follows immediately: it is not necessary to introduce some approximate equation of motion as is done usually. It does not matter that the underlying system is a particle storage ring, a periodic gutter in which a ball travels, or a pendulum undergoing forced oscillations at a given frequency. Incidently notice that I did not write $-\beta'/2$ for α since this substitution would in fact depend on the local Hamiltonian (see Equation (2.92)).

In the standard treatment of circular ring dynamics, the equations of motion and other details are introduced in parallel with the Courant-Snyder ansatz. However, it is important to realize that the ansatz would be of little use if Courant and Snyder *had not known in advance* that the motion is linearly stable and confined. Therefore, in a ring, the necessity of confinement and the obvious periodicity of the system leave no freedom to the theorist. *Independent* of the details of the system, his analysis of the motion will force him to invent the Courant-Snyder formalism and the Twiss parameters. We call this the global theory because its genesis depends only on very general statements about the ring: it is symplectic, it is nearly symplectic (electron), it is stable, etc...

Of course the ring is made out of magnets. And, in a simulation code, they must be represented. This can only be done accurately and easily if we concentrate on the details, i.e., the Hamiltonian of the magnet. The electron does not need the "Lagrangian of the Universe" and neither should your code. (The "Hamiltonian" is often used as a generic word for the equations of motion and the fields throughout this book.)

Consequently, it is natural to discuss the analysis of a periodic system independent from the details of its simulation. In contrast, traditional accelerator theory attempts to merge analysis (i.e., lattice function etc.) and simulation (i.e., $F = ma$ or tracking) in one integrated treatment.

It is the separation of the integration (local) and the analysis (global) which we call the global/local dichotomy.

1.1 DICHOTOMY

In many areas of physics it is often the case that there are two or more equivalent points of view which compete in the mind of the practitioners. One famous case is the Feynman-Schwinger description of quantum-electrodynamics. In the fall of 1947, Schwinger had succeeded in elucidating a string of famous problems using his formulation of QED. His formulation, based on a traditional set of equations, seemed understandable to most physicists. Feynman's summation over all possible paths was at best bizarre. It took the work of Freeman Dyson and a certain level of bravado to convince Bethe, Oppenheimer and others that Feynman's techniques were equivalent. In the case of QED, the "summing-over-all-path" ideas of Feynman replaced nearly completely Schwinger's approach. This happened because Feynman's methods were pedagogically and computationally superior.

Similar to QED, *given a ring*, either real or inside a simulation code, we must understand and analyze its properties. Because of clear experimental and computational constraints, we examine this motion at a finite number of locations. For example, simulation codes produce phase plots at certain locations around the ring and experiments are limited by the finite number of diagnostic tools. The theory should reflect this fact and not attempt to compute unmeasurable quantities. We describe this part of the theory with the epithet "global" because it looks at results which can only be comprehended if the ring exists as a complete conceptual or physical entity.

In great contrast, simulation requires us to look at the individual magnets as objects which have an existence of their own. Computationally, our most "reliable" map is still obtained via element by element tracking. Such a process must rely on the Hamiltonian of the magnet which is well defined[1.1] whether the magnet is sitting on a measuring bench or installed in the ring. Hence, simulation is a "local" process and must be described independent of global concepts such as closed orbit or beta functions.

Finally, precisely because rings do not come already made but are man made contraptions, their design requires a point of view which merges the local and global properties of a ring. Typically a simplified model is used to estimate the final relevant properties of a ring and to guide its design. Design must at a fundamental level mix the local and global properties of a ring. The effect of putting a quadrupole into a beam line is both local and global. Local because you should be able to compute the map for a quadrupole (i.e., write an integration routine), and global because unless you know the Twiss parameters (i.e., one-turn map) at that location you will have no idea of the ultimate effect of this quadrupole.

For this reason, design is algorithmically more complex to describe. In fact, it is still an art practiced successfully by few, and the author is not a member of this elite. Nevertheless, the global/local paradigm promoted in this book has led to modular and powerful design techniques exemplified by Irwin's work at SLAC[8].

Essentially, I believe that the most powerful view of circular ring dynamics (and clearly of single pass systems as existing software proves) requires a dichotomous and schizophrenous point of view. In fact, we find it easier to separate completely the local and the global treatments. They are logically almost separate: one can learn one without the other. They are interlinked mentally only when one must

---

[1.1] Theoretically the Hamiltonian of the magnet is *never* well defined: it is a metaphor. Only the grandiose Hamiltonian of the ring is. However, because magnetic fields die out fast in the fringe region, one can and should define the "Hamiltonian of a magnet." Remember that Hamiltonian "terrorists" do not: they view the magnet as a localized fluctuation in the time-like variable "s" requiring a whole slew of Fourier modes.

design a ring since the performance of the system (global) is affected by the actual choice of beam elements (local).

By contrast, traditional treatments try to find descriptions of the entire field which treat together the local and global aspects of the dynamics. This is an honorable goal which is doomed to failure in complex systems.

1.2 GLOBAL DYNAMICS

The very independence of the two aspects presents the first pedagogical problem. Should I introduce first the local techniques or the global ones? To the dismay of many accelerator physicists, I consider the global aspects of the dynamics to be conceptually the more important and warranting the leading rôle. Actually it is also the choice of introductory treatments- notably Edwards and Syphers: getting to the Courant-Snyder parameterization is more critical to them than describing the nuts and bolts of magnets.

To some, especially in the single pass community, this choice appears highly theoretical and counter to experience. Maybe. Personally, I asked the following question: how would I explain to my nonscientific friends what goes on in rings? Would I describe first the nuts and bolts of each magnet, then piece them together in arcs and interaction regions (or "I.R.'s") and finally discuss the purpose of all this hardware? In reality, we always start with its purpose which may be the production of light or new particles. And this purpose immediately implies stability. Particles must spend some time in the machine. Remarkably, in a ring, stability restricts considerably the form of the map and consequently the tools needed to describe it. If a ring had been brought to us by aliens, existing on its own like the hydrogen atom, we would first marvel at the lifetime of the beam. We would study it as we studied atoms. It is the global properties which would fascinate us primarily. And, just like Mendeleev discovering the periodic table before quantum mechanics, we would discover the Courant-Snyder theory before understanding the details of each magnet. At least this is how I learned accelerator physics! And consequently, as an iconoclast, I have recommended to students that they avoid reading any reference on basic accelerator physics. Presumably I failed once more if you are still reading!

The global aspects of the theory emerge by answering a Feynman-like question. Feynman once asked what are the most important physical facts that the survivors of a worldwide catastrophe should remember? He thought that the atomic nature of matter was a rich source of information about our universe. If we ask the same question about rings we conclude that the following facts are remarkable:

1. The motion is nearly stable: particles seem to spend a long time in the ring.

2. The motion is symplectic in hadron machines and nearly symplectic in electron machines (but also slightly stochastic).

The sections of this book on global properties are based on these facts. They are stripped of unnecessary information which usual treatments carry.

It is important to remember that the description of global properties takes a "contemplative" point of view: one first assumes that rings exist just as atoms do, then one investigates their properties. To contemplate the ring as if one did not know or care how to build it is somewhat disturbing to many accelerator physicists. I make no apologies for this because it is this novel point of view which provided the fertilizing agents (yes, some say "cut the crap!") for this dichotomous global/local approach. The importance I attach to global concepts is not the result of an "after the fact" theorizing. It is the result of my own experience which this book is supposed to describe.

1.3 LOCAL DYNAMICS

The description of the local dynamics is geared towards simulation. In contrast with the global theory, it seeks a theory of single particle dynamics where the individual magnets are functionally independent. The local description looks at each magnet individually and asks the following questions:

1. What is(are) the best frame(s) of reference(s) to compute the motion of a particle through a given magnet?

2. What should be the frames of reference at the entrance and at the exit of the said magnet? For example, if it is the purpose of a magnet to bend the trajectory by an angle α, the beam pipe at the entrance will make an angle α with the pipe at the exit. Hence we expect the frame of references to do so. Incidentally, this does not imply that it is necessary to use a cylindrical frame to integrate through this magnet as we will see shortly.

In question 1, we are preoccupied with the internal symmetries of the magnet. Generally, the frame of reference has to do with the shape of the laminations or the coils which produce the field. In the ALS (Berkeley Advanced Light Source), the bending magnets are shifted quadrupoles. Thus it is clear that the magnetic field is very simply expressed in Cartesian coordinates:

$$B_x = -ky$$
$$B_y = kx. \tag{1.2}$$

If the quadrupole is shifted so that a typical orbit enters at $x = x_0$, then the trajectory of a typical orbit will be close to an arc of circle of inverse bending radius $\frac{qkx_0}{p_0}$, where q is the charge of the particle and p_0 its momentum. The field inside the shifted quadrupole is always given by Equation (1.2) if Cartesian coordinates are used.

This example will force a nontrivial change of coordinates at both ends of the magnet. A typical particle will enter the magnet with an angle of approximately half the total bending angle. Somehow we need to rotate, in the horizontal plane, the frames of reference of the particle at both ends by this half angle. Traditionally accelerator physicists have insured this by the use of cylindrical coordinates when dealing with bending magnets, disregarding any internal symmetry of the field. In the case of the "shifted quadrupole" bend we can list immediately the difficulties which arise when trying to use cylindrical coordinates:

- The magnetic field becomes more complicated. A pure quadrupole in rectangular coordinates will contain nonlinear terms in cylindrical coordinates. Indeed Equation (1.2) is not a solution of Maxwell's equations in cylindrical coordinates.

- Fringe field effects are much simpler in Cartesian coordinates. For example, the boundary of the ideal magnet is expressed by the simple equation $z =$ constant.

In general the coordinates used in the body of a magnet will reflect the symmetry of this magnet. The reference frames at the ends will reflect the position of the magnet within the layout of the ring irrespective of the nature of the internal fields. This is what Hirata and I dubbed the "LEGO" approach to integration.[1.2] This approach neatly decouples the issues of questions 1 and 2.

[1.2] LEGO is a registered trade mark of Interlego A.G. It is a marvelous block set with which children of all ages should play.

The LEGO block approach presupposes that magnets are objects which can be moved around the layout of the ring with (almost) total impunity. It is somewhat true for the physical pieces of hardware but it is not so obvious with the mathematical maps needed by the local theory. The equations of motion for the rays are normally expressed with time as a parameter; the three positions and momenta of a particle are obtained as functions of time. This parametrization goes contrary to the local (LEGO) approach which propagates a particle from one end of a magnet to the other in perfect harmony with the magnet metaphor. In the LEGO approach it is necessary to view time as a dependent variable and to use a coordinate as the variable of integration. This particular coordinate will be a Cartesian variable or an angle depending on the internal symmetries of the magnetic field. In the local theory, this coordinate varies from magnet to magnet and is not necessarily a distance along a particular trajectory of the (ideal) system. In the case of the "shifted quadrupole" bend, the Cartesian coordinate along the axis of the bend cannot be a trajectory resembling that of a bend under any circumstances!

So, in the case of Cartesian symmetry, the local theory asks whether or not it is possible to replace the usual variables (x, p_x, y, p_y, z, p_z) by a new set (x, p_x, y, p_y, t, p_t) which is also canonical. It is clear from the chain rule that if $\frac{dz}{dt}$ does not change sign for particles within the "pipe," then we can re-express the equations of motion in terms of z. In other words, there is within that region of phase space a one-to-one correspondence between the time along a trajectory and a measure of length along the pipe. What is not so clear is the existence of a canonical set of variables. However, if we select p_t to be minus the Hamiltonian of the original system, then the variables (x, p_x, y, p_y, t, p_t) form a canonical set because their motion is generated by the Hamiltonian $-p_z$. This Hamiltonian is obtained by solving for the original momentum p_z in terms of the other variables. Let us do the trivial example of a non-relativistic free particle (drift). The Hamiltonian is given by

$$H = \frac{p_x^2 + p_y^2 + p_z^2}{2m}, \tag{1.3}$$

where m is the mass. We can solve for p_z in terms of the rest and obtain

$$
\begin{aligned}
K &= -p_z \\
&= -\sqrt{2mH - p_x^2 - p_y^2} \\
&= -\sqrt{-2mp_t - p_x^2 - p_y^2}.
\end{aligned}
\tag{1.4}
$$

This Hamiltonian describes a particle moving along the positive z direction. One notices that it is nonlinear.

Due to the conservation of p_z, the Hamiltonian K in Equation (1.4) will, for all time, generate the motion of (x, p_x, y, p_y, t, p_t). This is not true in general. If we complicate the situation by adding a potential

$$
H = \frac{p_x^2 + p_y^2 + p_z^2}{2m} + \frac{kz^2}{2}, \tag{1.5}
$$

then the Hamiltonian K of Equation (1.4) becomes

$$
\begin{aligned}
K &= -\sqrt{-2m(p_t + V) - p_x^2 - p_y^2} \\
\text{where } V &= \frac{kz^2}{2} .
\end{aligned}
\tag{1.6}
$$

In this case, the quantity under the square-root can vanish and in fact will do so if we wait long enough simply because the motion exhibits harmonic oscillations in the $z - p_z$ plane, as is evident from the form of the original Hamiltonian. The point in time when this happens coincides with $\frac{dz}{dt}$ changing sign. This is precisely when the chain rule breaks down. Physically the particle reverses direction within the magnet described locally by K. Needless to say, this does not happen in realistic design of circular rings. In fact, in circular rings it is often true that the transverse momenta (p_x, p_y) are so small that it is appropriate to expand the square-root in the Hamiltonian K and retain the lowest order terms. This approximation has become dangerous as technology pushes designs to the limit. It has led to the strictly false belief in the linearity of ideal quadrupoles and dipoles.

In conclusion, we can track particles with reasonable initial conditions from one end of a magnet to the other. As we will see, we can then move this magnet in free space, as if it were a LEGO block. It suffices to find how the elements of the Euclidean group (3 rotations and 3 translations) act on the map which transports the variables (x, p_x, y, p_y, t, p_t) from entrance frame to the exit frame. Not too surprisingly, they act nonlinearly! The central goal of all our discussions on locality is to elucidate these concepts. By the way, this question is meaningless if one talks in terms of a grandiose global Hamiltonian: the magnet metaphor is, as I said, more that just a mere verbal game!

Not too surprisingly the global/local paradigm which is advertised here can translate itself into the "object oriented" paradigm in computer science. Naïvely, one would think that the magnets themselves (i.e., the actual magnetic field configuration) are the fundamental objects of dynamics. The LEGO approach of Hirata and myself points out very clearly that the layout bend[1.3] is not defined in terms of its magnetic field configuration. The starting point is the layout or channel. Layouts for rings are always made of arcs of circles and straight lines. The so-called layout (or LEGO) block is an arc of circle equipped with two reference planes: one at the beginning and one at the end. Both planes are perpendicular to the arc of circle and their frames may be rotated by some angle.

Hence, when we build a complex ring we first proceed by gluing layout blocks together. The frames at the ends of the magnets must be smoothly joined. This operation is model independent. In fact, in the metaphorical language, we are digging the tunnel, and in anticipation of performing a dynamical simulation, we are installing local frames around the magnet slots. In the real ring the electron does not care about our frames of reference, but, in tracking codes, we are not so lucky.

The second step consists of literally lowering the magnets into their slots; mathematically we provide the map from one face of the block to the other. It is clear that they inherit some properties of the block. In fact, the entire action of the Euclidean group on the magnet is inherited from the block and has nothing to do with the actual field configuration and model of the magnet. The example of the shifted quadrupole typifies this fact. It is possible to imagine a quadrupole with a large aperture which we then move around the beam pipe. In such a case, from the point of view of the magnet designer, it is always a quadrupole. From the point of view of the alignment in the tunnel and of the simulation code, it can be a bend if it is purposely shifted as is the case of the ALS. Then it inherits from the layout the transformational properties of a bend block. This can be seen by performing a translation in the longitudinal direction of the quadrupole: it will affect the ideal closed orbit! It is the layout in the tunnel which determines the transformational properties of the

[1.3] I am using the language of Part III. A layout bend is a place where the beam pipe has a kink and presumably a bending magnet is located there.

magnet under Euclidean transformations.

In conclusion, I list critical points which must be addressed in a code implementation:

1. The tunnel, made up of layout slots or channels, must be created and handled by a layout routine. It is in this computer class that the Euclidean group is implemented.

2. Objects called magnets, which are members of a beam line class, produce the maps.[1.4] Internally, they know nothing about the layout. Their internal charts (frames) are determined by field considerations. Obviously, several types of magnets can exist. Therefore data structures of some complexity must exist which will include a description of the electro-magnetic fields.

3. Their charts must now match those of the block in which they sit. This is done by using Euclidean patches.

4. The magnets inherit properties from the layout once they are put in it. The Euclidean group is now operative on them.

5. The map at a minimum connects phase space at the entrance and exit faces of the block. But it can also include a Taylor series, moments, spin, beam envelope for radiation, etc...

It is remarkable that the local approach to particle simulations mimics the actual positioning of magnets and the beam pipe. And, in fact it should, if we expect our code to be a real "simulation" code. It has to imitate reality and therefore it cannot depend on concepts like closed orbit or tunes which may not exist in actual rings. (How many rings had a closed orbit when first turned on? Typically, the particles travel for awhile and die in the beam pipe!)

1.4 "LAYOUT OF THIS BOOK!"

Here is a short description of what lies ahead:

- Many readers would like to see examples of the abstract concepts before swallowing the theory. In fact this is how many of the concepts presented in this book emerged. They resulted from attempts to compute properties of actual designs or rings.

[1.4]Maps can be anything from tedious high order integration routines to approximate paraxial matrices.

For this reason, in Part I, we present a comic strip version of this book. We say "comic strip" because the examples are often purely imaginary and usually in one degree of freedom. They are chosen for their absolute simplicity: the reader can really "play with them." In fact, the reader is invited to come back to Part I to test his understanding of the rest of the book. Part I is divided into two chapters: one on global dynamics and one on local integration. It contains a lot of "advanced" materials which are lumped into "interludes." This material can be skimmed through and logically belongs to the more advanced chapters. Exercises in this book should always be read because they are part of the text and the answers are usually provided. I encourage the reader to acquire software with graphical interface to do the exercises in this book (see page xix).

- In Part II, we describe the ring in *global* terms. Starting with confinement and a few reasonable assumptions, we develop the so-called "Hamiltonian-Free" approach to a ring. We start with a classification of one-turn maps. From this classification follows generalizations of the Courant-Snyder theory in two directions:

  - We generalize to unstable linear maps with and without classical radiation damping.
  - We generalize to nonlinear maps with and without classical radiation.

  A mathematical definition of the ring as a finite collection of maps is presented. It is based on intuitive (experimental) needs. The Floquet ring is then defined and the usual perturbation theory is regained by taking the limit of a Floquet ring with an infinite number of maps. This limiting process leads to trivial derivations of standard accelerator results. But we always keep in mind that most "Hamiltonian" results usually come from unnecessary complexifications of problems that are actually solvable by a finite number of maps. The indispensable tool of the Lie operator enters in Part II in a fundamental way (the Courant-Snyder invariant is the Lie transform of the one-turn matrix in one degree of freedom. Voilà!).

- In Part III, we actually describe the *local* tools necessary for the realization of a simulation model. The Euclidean group, translations and rotations, is studied in detail. The central problem consists in finding an appropriate dynamical representation of this group: how does it act on the map for the magnet? Without such a representation the magnet metaphor is not theoretically grounded. We then discuss the map itself in the context of symplectic integration: the most important representation of a map in accelerator theory and simulation.

In traditional treatments, one assumes that it is possible to write a global differential equation for the entire ring thus abandoning the magnet/metaphor. For this equation to be at all manageable, one assumes that it possible to write it using variables expressing displacements from some ideal reference curve. Usually, the set chosen is orthogonal to the reference orbit—a global Frenet-Serrat coordinate frame made of the tangent, normal and binormal vectors. This is done to facilitate simultaneously the analysis and the simulation. Indeed, standard Hamiltonian perturbation theory (analysis), if possible at all, can only be performed on a smooth Hamiltonian. However, in general, there is no guarantee that a frame of reference which facilitates perturbation theory is the best one to use for integration—a frame of reference attached to the ideal closed orbit is not automatically ideal for the local purpose of particle tracking. A good example of this is our ALS rectangular gradient dipole which is nothing but a displaced quadrupole. Common sense dictates the use of a rectangular coordinate frame even though the actual orbit is close to a circle. To avoid this problem, we introduce the concepts of local coordinates and coordinate patches. Here, the emphasis is on the smallest building blocks necessary to model an accelerator. These blocks exist on their own whether or not the beam line is a circular ring, an electron microscope or a linac. Our treatment reflects this fact.

The idea of a global Frenet-Serrat coordinate frame is entirely and permanently discarded since it is based on a global concept, i.e., some reference closed orbit.

Go ahead and enjoy the ride if at all possible...

PART I
A Pictorial View in One Degree of Freedom

CHAPTER 2
FROM THE MAP TO THE HAMILTONIAN

Imagine that you just got a job as an accelerator physicist and you are told to run a simulation code. You are given a manual in the English language (or something like it) and the description of a lattice for some circular ring. You don't have to design it, you don't have to know anything about the code, you just have to "track" which is a word for simulating the motion of a particle around a ring. (You can laugh, but this is a typical assignment for a beginner!). If you don't like this example, try this one: you are sitting next to a circular frictionless gutter with some arbitrary cross-section. Now throw a marble down the gutter with a given speed. This problem and the one-dimensional restriction of certain accelerators are equivalent problems, i.e., accelerators with mid-plane symmetry.

2.1 STABILITY: EXISTENCE OF A CLOSED ORBIT

The first thing you will notice is whether the marble stays in the gutter or goes out of the gutter. If it stays in for many turns, you may want to ask the following question: is there a closed orbit to this gutter? After all, in a gutter with rotational invariance, the closed orbit at a given energy is a circle. You simply match the centripetal, gravitational, and normal forces. But even in general, the answer to the closed orbit question is yes. In the following "game", we assume that a decently designed accelerator must have a closed orbit. So, imagine yourself in front of a computer terminal watching the phase space of an accelerator or a gutter (your pick). In the following set of exercises, you are invited to play a "let's pretend game" along these lines.

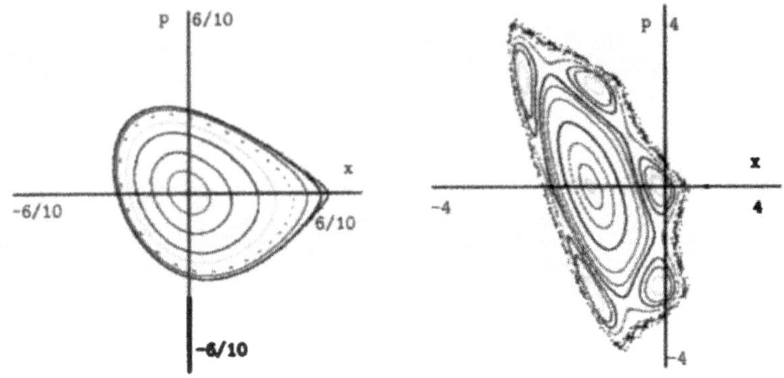

Figures 2.1 and 2.2: $d = 0$ and $d = -2.3$.

Exercise 2-1

Consider the following one degree of freedom area preserving (symplectic) map:

$$\overline{x} = x + 0.5p$$

$$\overline{p} = p - 0.5\overline{x} + d + \overline{x}^2 \tag{2.1}$$

Think of d as some knob which you are allowed to turn. First, before doing anything, check that (2.1) is area preserving. In one degree of freedom, prove that this is equivalent to a Jacobian with unit determinant for all x and p.

1. *Find its fixed points as a function of d. Show that the solution is given by:*

$$(x, p) = (f, 0) \text{ where } f = \frac{0.5 \pm \sqrt{0.25 - 4d}}{2}. \tag{2.2}$$

   Examine the stability of the fixed points.

2. *Investigate the nature of phase space for $d = 0$ by plotting a few trajectories. You should get what appears to be a "fish" if you bother to plot the separatrix. (Actually the sepatrix itself does not exist. What you are seeing are the stable and unstable manifolds that intersect homoclinically with a small angle.)*

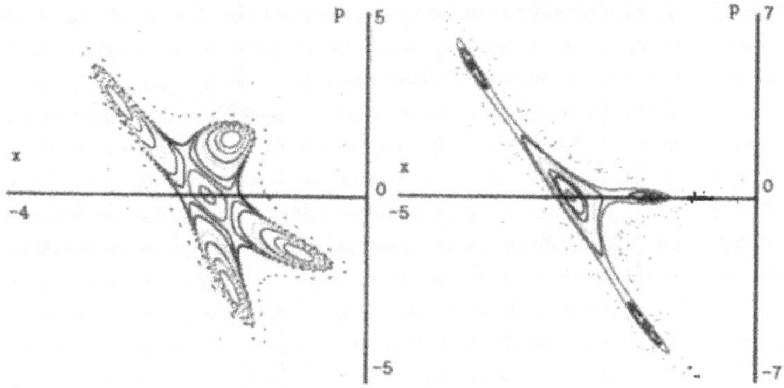

Figures 2.3 and 2.4: $d = -4$ and $d = -8$.

3. *Do the same around the stable fixed point for* $d = -2.3$. *You should get an ugly fish with some ugly warts on it (islands).*

4. *Do the same for* $d = -4$. *You should get an ugly fish with more ugly warts on it (islands).*

5. *Do the same for* $d = -8$. *Same story!*

2.2 MAP AROUND THE FIXED POINT

We have stated in section Sect. 2.1 through Exercise 2-1 that it is necessary to have a fixed point for stability. Moreover not all fixed points are linearly stable.

We now examine the motion around a stable fixed point.

Exercise 2-2

1. *Show that the map of Equation* (2.1), *when expressed in coordinates measuring deviation from its fixed points (i.e.,* $x_{\text{old}} = x + f$*), is given by*

$$\overline{x} = x + 0.5p$$

$$\overline{p} = (0.75 + f)p + (2f - 0.5)x + x^2 + xp + 0.25p^2. \qquad (2.3)$$

2. *Show that the eigenvalues of the linear map are given by*

$$\lambda_\pm = \cos(\mu) \pm i\ \sin(\mu)$$
$$\cos(\mu) = \frac{1.75 + f_\pm}{2} = 1 \pm \frac{\sqrt{0.25 - 4d}}{4}. \tag{2.4}$$

Notice that one fixed point leads to linearly stable motion (elliptic fixed point) while the other one is linearly unstable (hyperbolic).

3. *Determine* α, β, γ *such that*

$$\varepsilon = \gamma x^2 + 2\alpha x p + \beta p^2 \text{ is an invariant for the linear map} \tag{2.5}$$
$$\text{assuming } \beta\gamma = 1 + \alpha^2.$$

We will come back to this problem of invariants because it is central to the entire theory.[2.1]

4. *Compute* $\nu = \frac{\mu}{2\pi}$, α, β *and* γ *for the examples of Exercise* 2-1.

Answers:

TABLE I. Map parameters for various values of d.

d	f	ν	β	α	γ
0	0	$8.043\ 10^{-2}$	1.0328	0.2582	1.0328
−2.3	−1.287	0.2128	0.5140	0.7900	3.1600
−4	−1.766	0.2512	0.5000	1.0078	4.031
−8	−2.590	0.3189	0.5509	1.5642	6.2567

Now that you have played enough with this map, what can you conclude about the four cases $d = 0, -2.3, -4, -8$? Clearly, the islands cannot be due to the linear problem alone. Can you identify one property which is changing with d and might be responsible for the most obvious nonlinear structure you are seeing? Because the nonlinear term is left unchanged (see Equation (2.9)), something dramatic happens in the linear map to induce the nonlinear term to produce

[2.1] Notice that if ε is an invariant, any function of it is also an invariant. We chose a quadratic invariant because linear maps are generated by quadratic Hamiltonians. Therefore we will see that our choice is related to the Lie operator representation of the map.

these islands. If you still have no clue, then look at the tunes $\nu = \frac{\mu}{2\pi}$ which you computed in Exercise 2-2.4. To be more precise, look at the inverses.

TABLE II. Inverses of the tunes for various values of d.

d	ν	ν^{-1}
0	$8.043\ 10^{-2}$	12.43
−2.3	0.2128	4.70
−4	0.2512	3.98
−8	0.3189	3.14

As the particle goes around and around, it comes back to its original location every $\frac{1}{\nu}$ turns if $\frac{1}{\nu}$ is an integer. The question is whether or not there is something in the nonlinear part of the map which can resonate with a frequency $m\nu$ where m is an integer. Obviously, our particular map resonates very well near $\nu = \frac{1}{3}, \frac{1}{4}$, and $\frac{1}{5}$. For some reason, the ellipse parameters α, β, and γ seem to be more relevant to the actual scaling of the map (i.e., units so to speak) than to its dynamical properties. In other words, if one changes slightly the shape of the ellipse (i.e., α, β, γ) but not the tune, the map will have a similar topology. The islands will still be there, but deformed. Just to illustrate again the importance of the tune ν, I now invite the reader to simulate a simpler map in Exercise 2-3. Notice that this map has a topology remarkably similar to the map of Equation (2.3).

Exercise 2-3

Look at the phase space of the following map for the four values of ν of Exercise 2-2 and for $\delta = 0$:

$$(x, p) \longmapsto (\overline{x}, \overline{p})$$

$$\begin{aligned}
\widetilde{x} &= \cos(\mu)x + \sin(\mu)p \\
\widetilde{p} &= \cos(\mu)p - \sin(\mu)x \\
\overline{x} &= \widetilde{x} \\
\overline{p} &= \widetilde{p} + \widetilde{x}^2 + \delta\widetilde{x}^3.
\end{aligned} \tag{2.6}$$

You should see the same islands again. Be careful however with the $m = 3$ case. The islands almost vanish. Clearly, we have a qualitatively similar map, but not a quantitatively identical map. Actually the reader might have noticed a swindle: under a suitable change of *coordinates, the quadratic map of Exercise* 2-2 *is rigorously equivalent to the map* (2.6) *if* $\delta = 0$. *The reader is invited to change the value of* δ.

The reader should conclude from Exercises 2-1, 2-2, and 2-3 that a quadratic term in the map is good at generating islands of all sorts.

Exercise 2-4
Using the map of Equation (2.6), *study the behavior of phase space as you travel in tunes from* $\nu = 0.30$ *to* 0.36 *in small steps. You are now numerically studying third order extraction! The area of mostly linear motion slowly diminishes as you approach* $\nu = \frac{1}{3}$.

The reader has learned already why designers of accelerators try to avoid quadratic terms. They are good at destroying the area of stability. What can we say about a map with cubic terms only? This is the topic of Exercise 2-5.

Exercise 2-5
Repeat the studies of Exercise 2-3 *with the map*

$$(x, p) \longmapsto (\overline{x}, \overline{p})$$

$$\begin{aligned} \widetilde{x} &= \cos(\mu)x + \sin(\mu)p \\ \widetilde{p} &= \cos(\mu)p - \sin(\mu)x \\ \overline{x} &= \widetilde{x} \\ \overline{p} &= \widetilde{p} + \widetilde{x}^3 \ . \end{aligned} \tag{2.7}$$

The reader must be tired of simulations. The question one should ask is whether or not there exists a mathematical theory capable of computing some of the features which appeared on the phase plots. We have promised the reader that our approach extends naturally from the linear to the nonlinear—quite naturally the next section is about the tools necessary for the computation of (non)linear invariants. This is where Lie methods appear in a fundamental way in circular ring dynamics.

2.3 LIE MAPS: SEARCH FOR INVARIANTS

2.3.1 Definition of the Compositional (Lie) Map

From our discussion of the linearized problem, we know that a certain ellipse is left invariant by the map. Mathematically, this is a bit ill-defined. The map itself, as we defined it, transforms phase space into phase space. Specifically, in Equation (2.3) the map which we will now denote by $\vec{\zeta}$ is a transformation of $\mathbb{R}^2$:

$$\mathbb{R}^2 \xrightarrow{\vec{\zeta}} \mathbb{R}^2$$

$$\vec{z} = (x, p) \longmapsto \vec{\zeta}(\vec{z}) \,, \tag{2.8}$$

which for the case of Equation (2.3) is just

$$\zeta_1(\vec{z}) = z_1 + 0.5 z_2$$

$$\zeta_2(\vec{z}) = (0.75 + f) z_2 + (2f - 0.5) z_1 + z_1^2 + 0.25 z_2^2 + z_1 z_2 \tag{2.9}$$

Such a map is often called a transfer map. Now we can state precisely what is meant by an invariant function ε. If we have

$$\mathbb{R}^2 \xrightarrow{\varepsilon} \mathbb{R}$$

$$\vec{z} = (x, p) \longmapsto \varepsilon(\vec{z}), \tag{2.10}$$

where, for example, $\varepsilon(\vec{z}) = \gamma x^2 + 2\alpha x p + \beta p^2$ as in (2.5), then ε is an invariant of $\vec{\zeta}$ means

$$\forall \vec{z} \; \varepsilon(\vec{z}) = \varepsilon\left(\vec{\zeta}(\vec{z})\right)$$

or

$$\varepsilon = \varepsilon \circ \vec{\zeta} \tag{2.11}$$

In mathematics, the notation $f \circ g$ denotes the composition of two functions: "plugging" g into f. We are now ready to introduce the concept of a compositional map for lack of a better word. This new type of map will allow us to extend the linear theory to the nonlinear régime without any new formal construction. It is very important that the reader grasps this small but crucial increment in abstraction. The reason is simple: all the Lie operators, the Poisson brackets, the Lie brackets, the commutators, etc., which are essential in a map approach to classical dynamics live in the world of compositional maps.

To define them, we get our inspiration from Equation (2.11). To every ordinary transfer map of phase space such as $\vec{\zeta}$, we associate a compositional map $\mathcal{M}_{\vec{\zeta}}$:

$$\mathcal{M}_{\vec{\zeta}} f = f \circ \vec{\zeta}\,. \tag{2.12}$$

Therefore, in this language, the invariance of the Courant-Snyder function ε is expressed by the relation

$$\mathcal{M}_{\vec{\zeta}} \epsilon = \varepsilon. \text{ i.e. } \mathcal{M}_{\vec{\zeta}} \epsilon = \varepsilon \circ \vec{\zeta} = \varepsilon. \tag{2.13}$$

Notice that in theory we say nothing about the nature of $\mathcal{M}$ (whenever obvious I may drop the subscript referring to $\vec{\zeta}$). In particular, the map $\vec{\zeta}$ which gave rise to it may be linear or nonlinear—symplectic or arbitrary. It is therefore true that our definition of $\mathcal{M}$ does not preclude the formulation of a nonlinear general theory. If fact our definition of $\mathcal{M}$ is related to a process called "Koopmanism" in nonlinear dynamics: the replacement of a nonlinear problem of finite dimension by a linear problem of infinite dimension. The operator $\mathcal{M}$ is sometimes called the Koopman operator associated with $\vec{\zeta}$; in Chapter 15 we connect it to the Frobenius-Perron operator which acts on distributions and is actually dual to the Koopman. But to see how this comes about, we will first emphasize the linear symplectic result—the existence of an invariant ε for a linear map. This brings us to Poisson brackets and Lie operators.

2.3.2 Representing Compositional Maps: Lie Operators

I will first define the derivative of a function in the usual way

$$\begin{aligned} & f \longmapsto \partial_k f \\ & \forall \vec{z}\ \ (\partial_k f)(\vec{z}) = \lim_{\delta \to 0} \frac{f(\cdots, z_k + \delta, \cdots) - f(\cdots, z_k, \cdots)}{\delta}. \end{aligned} \tag{2.14}$$

The one-degree of freedom Poisson bracket is defined as follows:

$$[f, g] = \partial_1 f\, \partial_2 g \ - \ \partial_1 g\, \partial_2 f\,. \tag{2.15}$$

The first thing to compute is the Poisson bracket of "x" and "p". Why do I put these two coordinates in quotation marks? The mathematically inclined reader has probably noticed (before the author) that I did not define the functions "x" and "p", and therefore, strictly speaking, we are incapable of taking their Poisson bracket. So, let us define them first for the sake of completeness and precision:

$$\forall \vec{z} \quad x(\vec{z}) \stackrel{\text{def}}{=} z_1$$
$$p(\vec{z}) \stackrel{\text{def}}{=} z_2 \ . \tag{2.16}$$

This definition says that if you take a point in the plane $\mathbb{R}^2$, the function "x" will project out the first component of this point while the function "p" projects out the second component. With this trivial confirmation of what we mean, it is clear that the Poisson bracket of x and p, when viewed as a function, is equal to "unity." Area preserving maps, it is said, preserve the Poisson bracket. In other words, the equation $[\zeta_1, \zeta_2] =$"1" is satisfied[2.2] by a map $\vec{\zeta}$ as defined in Exercise 2-1 or 2-3. The reader will notice that preservation of the Poisson bracket in one degree of freedom is equivalent to a Jacobian with unit determinant (see Exercise 2-1).

To go any further, we need to introduce a new operator concocted with the Poisson bracket:

$$\text{Given } f \text{ then } \forall g \quad :f:g \stackrel{\text{def}}{=} [f,g] \ . \tag{2.17}$$

This operator lives in the same space as $\mathcal{M}_{\vec{\zeta}}$ because it acts on functions. Now, a miracle of nature will occur before your very eyes. We will create a map which not only acts in the same space as $\mathcal{M}_{\vec{\zeta}}$, but behaves very similarly. Consider the following map:

$$\mathcal{N} = \exp\ (: -\mu\varepsilon :)$$
$$\text{and}$$
$$\vec{\zeta}_{\mathcal{N}} = \exp\ (: -\mu\varepsilon :)\,(x,p) \ . \tag{2.18}$$

The exponential is defined in the usual way:

$$\exp\,(:f:)\,g \stackrel{\text{def}}{=} \sum_{i=0}^{\infty} \frac{:f:^i}{i!} g$$
$$\text{with } :f:^0 \stackrel{\text{def}}{=} \text{identity} = \mathcal{I} \ . \tag{2.19}$$

[2.2]The quotation mark is there for the last time: reader please remember that the Poisson bracket is a function. Technically $[\zeta_1, \zeta_2] = 1$ is meaningless while $[\zeta_1, \zeta_2]\,(\vec{z}) = 1$ is meaningful.

Incidentally, the reader notices that the couple of functions (x,p) in Equation (2.18) is actually the identity transfer map in $\mathbf{R}^2$:

$$\begin{array}{ccc} \mathbf{R}^2 & \xrightarrow{\vec{I} \text{ or } (x,p)} & \mathbf{R}^2 \\ \vec{z} & \longmapsto & \vec{I}(\vec{z}) = \vec{z}\,. \end{array} \tag{2.20}$$

We now list the few minor miracles coming from this definition of the map $\mathcal{N}$.

List of Miracles

1. The map $\mathcal{N}$ leaves the function ε invariant:

$$\begin{aligned} \mathcal{N}\varepsilon &= \exp\ (:-\mu\varepsilon:)\varepsilon = \left\{1 + \ :-\mu\varepsilon: + \frac{:-\mu\varepsilon:^2}{2} + \cdots\right\}\varepsilon \\ &= \varepsilon + [-\mu\varepsilon, \varepsilon] + \frac{[-\mu\varepsilon, [-\mu\varepsilon, \varepsilon]]}{2} + \cdots \\ &= \varepsilon \quad \text{because } [-\mu\varepsilon, \varepsilon] = 0\ . \end{aligned} \tag{2.21}$$

   By the way, notice that this is true for all functions ε, not just the invariant of Exercise 2-2.

2. Because the operator $:f:$ obeys the Leibnitz rule $:f:(gh) = (:f:g)h + g(:f:h)$, then

$$\mathcal{N}f = \exp\ (:-\mu\varepsilon:)f \ = \ f \circ \vec{\zeta}_{\mathcal{N}}\,. \tag{2.22}$$

   In other words, we have created a compositional map $\mathcal{N}$ with the help of a function $-\mu\varepsilon$. By letting this map act on the identity transformation in the plane, $\mathbf{R}^2$, we created a coordinate transformation of the plane, i.e., a map in the regular sense of the word. We denoted this map by $\vec{\zeta}_{\mathcal{N}}$. There is no reason to expect that if we turn around and use this map $\vec{\zeta}_{\mathcal{N}}$ to create a compositional map using Equation (2.12), that we will recover the map $\mathcal{N}$. But, in this case we do! This property will be proved in Interlude II and illustrated in Exercise 2-6.

3. And now, the greatest miracle of all, the map $\vec{\zeta}_{\mathcal{N}}$ is a symplectic map for all possible functions ε. In other words, it preserves the Poisson bracket.

What follows next are two interludes which the reader may want to read at least diagonally first. In Interlude I, I derive the differential equation for the compositional map. This is where the Poisson bracket operator first appears. This interlude is immediately followed by a proof of the "miracles" mentioned in this section.

INTERLUDE I
DIFFERENTIAL EQUATION FOR COMPOSITIONAL MAPS

The real physical object as well as its mathematical equivalent is a continuous object. Therefore, all the maps we talked about are generated by some differential equation. At this stage, we make the connection between the compositional map and its generator. This topic is not difficult, but because of its abstraction, I would have preferred to relegate it to Parts II or III.

We start with an arbitrary differential equation for a map

$$\begin{aligned}\frac{d}{ds}\vec{\zeta}_{s_0\to s_0+s} &= \vec{F}_s(\vec{\zeta}_{s_0\to s_0+s})\\ &\overset{\text{more rigorously}}{=} \vec{F}_s\circ\vec{\zeta}_{s_0\to s_0+s}\,. \end{aligned}\tag{2.23}$$

The vector function $\vec{F}_s$ represents the force on the particle as we go around our machine. It is, of course, s-dependent. Notice Equation (2.23) does not assume the Hamiltonian nature of the flow. Our results will apply to any flow representable by Equation (2.23). I now proceed to use Equation (2.23) to find an equation of motion for the map $\mathcal{M}_{s_0\to s_0+s}$ by letting it act on an arbitrary function f which can depend intrinsically on the parameter s:

$$\begin{aligned}\frac{d}{ds}\left(\mathcal{M}_{s_0\to s_0+s}f\right) &\overset{(2.12)}{=} \frac{d}{ds}\left(f\circ\vec{\zeta}_{s_0\to s_0+s}\right)\\ \left(\frac{d}{ds}\mathcal{M}_{s_0\to s_0+s}\right)f &\overset{\text{chain rule}}{=} \sum_k\left(\partial_k f\circ\vec{\zeta}_{s_0\to s_0+s}\right)\frac{d\zeta_{s_0\to s_0+s;k}}{ds}\\ +\mathcal{M}_{s_0\to s_0+s}\partial_s f &\qquad\qquad +\partial_s f\circ\vec{\zeta}_{s_0\to s_0+s}\\ &\Downarrow\\ \left(\frac{d}{ds}\mathcal{M}_{s_0\to s_0+s}\right)f &\overset{(2.23)}{=} \sum_k\left(\partial_k f\circ\vec{\zeta}_{s_0\to s_0+s}\right)F_k\circ\vec{\zeta}_{s_0\to s_0+s}\end{aligned}$$

$$
\begin{aligned}
&\overset{\text{Def. of }\circ}{=} \quad \sum_k (F_k\partial_k f)\circ \vec{\zeta}_{s_0\to s_0+s} \\
\frac{d}{ds}\mathcal{M}_{s_0\to s_0+s} f \quad &\overset{(2.12)}{=} \quad \mathcal{M}_{s_0\to s_0+s}\sum_k F_k\partial_k f \\
\frac{d}{ds}\mathcal{M}_{s_0\to s_0+s} \quad &\overset{f \text{ is arbitrary}}{=} \quad \mathcal{M}_{s_0\to s_0+s}\vec{F}\cdot\vec{\nabla}
\end{aligned} \tag{2.24}
$$

In the case of the Hamiltonian treatment, the force $\vec{F}_s$ is given by $\vec{F}_s\cdot\vec{\nabla} = \mathrm{J}\vec{\nabla}H_s$. Therefore the operator $\vec{F}_s\cdot\vec{\nabla}$ is just:

$$
\begin{aligned}
\vec{F}_s\cdot\vec{\nabla}\Big|_a &= \mathrm{J}\vec{\nabla}H_s\cdot\vec{\nabla}\Big|_a \\
&= \mathrm{J}_{ab}\,(\partial_b H_s)\,\partial_a \;=\; -\mathrm{J}_{ba}\,(\partial_b H_s)\,\partial_a \\
&\Rightarrow \vec{F}_s\cdot\vec{\nabla} =: -H_s : \ .
\end{aligned} \tag{2.25}
$$

Here J is the antisymmetric matrix defining the Poisson bracket

$$
\mathrm{J} = \begin{pmatrix} 0 & 1 & & & \\ -1 & 0 & & & \\ & & \ddots & & \\ & & & 0 & 1 \\ & & & -1 & 0 \end{pmatrix}. \tag{2.26}
$$

We get the symplectic form for the compositional map equation:

$$
\begin{aligned}
\frac{d}{ds}\mathcal{M}_{s_0\to s_0+s} &= \mathcal{M}_{s_0\to s_0+s} \; :-H_s: \\
\mathcal{M}_{s_0\to s_0} &= \mathcal{I}\ .
\end{aligned} \tag{2.27}
$$

Equation (2.27) provides a reason for exploring maps which are exponentials of Poisson bracket operators. If the Hamiltonian is independent of s, Equation (2.27) admits the solution $\mathcal{M}_{s_0\to s_0+s} = \exp\left(s : -H :\right)$. This implies that exponentials of Poisson brackets are Hamiltonian maps and, as we will prove in Interlude II, preserve the Poisson bracket.
END OF INTERLUDE

While we are talking about symplectic maps, let us recall the definition of a symplectic map of phase space. The map $\vec{\zeta}$ of $\mathbb{R}^{2N}$ will be called symplectic in this book if and only if there exists a set of canonical variables where the Jacobian $D\vec{\zeta}$ of $\vec{\zeta}$ obeys the relation

$$
\forall \vec{z}\in\mathbb{R}^{2N} \quad \left(D\vec{\zeta}^{\,\mathsf{T}}\mathrm{J}D\vec{\zeta}\right)(\vec{z}) = \mathrm{J}. \tag{2.28}
$$

Having justified the existence of these Poisson bracket operators and introduced symplectic maps, we are ready to prove the miracles.

INTERLUDE II
PROOF OF THE MIRACLES

This topic really belongs to the later chapters, but the reader may be annoyed by my introduction of Poisson bracket operators from nowhere. So, I throw in a theoretical interlude which the reader may choose to skip.

First, the invariance of f under the map $\exp(:f:)$ is immediate from the antisymmetry of the Poisson bracket, i.e., $[f, f] = 0$. Next we jump to "miracle" number 2. The proof presented here is due to Dragt. Actually since I introduced the compositional map first, unlike Dragt, miracle 2 can be deduced immediately from the compositional map equation derived in Interlude I. Following Dragt, consider the special case:

$$\left(e^{\sigma:f:}g\right)\ \left(e^{\sigma:f:}h\right) \stackrel{?}{=} e^{\sigma:f:}(gh). \tag{2.29}$$

The proof follows from taking the σ derivative of Equation (2.29). The derivative of the left-hand side is:

$$\begin{aligned}\frac{d}{d\sigma}\left\{\left(e^{\sigma:f:}g\right)\ \left(e^{\sigma:f:}h\right)\right\} &= \left(:f:\ e^{\sigma:f:}g\right)\ \left(e^{\sigma:f:}h\right) \\ &\quad + \left(e^{\sigma:f:}g\right)\ \left(:f:\ e^{\sigma:f:}h\right) \\ &\stackrel{\text{Leibnitz}}{=} :f:\left\{\left(e^{\sigma:f:}g\right)\ \left(e^{\sigma:f:}h\right)\right\}.\end{aligned} \tag{2.30}$$

But, the derivative of the right-hand side is given by:

$$\frac{d}{d\sigma}\exp\left(\sigma:f:\right)gh \;=:f:\exp\left(\sigma:f:\right)gh\,. \tag{2.31}$$

By induction on Equations (2.30) and (2.31), we conclude that

$$\begin{gathered}\frac{d^k}{d\sigma^k}\exp\left(\sigma:f:\right)gh \;=:f:^k\exp\left(\sigma:f:\right)gh \\ \text{and} \\ \frac{d^k}{d\sigma^k}\left(e^{\sigma:f:}g\right)\left(e^{\sigma:f:}h\right) =:f:^k\left(e^{\sigma:f:}g\right)\left(e^{\sigma:f:}h\right)\,.\end{gathered} \tag{2.32}$$

Therefore all the derivatives and the values of the two functions at $\sigma = 0$ are equal which implies that the two expressions are identical for all σ. To get the general result, we apply this special result inductively on the space of polynomials. Finally, we use the fact that polynomials of arbitrary degree form a dense set in the space of functions, hence the theorem applies to an arbitrary function. Next, we are left with the preservation of the Poisson bracket: the third "miracle." Again we compare the σ derivative of two expressions:

$$\begin{aligned}
\frac{d}{d\sigma} e^{\sigma:f:}[g,h] \quad &\stackrel{?}{=} \quad \frac{d}{d\sigma}\left[e^{\sigma:f:}g, e^{\sigma:f:}h\right] \\
:f: e^{\sigma:f:}[g,h] \quad &\stackrel{?}{=} \quad \left[:f: e^{\sigma:f:}g, e^{\sigma:f:}h\right] + \left[e^{\sigma:f:}g, :f: e^{\sigma:f:}h\right] \\
&\stackrel{\text{Jacobi Identity}}{=} \; :f: \; \left[e^{\sigma:f:}g, e^{\sigma:f:}h\right] \, . \qquad (2.33)
\end{aligned}$$

We used the Jacobi identity, $[[A,B],C]+[[C,A],B]+[[B,C],A]=0$, which is obeyed by any Lie algebra: Poisson brackets, commutators, etc... (The reader should check that commutators and Poisson brackets form a Lie Algebra.) The rest of the demonstration is done inductively: all the σ derivatives and the functions at $\sigma = 0$ are equal once more and therefore we conclude that the functions are equal for all σ. Notice that, to prove our results, we used only the Jacobi identity on the bracket algebra and the Leibnitz rule.
END OF INTERLUDE

To illustrate all this abstraction, a few exercises are in order.

Exercise 2-6

1. *Consider the function f whose functional form is given by $f = -L\frac{p^2}{2}$. Show that*

$$\begin{aligned}
&1) \quad \vec{\zeta} = \exp\,(:f:)\,(x,p) \; = \; (x + Lp, p) \\
&2) \quad \exp\,(:f:)\,x^3 = x^3 + 3x^2Lp + 3xL^2p^2 + L^3p^3 \\
&3) \quad x^3 \circ \vec{\zeta} \; = \; x^3 + 3x^2Lp + 3xL^2p^2 + L^3p^3 \, . \qquad (2.34)
\end{aligned}$$

2. *The next step is a counter-example: not every useful exponential operator is a compositional operator. In the Fokker-Planck equation, we have a diffusion operator of the form $a\partial_1^2$. Essentially, repeat the same exercise.*

$$\begin{aligned}
&1) \quad \vec{\zeta} = \exp\left(a\partial_1^2\right)(x,p) \; = \; (x,p) \\
&2) \quad \exp\left(a\partial_1^2\right)x^3 = x^3 + 6ax \\
&3) \quad x^3 \circ \vec{\zeta} \; = \; x^3 \, . \qquad (2.35)
\end{aligned}$$

3. *The resulting transfer map is area-preserving because we insisted on using a Poisson bracket operator. However, the self-consistency (the first two miracles), is a consequence of the Leibnitz rule alone. Consider the following nonsymplectic example. Show that*

$$\begin{aligned}
&1)\quad \vec{\zeta} = \exp\left(a(x\partial_1 + p\partial_2)\right)(x,p) = e^{a}(x,p) \\
&2)\quad \exp\left(a(x\partial_1 + p\partial_2)\right)x^3 = e^{3a}x^3 \\
&3)\quad x^3 \circ \vec{\zeta} = e^{3a}x^3 \,. \qquad (2.36)
\end{aligned}$$

Summary: We can summarize this section on compositional maps as follows:

- In Sects. 2.1 and 2.2, we saw phase plots on which particles seem to move approximately on invariants. To find these invariants in functional form, we are tempted to try to "plug in" the transfer map into a function to see if it changes.
- This operation seems very fundamental. After all, in some of the phase plots of Exercises 2-1 and 2-3, the orbits, at least near the fixed point, seem to move on some invariant curve. Therefore, in Sect. 2.3.1, we upgraded the "plug in" operation to a new type of map which acts on functions.
- In Sect. 2.3.2, we create a map out of the Poisson bracket since it also acts on functions.
- Miracle! The Poisson bracket map is a compositional map which creates symplectic transfer maps when it acts on the identity (x,p). We got ourselves a factory for making symplectic compositional maps. Better yet, the exponent of this Poisson bracket map is connected directly to the invariant of the associated transfer map. See Interlude I for a justification.

So you now have a goal in life: take the transfer map you are studying and try to write its associated compositional map (i.e., "plug in map") using a single Poisson bracket operator. In the next section, we explore how to compute this operator in the linear case. We also touch on the nonlinear case.

2.4 PURPOSE OF LIE OPERATORS

We will first look at the eigenvectors of the linear compositional map from which we can construct the quadratic invariant—the Courant-Snyder invariant. We also show why this invariant must be the Lie polynomial of this map.

2.4.1 Resonance Eigenbasis: the Invariant

I will start with the linear case. We first consider a linear transfer map of $\mathbb{R}^2$ which we assume to be stable:

$$\begin{aligned}\vec{\zeta}(\vec{z}) &= (az_1 + bz_2, cz_1 + dz_2) \\ &= \begin{pmatrix} a & b \\ c & d \end{pmatrix} \begin{pmatrix} z_1 \\ z_2 \end{pmatrix} = M\vec{z}.\end{aligned} \tag{2.37}$$

We define the compositional map $\mathcal{M}_{\vec{\zeta}}$ in the usual way. Now, because the map $\vec{\zeta}$ is a linear map, we notice that $\mathcal{M}_{\vec{\zeta}}$ leaves the space of homogeneous polynomials globally invariant. For example, take the monomial xp^2. The action of $\mathcal{M}$ on it is just:

$$\begin{aligned}\mathcal{M}xp^2 = xp^2 \circ \vec{\zeta} &= (x \circ \vec{\zeta})(p \circ \vec{\zeta})^2 \\ &= (ax + bp)\,(cx + dp)^2\,.\end{aligned} \tag{2.38}$$

The result is clearly a homogeneous polynomial of degree 3. So, let us consider the effect of $\mathcal{M}$ on an arbitrary linear polynomial f. The most general linear polynomial f is just $f = v_1 x + v_2 p$:

$$\begin{aligned}\overline{f} = \mathcal{M}\, f = \mathcal{M}\,(v_1 x + v_2 p) &= v_1 x \circ \vec{\zeta} + v_2 p \circ \vec{\zeta} \\ &= v_1 ax + v_1 bp + v_2 cx + v_2 dp \\ &= \overline{v}_1 x + \overline{v}_2 p \end{aligned} \tag{2.39}$$

$$\text{where } \begin{pmatrix} \overline{v}_1 \\ \overline{v}_2 \end{pmatrix} = \begin{pmatrix} a & c \\ b & d \end{pmatrix} \begin{pmatrix} v_1 \\ v_2 \end{pmatrix}.$$

It is important to notice that, in the vector space of linear homogeneous functions, the doublet $\mathbf{v}$ contains the components of f. The components of the transformed function $\overline{f}$ are given by $\overline{\mathbf{v}}$. Finally, we also notice that these components transform according to the transpose of the matrix governing the original transfer map $\vec{\zeta}$. This is a feature of compositional maps: they act in the reverse order from ordinary map composition. We see why here: they act on the functions of phase space instead of acting on phase space itself.

The search for the invariant starts with the diagonalization of $\mathcal{M}$ in the linear restriction, i.e., the diagonalization of matrix $M^{\mathsf{T}} = \begin{pmatrix} a & c \\ b & d \end{pmatrix}$.

Exercise 2-7

1. *Assuming area preservation, show that the eigenvalues of M^{T} are*

$$\lambda_{\pm} = \frac{a+d}{2} \pm \sqrt{\left(\frac{a+d}{2}\right)^2 - 1} . \qquad (2.40)$$

Notice that the real solution is unstable and that the eigenvalues are reciprocals of each other. This is true for all symplectic matrices[9].

2. *Show that the stable solution is on the unit circle, i.e., $\lambda_{\pm} = \cos(\mu) \mp i \sin(\mu)$. Remark: In a complete diagonalization of a complex matrix, the tune μ is defined within a sign. If we normalized the matrix into a rotation it will fix uniquely the sign of μ. For consistency with the rest of the book, we associate the eigenvector h_+ with the eigenvalue $\exp(-i\mu)$ where μ is the angle of this still unknown rotation. See Sect. 2.4.2.*

3. *Show that the two linear eigenfunctions of $\mathcal{M}$ associated with $\lambda_{\pm}$ are*

$$h_{\pm} = K \left(cx - (\frac{a-d}{2})p \pm i \sqrt{1 - (\frac{a+d}{2})^2 p}\right)$$
$$\text{where } K \in \mathbf{C} . \qquad (2.41)$$

We have succeeded in finding two linear eigenfunctions of $\mathcal{M}$. Now comes the crucial point: all the monomials of the eigenfunctions h_+ and h_- are eigenfunctions of $\mathcal{M}$:

$$\begin{aligned} \mathcal{M} h_+^m h_-^n &= h_+^m h_-^n \circ \vec{\zeta} = \left(h_+ \circ \vec{\zeta}\right)^m \left(h_- \circ \vec{\zeta}\right)^n \\ &= (\mathcal{M} h_+)^m (\mathcal{M} h_-)^n \\ &= (\lambda_+^m h_+^m)(\lambda_-^n h_-^n) \\ &= \exp\left(i(n-m)\mu\right) h_+^m h_-^n . \qquad (2.42) \end{aligned}$$

For tunes away from linear resonances, the construction of invariants is now immediate using Equation (2.42): select $m = n$. Notice that all the monomial invariants are powers of the quadratic invariant h_+h_-. At this point, we are in a position to express the map $\mathcal{M}$ with a single Lie exponent. We know from the property of the Poisson bracket that it must have the form $\mathcal{M} = \exp(k : h_+h_- :)$ where $k \in \mathbb{R}$. *Why?* Clearly, the map must leave h_+h_- invariant, and therefore, the function associated with its Lie exponent (i.e., f as in $\mathcal{M} = \exp\ (: f :)$) must be a function of h_+h_- only. However, we know that if $\mathcal{M}$ acts on the identity map (x, p) it must produce a linear map. In fact, it must reproduce Equation (2.37). The reader can convince himself (through examples if necessary), that only a quadratic homogeneous polynomial can produce a map of the form of Equation (2.37). To get the actual value of k elegantly, it is best to stop here and to introduce the concept of a canonical transformation.

INTERLUDE III
ELEGANCE

What do I mean by elegance? In mathematics, for example, it might refer to a very abstract demonstration which encapsulates in a few lines what takes pages using a more primitive approach. The mathematician is more a poet than a writer of prose. Pragmatism and practicality is of little consequence to the mathematician. This is not our definition of elegance. My primary reason for introducing "elegant" techniques is very pragmatic. While the reader can sit and grind at the linear two-dimensional problem and get the Lie exponent of the map, the problem of linear coupling and nonlinear effects will be intractable without an investment in more powerful tools. The purpose of this book is to introduce elegant techniques: tools and methods that have obvious extensions to more complicated situations.
END OF INTERLUDE

2.4.2 The Canonical Transformation $\mathcal{A}$: Normalization

Here, we get our inspiration from the map of Exercise 2-5. The linear part of this map is a rotation. The reader can check quite trivially that the compositional map of a rotation is given by:

$$\text{if } \vec{\rho}(\vec{z}) = \begin{pmatrix} \cos(\mu) & \sin(\mu) \\ -\sin(\mu) & \cos(\mu) \end{pmatrix} \begin{pmatrix} z_1 \\ z_2 \end{pmatrix}$$

$$\Downarrow$$
$$\mathcal{R}_{\vec{\rho}} = \exp\ (: -\mu J :) \tag{2.43}$$

where $J = \frac{x^2+p^2}{2}$ (this function J should not be confused with the matrix J of Equation (2.26).). The sign of μ is now fixed by the form of Equation (2.43) and our desire to associate $\exp(-i\mu)$ with the eigenfunction h_+. We will now see how it all comes about!

Exercise 2-8

Using the resonance eigenbasis of Exercise 2-7, prove the statement of Equation (2.43).

Let us assume that, by some procedure, we succeed in finding a compositional map $\mathcal{A}$ which transforms $\mathcal{M}$ into a rotation. In other words, we have:

$$\begin{aligned}
\vec{\rho} &= \vec{\alpha}^{-1} \circ \vec{\zeta} \circ \vec{\alpha} \\
&\Downarrow \text{ where } \vec{\zeta}(\vec{z}) = \begin{pmatrix} a & b \\ c & d \end{pmatrix} \begin{pmatrix} z_1 \\ z_2 \end{pmatrix} \\
\mathcal{R} &= \mathcal{A}_{\vec{\alpha}} \mathcal{M} \mathcal{A}_{\vec{\alpha}}^{-1} \ .
\end{aligned} \tag{2.44}$$

From Equation (2.44), it appears that we can immediately get our invariant as well as the single Lie exponent representation:

$$\begin{aligned}
\mathcal{R} &= \mathcal{A}\mathcal{M}\mathcal{A}^{-1} \\
&\Downarrow \\
\mathcal{M} &= \mathcal{A}^{-1}\mathcal{R}\mathcal{A} = \mathcal{A}^{-1}\exp(: -\mu J :)\mathcal{A} \\
&= \exp(\mathcal{A}^{-1} : -\mu J : \mathcal{A}) \\
&\overset{\text{See Exercise 2-9}}{=} \exp(: -\mu \mathcal{A}^{-1} J :) = \exp(: -\mu\ J \circ \vec{\alpha}^{-1}\ :)\ .
\end{aligned} \tag{2.45}$$

From Equation (2.45), we deduce that the invariant of $\mathcal{M}$ is just $\mathcal{A}^{-1}J = J \circ \vec{\alpha}^{-1}$.

Exercise 2-9

This exercise is an attempt to justify the last line of Equation (2.45)*: the transform of an operator* $: f :$ *associated with a function* f, *i.e.,* $\mathcal{A} : f : \mathcal{A}^{-1}$, *is the operator of the transformed function* $: f \circ \vec{\alpha} :$. *Prove this result using the invariance of the Poisson bracket which was proved in Interlude* II.

The only thing left is to construct the map $\mathcal{A}$ using the information we have gathered so far. To do this, we go back to the two eigenvectors $h_{\pm}$. We first write h_+ as

$$\begin{aligned} h_+ &= \Re h_+ + i\Im h_+ \\ &= \underbrace{(\Re_x x + \Re_p p)}_{\Re h_+} + i \underbrace{(\Im_x x + \Im_p p)}_{\Im h_+}. \end{aligned} \tag{2.46}$$

The quantities $\Re_x$, $\Re_p$, $\Im_x$, and $\Im_p$ are all real numbers. We then normalize the real and imaginary part of h_+ by their Poisson bracket, which one can show cannot be zero. Indeed because the initial bracket $[x,p]$ is equal to one and $\{h_+, h_-\}$ forms a basis for the space of complex homogeneous polynomial of degree one, then the bracket of the real and imaginary parts cannot vanish. Hence

$$\begin{aligned} h_+ &= \frac{1}{\left|\left[\Re h_+^{old}, \Im h_+^{old}\right]\right|^{1/2}} h_+^{old} \\ \text{where} \quad & [\Re h_+, \Im h_+] = \Re_x \Im_p - \Re_p \Im_x. \end{aligned} \tag{2.47}$$

Using this normalized eigenfunction, we define two new functions x_{new} and p_{new} as follows:

$$\begin{aligned} x_{new} &= \Re h_+ = \frac{h_+ + h_-}{2} \\ p_{new} &= \epsilon \Im h_+ = \frac{h_+ - h_-}{2i\epsilon} \\ \text{where } \epsilon &= \operatorname{sign}([\Re h_+, \Im h_+]). \end{aligned} \tag{2.48}$$

Notice that by construction the Poisson bracket $[x_{new}, p_{new}]$ is equal to one. Therefore this new transformation is symplectic. But, what is more exciting is the fact that the quantity $J_{new} = \frac{x_{new}^2 + p_{new}^2}{2}$ is precisely the invariant $h_+ h_-$. Now, we are ready for the big result: $\mathcal{AMA}^{-1}$ is a rotation! If we choose

$$(\alpha_1^{-1}, \alpha_2^{-1}) = (x_{new}, p_{new}),$$

then it follows that

$$\begin{aligned} \vec{\rho} &= \mathcal{AMA}^{-1}(x,p) \\ & \quad \text{here } (x,p) = \vec{I} = \text{identity map}, \\ &= \mathcal{AM}(x_{new}, p_{new}) \end{aligned}$$

$$
\begin{aligned}
&= \mathcal{AM}\left(\frac{h_+ + h_-}{2}, \frac{h_+ - h_-}{2i\epsilon}\right) \\
&= \mathcal{A}\left(\frac{\mathcal{M}h_+ + \mathcal{M}h_-}{2}, \frac{\mathcal{M}h_+ - \mathcal{M}h_-}{2i\epsilon}\right) \\
&= \mathcal{A}\left(\frac{e^{-i\mu}h_+ + e^{i\mu}h_-}{2}, \frac{e^{-i\mu}h_+ - e^{i\mu}h_-}{2i\epsilon}\right)
\end{aligned}
$$

$$
\begin{aligned}
&= \mathcal{A}\left(\cos(\mu)x_{new} + \sin(\mu)\epsilon p_{new}, \frac{\cos(\mu)\epsilon p_{new} - \sin(\mu)x_{new}}{\epsilon}\right) \\
&= (\cos(\mu)\mathcal{A}x_{new} + \sin(\mu)\epsilon\mathcal{A}p_{new}, \cos(\mu)\mathcal{A}p_{new} - \epsilon\sin(\mu)\mathcal{A}x_{new}) \\
&= (\cos(\mu)x + \epsilon\sin(\mu)p, \cos(\mu)p - \epsilon\sin(\mu)x) \ . \qquad (2.49)
\end{aligned}
$$

From the last line of Equation (2.49), we conclude that the map $\mathcal{AMA}^{-1}$ is associated with the rotation $\vec{\rho}$ as defined by Equation (2.43). The reader notices that ϵ can be minus one. In such a case, we conclude that the tune should have been $\tilde{\mu} = -\mu$ instead of μ. This initial ambiguity comes from the complex nature of the eigenvalues. Since, we are interested in real similarity transformations only, our normalized matrix cannot be diagonal, but a two by two rotation. This resolves the ambiguity in the tune, but only at the end of the calculation when the rotation is calculated. In higher dimensionality a similar result applies: we can turn the map into a series of two by two rotations. The resulting tunes are unique but there is a freedom in deciding which tune corresponds to the first degree of freedom, which one associates to the second degree of freedom and so on. Of course, any properly formulated problem in terms of the real physical observables *cannot* depend on choices affecting the normalized variables.

The reader may have noticed that our argument works perfectly well in the case of a fully coupled oscillator in 2 or 3 degrees of freedom. It suffices to have access to a library capable of finding the complex eigenvectors and eigenvalues of M^{T} for the entire diagonalization procedure of Equation (2.48) to work.

2.4.3 From Linear $\mathcal{A}$ to Nonlinear $\mathcal{A}$

The reader has heard us mention more than once that the theory extends to nonlinear effects formally (i.e., matters of convergence ignored). This section as well as the next will illustrate this fact. First we define in Exercise 2-10 the concept of a nonlinear rotation.

Exercise 2-10

Show that the map $\vec{\rho} = \mathcal{R}\vec{I} = \exp(: f :)\vec{I}$, where $\vec{I}$ is our vector notation for the identity map (x,p) and f is a function of J only, is given by:

$$\rho_1 = \cos(\mu)x + \sin(\mu)p$$
$$\rho_2 = \cos(\mu)p - \sin(\mu)x$$
$$\mu = -\frac{df}{dJ} \equiv -f' \circ J \tag{2.50}$$

The expression $\frac{df}{dJ}$ is to be interpreted in the usual nonrigorous manner as the derivative of a single valued function in which we "plug in" the function $J = \frac{x^2+p^2}{2}$. In Interlude IV, Exercise 2-13, we will define an action-angle canonical transformation which formalizes this substitution.

A nonlinear rotation rotates the ray by a constant angle which depends on the radius $x^2 + p^2$. It is quite natural to ask if nonlinear maps can be transformed into nonlinear rotations by a canonical transformation $\mathcal{A} = \exp(: F :)$. We now explore this in the next exercise.

Exercise 2-11

1. *Show that the compositional map in Exercise 2-5 can be written as follows:*

$$\mathcal{M} = \exp(: -\mu J :)\exp\left(: \frac{x^4}{4} :\right) . \tag{2.51}$$

This is the so-called Dragt-Finn factorization.

2. *Show numerical evidence that the map of Equation (2.51) can be written as follows:*

$$\mathcal{M}\vec{I} = \underbrace{\exp(- : F :)\ \exp(: -\mu J + \frac{3}{8}J^2 :)\ \exp(: F :)\vec{I}}_{\mathcal{M}_4^{KAM}} + \cdots O(x^5)$$

$$F = -\frac{1}{64}\Big\{ - 5x^4 + 3p^4 + 6x^2p^2 + (8\cot(\mu) + 4\cot(2\mu))x^3p$$
$$+(8\cot(\mu) - 4\cot(2\mu))p^3x \Big\} . \tag{2.52}$$

Hint: Evaluate $\left(\mathcal{M}\vec{I} - \mathcal{M}_4^{KAM}\vec{I}\right)(\lambda\vec{z})$ for different λ's and a fixed $\vec{z}$ near the origin. What should you see?

3. *What is the invariant of* $\mathcal{M}_4^{KAM}$*?*

4. *Does the real map* $\mathcal{M}$ *have an invariant? Hint: Look at the phase plot. What do you conclude about the convergence of the procedure which generates the nonlinear canonical transformation that turns* $\mathcal{M}$ *into a rotation?*

The complexity of Equation (2.52), which was computed with pencil and paper, suggests that the author used some powerful tools in computing the normalizing map. Here is a sketch of the manipulations where f stands for $\frac{x^4}{4}$:

$$
\begin{aligned}
e^{:F:}\mathcal{M}e^{:-F:} &= \exp(:F:)e^{:-\mu J:}\exp(:f:)\exp(:-F:) \\
&\overset{\text{Inserting Identity as } e^{:\mu J:}e^{:-\mu J:}}{=} \exp(:F:)\exp\left(e^{:-\mu J:}:f:e^{:\mu J:}\right) \\
&\qquad\qquad \times\exp(e^{:-\mu J:}:-F:e^{:\mu J:})e^{:-\mu J:} \\
&\overset{\text{exercise 2-9}}{=} e^{:F:}\exp\left(:e^{:-\mu J:}f:\right)\exp\left(:-e^{:-\mu J:}F:\right)e^{:-\mu J:} \\
&\approx \exp(:(\mathcal{I}-e^{:-\mu J:})F+e^{:-\mu J:}f:)e^{:-\mu J:} \\
\Rightarrow\quad & (\mathcal{I}-e^{:-\mu J:})F+e^{:-\mu J:}f \\
& \text{depends on } J \text{ only}\,.
\end{aligned}
\tag{2.53}
$$

In Equation (2.53) we only did the calculation to leading order in the quartic perturbation.

Exercise 2-12 (Instructive: Do it and Remember it!)

1. *Show that the eigenfunctions of* $\exp(:-\mu J:)$ *are* $r_\pm = x \pm i\,p$.

2. *By using these eigenfunctions, try to solve for* F *using the last line of Equation* (2.53).

3. *Do the same thing on the map* $\mathcal{M}_\epsilon = \exp(:-\mu J:)\exp\left(:\epsilon\,\frac{x^2}{2}:\right)$ *to first order in* ϵ*. Notice that the map* $\mathcal{M}_\epsilon$ *is linear. So, here is your chance to perform the calculations two different ways. The exact way using the methods of Equation* (2.48) *and the approximate way (first order in* ϵ *will suffice) using the transformation* $\exp(:\epsilon F:)$.

Exercise 2-12 should be done by the reader at some point during his travel through this book. It is central to the theme of this book—the introduction of methods that work whether we are doing linear perturbation theory or nonlinear perturbation theory. For example, when analyzing a skew quadrupole effect to second order or a sextupole effect to second order, there is no reason to jump around from one method to another as is traditionally required. The trick of perturbation theory with Lie methods is to select F so that the remaining exponents of the map depend only on J. In fact, if the reader has worked out Exercise 2-11 or 2-12, he knows by now that the rotation terms cannot be removed. Later, in the book, we will discuss the case where one resonance is left in the map. This is used when perturbation theory attempts to explain the islands appearing in the figures of Exercise 2-1. Mathematically, in such a case, we leave in the map not only terms in J, but also terms which have the symmetry of the islands. For example, in a four-island problem, we must leave terms such as $4J^2\cos(4\phi)$ and $4J^2\sin(4\phi)$ (these are the real and imaginary parts of r_+^4 written as $r_+^4 = \left(\sqrt{2J}e^{-i\phi}\right)^4$, see Exercise 2-12.1 for a definition of r_+.). The reader can try to untangle this by himself or look into the future chapters. But, there is no crime in concentrating on the simpler transformation to a rotation for the time being. It is in fact a more important concept for the entire theory. Anyway, here is an interlude on the rigorous definition of the action-angle variable (ϕ, J) viewed as a canonical transformation.

INTERLUDE IV
ACTION ANGLE COORDINATES

In a total normalization of a map (linear or nonlinear), it is clear that the particle travels on a circle. Hence, the use of cylindrical coordinates appears appropriate because the radius (i.e., $\sqrt{2J}$) will stay invariant and the phase will advance by a constant amount (which is usually radius-dependent in the nonlinear case). For this reason, I introduce rigorously the action-angle coordinate transformation. The reader notices that this transformation is singular at the origin.

Exercise 2-13 (Rigorous Definition of Action-Angles)

1. *Show that*

$$\vec{z} \overset{\vec{\varpi}}{\longmapsto} \vec{\varpi}(\vec{z}) = \left(\sqrt{2z_2}cos(z_1), -\sqrt{2z_2}sin(z_1)\right) \quad (2.54)$$

is a canonical transformation. It defines the so-called action-angle variables ($z_1 = \phi$ and $z_2 = J$).

2. *Convince yourself that the rigorous expression for the last line of Equation* (2.50) *in Exercise* 2-10 *is*

$$\mu = \partial_2(f \circ \vec{\varpi}) \circ \vec{\varpi}^{-1} \text{ where } \partial_1(f \circ \vec{\varpi}) = 0 \quad . \quad (2.55)$$

END OF INTERLUDE

2.4.4 Freedom in Choosing $\mathcal{A}$: Gauge Dependence

Without going too deep into mathematical details, it is clear that the canonical transformation $\mathcal{A}$ is not unique. Even if the final form of the map is a unique rotation, there are infinitely many ways to get there. This is what we mean by "gauge freedom," it results from the "freedom" in the choice of a canonical transformation. From our construction algorithm, which uses the real and imaginary parts of h_+, we can have an arbitrary phase factor K as defined by Equation (2.41). A simple "grinding" exercise reveals the freedom in choosing $\mathcal{A}$.

Exercise 2-14 (Do it and Forget it!)
By multiplying h_+ in Equation (2.48) by an arbitrary complex number $re^{i\theta}$, show that the definition of $\mathcal{A}$ changes by a rotation of angle θ.

Exercise 2-14 is a bad idea because it violates our idea of elegance: the method only works in the linear case (it does work with coupling however). Therefore the next step is to imagine that the map $\mathcal{M}$ is nonlinear:

$$\begin{aligned} \mathcal{M} &= \mathcal{A}^{-1}\mathcal{R}\mathcal{A} \\ &\text{where} \\ \mathcal{R} &= \exp\ (:f:) \rightarrow f \text{ is a function of } J \text{ only i.e. } \frac{x^2+p^2}{2} \\ &\text{and } \mathcal{A} \text{ is nonlinear} \ . \end{aligned} \quad (2.56)$$

Whenever Equation (2.56) is true, it is possible to rewrite the map $\mathcal{M}$ as follows:

$$\begin{aligned} \mathcal{M} &= \mathcal{A}^{-1}\mathcal{R}\mathcal{A} = \mathcal{A}^{-1}\mathcal{F}^{-1}\mathcal{R}\mathcal{F}\mathcal{A} \\ &\text{where} \\ \mathcal{F} &= \exp(:\Phi:)\,, \quad \Phi \text{ is also a function of } J\,. \end{aligned} \tag{2.57}$$

Both the maps $\mathcal{R}$ and $\mathcal{F}$ are nonlinear rotations and commute with each other (i.e., $\mathcal{F}^{-1}\mathcal{R}\mathcal{F} = \mathcal{R}\mathcal{F}^{-1}\mathcal{F} = \mathcal{R}$). This freedom is a bit puzzling because it appears that different persons could get different answers for the new normalized or Floquet[2.3] coordinates (x_{new}, p_{new}). Indeed, there is an infinite set of possible choices for (x_{new}, p_{new}). The reader is still puzzled only if he has forgotten that we live in the physical world not the Floquet world. Therefore we should only ask physical questions and only expect questions about physical quantities to have unique answers. As the reader becomes familiar with accelerator physics jargon (assuming he is not already), he will hear statements such as

- How does one generalize the beta function when there is coupling?
- How does one generalize the phase advance? (Sorry, a concept we have not yet defined, but is intimately connected to the phase freedom $\mathcal{F}$ in Equation (2.57).)
- What is the proper form of the canonical transformation when there is coupling?
- What is, let's say, $\frac{\partial \beta}{\partial \delta}$ where δ is the energy of the particle.
- etc...

These statements have one property in common: they are ill-defined questions. They might mean something to people who are working on exactly the same problem using the same tools and coordinates: for example, $\frac{\partial \beta}{\partial \delta}$. But, lofty questions, such as worrying about how to generalize the beta function to coupled systems, are meaningless.

[2.3]In this book the term "Floquet" refers to a space in which the motion is approximately circular when projected in the position-momentum plane. More precisely the motion is exactly circular up to a given degree in the Taylor series expansion of the map with respect to a set of parameters.

This is because they are not posed in terms of physical observables[2.4]. As an observer of the field, I think that this kind of confusion arises in part because the traditional treatment requires so many approximations that its extension to a complicated situation is not transparent. One starts to lose sight of what is fundamental (usually the one turn map) and slowly sinks into a morass of irrelevant minutiae (i.e., what is the new Hamiltonian, how can I find it, how can I express it in Frenet-Serrat coordinates, etc, etc, etc...). In this section, we would like to show for the simple example of the "beta" function how a well-posed problem leads to well-defined *physical* solutions.

2.4.5 Measurable Quantities: Gauge Independence

We first start with the linear one degree of freedom problem and ask a few physical questions. The first question we ask is how to compute the invariant in terms of the canonical transformation $\vec{\alpha}^{-1}$ as defined by Equation (2.44). The starting point consists of the matrices for $\vec{\alpha}$ and $\vec{\alpha}^{-1}$:

$$A = \begin{pmatrix} A_{11} & A_{12} \\ A_{21} & A_{22} \end{pmatrix} \overset{\text{Symplectic}}{\longrightarrow} A^{-1} = \begin{pmatrix} A_{22} & -A_{12} \\ -A_{21} & A_{11} \end{pmatrix} . \quad (2.58)$$

Now, we use the fact that the invariant of $\mathcal{M}$ is just $J \circ \vec{\alpha}^{-1}$:

$$\begin{aligned} \epsilon \overset{\text{def}}{=} 2J \circ \vec{\alpha}^{-1} &= (A_{22}\, x - A_{12} p)^2 + (-A_{21}\, x + A_{11} p)^2 \\ &= \underbrace{(A_{22}^2 + A_{21}^2)}_{\gamma} x^2 + 2\underbrace{(-A_{22}A_{12} - A_{21}A_{11})}_{\alpha} xp \\ &\quad + \underbrace{(A_{11}^2 + A_{12}^2)}_{\beta} p^2 . \end{aligned} \quad (2.59)$$

Here ϵ is the Courant-Snyder invariant.

Exercise 2-15 (Trivial but "Deep")
Using the symplectic condition again, i.e., $A_{11}A_{22} - A_{21}A_{12} = 1$, *show that* $1 + \alpha^2 = \beta\gamma$.

This exercise teaches an important lesson. The relation between the parameters α, β *and* γ *is important because it insures stability. Needless to say that it is more complicated for a six by six fully coupled*

[2.4] Here we mean a generalized physical observable: we include quantities which may require perfect experimental set-up or gedanken experiments.

matrix. However, by assuming that our matrix is related to a rotation by a symplectic similarity transformation, we need not compute such a relation. It is built-in automatically. Even better, in nonlinear systems the analogous more complicated relations will be built in as well.

Next we define three physical quantities. Suppose that you observe the particle over time, turn after turn, and that you compute the multiple turn averages $\langle x^2 \rangle$, $\langle p^2 \rangle$, and $\langle xp \rangle$. What are these quantities? Before, we attempt to compute them, let us mention that they must be functions of the invariants only. In other words, they depend on the initial conditions through the invariants. If you know on what invariants the particle sits, then you should be able to find the average of any quantity. Indeed, the multiple turn average will sample all the initial conditions on the invariant if the tunes are irrational amongst each other, i.e., the tunes are not resonant. This condition in one dimension means that the tune is not rational. Let us look at an arbitrary function f. Its multiple turn average is defined as:

$$\langle f \rangle = \lim_{n \to \infty} \frac{1}{n} \sum_{k=1}^{n} \mathcal{M}^k f \,. \tag{2.60}$$

Therefore, it is of paramount importance to compute the quantity $\mathcal{M}^k f = f \circ \underbrace{\vec{\zeta} \circ \cdots \circ \vec{\zeta}}_{\text{k times}}$. To compute this quantity, we assume the existence of the canonical transformation $\mathcal{A}$:

$$\mathcal{M}^k f = \mathcal{A}^{-1} \mathcal{R}^k \mathcal{A} f \,. \tag{2.61}$$

We now expand the function $\mathcal{A}f$ using the eigenvectors r_+ and r_- of the rotation $\mathcal{R}$ (see Exercise 2-12.1). The readers familiar with ordinary perturbation theory will see this as a Fourier transform. Those familiar with quantum mechanics can view r_+ and r_- as raising and lowering operators of the harmonic oscillators. Anyway, we just proceed with the multidimensional and nonlinear case immediately. We denote the monomial $r_{1+}^{n_1^+} r_{1-}^{n_1^-} \cdots r_{N+}^{n_N^+}, r_{N-}^{n_N^-}$ by the notation $r_+^{\vec{n}^+} r_-^{\vec{n}^-}$ where $\vec{r}_\pm = (r_{1\pm}, \cdots, r_{N\pm}) = (x_1 \pm ip_1, \cdots, x_N \pm ip_N)$. The calculation goes as follows:

$$\text{if } \mathcal{A}f = \sum F_{\vec{n}^+,\vec{n}^-} r_+^{\vec{n}^+} r_-^{\vec{n}^-}$$

$$\Downarrow$$

$$
\begin{aligned}
\mathcal{M}^k f &= \mathcal{A}^{-1}\mathcal{R}^k \sum F_{\vec{n}^+,\vec{n}^-} r_+^{\vec{n}^+} r_-^{\vec{n}^-} \\
&= \mathcal{A}^{-1} \sum F_{\vec{n}^+,\vec{n}^-} \exp\left(-ik(\vec{n}^+ - \vec{n}^-)\cdot\vec{\mu}\right) r_+^{\vec{n}^+} r_-^{\vec{n}^-} \\
&= \sum F_{\vec{n}^+,\vec{n}^-} \exp\left(-ik(\vec{n}^+ - \vec{n}^-)\cdot(\mathcal{A}^{-1}\vec{\mu})\right) \\
&\quad \times(\mathcal{A}^{-1} r_+^{\vec{n}^+})(\mathcal{A}^{-1} r_-^{\vec{n}^-}) \, .
\end{aligned}
\tag{2.62}
$$

The notation $\mathcal{A}^{-1}r_+^{\vec{n}^+}$ simply means $(\mathcal{A}^{-1}r_{1+}^{n_1^+})\cdots(\mathcal{A}^{-1}r_{N+}^{n_N^+})$. First, we notice that our derivation includes nonlinear terms since nothing was said about the angle $\vec{\mu}$ of the rotation. The angle $\vec{\mu}$ is only a function of $\vec{J}$ as defined by Equation (2.50) and rigorously in Equation (2.55). Hence, the quantity $\mathcal{A}^{-1}\vec{\mu} = \vec{\mu}\circ\vec{\alpha}^{-1}$ is also only a function of the invariants of $\mathcal{M}$, i.e., $\mathcal{A}^{-1}\vec{J} = \vec{J}\circ\vec{\alpha}^{-1}$. If we take the limit as n goes to infinity only the invariant terms survive (i.e., $\vec{n}^+ = \vec{n}^-$):

$$
\begin{aligned}
\langle f\rangle &= \sum_{\vec{n}} F_{\vec{n},\vec{n}}\, \mathcal{A}^{-1}(r_+^{\vec{n}} r_-^{\vec{n}}) \\
&= \sum_{\vec{n}} F_{\vec{n},\vec{n}} 2^{|\vec{n}|}(\mathcal{A}^{-1}\vec{J})^{\vec{n}} \\
\text{where}\quad |\vec{n}| &= \sum_{i=1}^{N} |n_i| = \sum_{i=1}^{N} n_i \, .
\end{aligned}
\tag{2.63}
$$

Our first application is the case $\langle x^2\rangle$ for a one degree of freedom linear map. The function f is just x^2 and the map $\mathcal{A}$ is obtained from Equation (2.58). We first compute $\mathcal{A}x^2$:

$$
\begin{aligned}
\mathcal{A}x^2 &= (A_{11}x + A_{12}p)^2 \\
&= \left(A_{11}\frac{r_+ + r_-}{2} + A_{12}\frac{r_+ - r_-}{2i}\right)^2 \\
&= \frac{1}{4}(A_{11} - i\,A_{12})^2 r_+^2 + \frac{1}{4}(A_{11} + i\,A_{12})^2 r_-^2 \\
&\quad + \frac{1}{2}(A_{11}^2 + A_{12}^2) r_+ r_- \, .
\end{aligned}
\tag{2.64}
$$

So, we identify immediately the coefficient $F_{(1)(1)}$ as

$$
\begin{aligned}
\langle x^2\rangle &= (A_{11}^2 + A_{12}^2)\mathcal{A}^{-1}J \\
&= \beta\mathcal{A}^{-1}J \overset{\text{Eq. (2.59)}}{=} \frac{1}{2}\beta\varepsilon \, .
\end{aligned}
\tag{2.65}
$$

Exercise 2-16

1. *Show that* $\langle p^2 \rangle$ *is just* $\frac{1}{2}\gamma\varepsilon$ *where* $\gamma = A_{22}^2 + A_{21}^2$.

2. *Show that* $\langle xp \rangle$ *is given by* $-\frac{1}{2}\alpha\varepsilon$ *where* $\alpha = -A_{22}A_{12} - A_{21}A_{11}$.

3. *Consider the case of a two degree of freedom problem. Assuming that the problem is linear and stable, we know that there exists a linear canonical transformation, whose matrix we denote by A, which turns the map into a rotation. Using formula* (2.63), *show that* $\langle x_1^2 \rangle$ *is given by:*

$$\langle x_1^2 \rangle = (A_{11}^2 + A_{12}^2)\mathcal{A}^{-1}J_1 + (A_{13}^2 + A_{14}^2)\mathcal{A}^{-1}J_2 \,. \quad (2.66)$$

   Notice that there are two "beta functions" in the case of coupled oscillations in two degrees of freedom. Remark: It is a good place to point out to the reader that the Lie transformation takes the "passive" point of view. Therefore, it is natural to define J_i *to be the function equal to half the radius in the* i^{th} *plane.*

4. *Define the quantity* $\widehat{x_1}$ *as the extremum value reached by the variable* x_1. *Show that it is given by:*

$$\begin{aligned}\widehat{x_1} &= \pm\sqrt{2(A_{11}^2 + A_{12}^2)\mathcal{A}^{-1}J_1} \pm \sqrt{2(A_{13}^2 + A_{14}^2)\mathcal{A}^{-1}J_2} \\ &= \pm\sqrt{(A_{11}^2 + A_{12}^2)\varepsilon_1} \pm \sqrt{(A_{13}^2 + A_{14}^2)\varepsilon_2} \,. \quad (2.67)\end{aligned}$$

   Here, ε_1 *and* ε_2 *are generalizations of the Courant-Snyder invariant in two degrees of freedom—the map* $\mathcal{M}$ *is just* $\exp(-:\mu_1\varepsilon_1 + \mu_2\varepsilon_2:)$. *The reader should convince himself that the* $\pm$ *sign in Equation* (2.67) *produces the inner and outer boundaries of the donut shape typical of coupled motion[10]. Remark: I would like to point out that in reference [10] the authors, after a lot of algebra leading to the discovery of Equation* (2.67), *rhetorically ask the readers: "What is the meaning of these lattice functions?" This is totally antipodal to our way of thinking. Here we start with a well-defined phase plot and a simple question, and then we derive* (2.67) *in a couple of lines!*

5. *Using the map* $e^{:\epsilon x_1 x_3:}e^{-:\mu_1 J_1 + \mu_2 J_2:}e^{-:\epsilon x_1 x_3:}$ *check numerically the various assertions of these exercises.*

6. *Using the nonlinear map of Exercise 2-11, show that $\langle x^2 \rangle$ is given by:*

$$\langle x^2 \rangle = \frac{1}{2}\varepsilon + \frac{3}{16}\cot(\mu)\varepsilon^2$$
$$\textit{where } \varepsilon = \exp\left(: -F :\right)\, r_+ r_- \ . \tag{2.68}$$

Check this numerically with scaling tests (see Exercise 2-11).

In this section, we have seen that if one asks physical questions, one gets answers which only depend on the invariants and quantities which are invariant under a rotation. In other words, if a problem is well-posed, the answer does not depend on methods and gauges. We will emphasize this when we discuss the concept of "phase advance" in the next section.

2.5 THE RING

The purpose of Chapter 2 has been to convince the reader that the one-turn map of the ring contains a lot of information about our system; that only a few reasonable assumptions were necessary to understand the nature of this map; that stability insures the existence of a fixed point—etc... as was pointed out at the beginning of this chapter.

The reader remembers that this chapter is supposed to be a "comic book" version of Parts II and III. In this chapter, we have introduced new concepts as we went along. To some extent, it is a coloring book, also. The reader is invited to learn through exercises. Ideally, the reader should, up to a point, be able to fight with the author and rewrite this chapter as he goes along. In other words, this should be not just a coloring book, but a book with blank pages. I mention this now because it is probable that a few readers are wishing desperately to see a magnet or some sort of Hamiltonian for this ring. Some must wonder if they will *ever* see a magnet or a Hamiltonian. There is a great temptation for the author to give in and to start writing down some Hamiltonian. But, we will keep ourselves faithful to the logical division between the local and global aspects of our description. Needless to say that the reader can cheat and jump to section 3 of this chapter. Bolder readers can also immediately read Part III if they want.

So, true to ourselves, we will descend towards the Hamiltonian in the abstract limit of a ring with an infinite number of maps. In the comical Chapter 3 of Part I or in the "serious" Part III, we will do the reverse: we will start in the hellish world of local dynamics and show how one emerges from it with a simulation code and, therefore, with a one-turn map.

2.5.1 Why More than One Map?

We have spent a lot of effort to convince the reader of the importance of the one-turn map. Of course, in practice there are a lot of problems which require the examination of several locations around a ring. So, now we will create a fictitious ring giving rise to a map closely analogous to that of Equation (2.51). Consider the following three transformations:

$$\begin{aligned} \mathcal{D}_L \begin{pmatrix} x \\ p \end{pmatrix} &= \begin{pmatrix} x + Lp \\ p \end{pmatrix} \Rightarrow \mathcal{D}_L = \exp\left(: - L\,\frac{p^2}{2} : \right) \\ \mathcal{Q}_{k_2} \begin{pmatrix} x \\ p \end{pmatrix} &= \begin{pmatrix} x \\ p - k_2 x \end{pmatrix} \Rightarrow \mathcal{Q}_{k_2} = \exp\left(: - k_2\,\frac{x^2}{2} : \right) \\ \mathcal{O}_{k_4} \begin{pmatrix} x \\ p \end{pmatrix} &= \begin{pmatrix} x \\ p - k_4 x^3 \end{pmatrix} \Rightarrow \mathcal{O}_{k_4} = \exp\left(: - k_4\,\frac{x^4}{4} : \right) . \end{aligned} \tag{2.69}$$

We will call the map $\mathcal{D}_L$ a drift of length L, $\mathcal{Q}_k$ a quadrupole of strength k_2, and $\mathcal{O}_k$ an octupole of strength k_4. The readers who know too much are probably complaining that these formulas are approximate for all sorts of reasons—true. At the same time, it is fair to say that these approximate maps are not devoid of physical significance. Anyway, let us use them as "gifts from God," without questioning their accuracy (the angry reader can take a look at Part III to satisfy his thirst for truth and exactness).

Let us consider the following ordered set[2.5]

$$Ring_4 = \left\{ \mathcal{D}_{\tan(\frac{\mu}{2})}, \mathcal{Q}_{\sin(\mu)}, \mathcal{O}_{k=-1}, \mathcal{D}_{\tan(\frac{\mu}{2})} \right\} . \tag{2.70}$$

In an ordered set, we have a first map, second map, third map, and fourth map:

$$\begin{aligned} \mathcal{M}_{0\to 1} &= \mathcal{D}_{\tan(\frac{\mu}{2})} \\ \mathcal{M}_{1\to 2} &= \mathcal{Q}_{\sin(\mu)} \end{aligned}$$

[2.5] I am sure that some readers feel that the ring of Equation (2.71) cannot close because it has no dipoles! Remember that we are describing the properties of a ring around the closed orbit. Thus it is quite possible that its quadrupoles or drifts may in fact be dipoles.

$$\begin{aligned}\mathcal{M}_{2\to 3} &= \mathcal{O}_{k=-1}\\ \mathcal{M}_{3\to 4} &= \mathcal{D}_{\tan(\frac{\mu}{2})}\\ &\text{where } 4=0 \quad \text{modulo } 4\,.\end{aligned} \tag{2.71}$$

Now, let us give a "physical meaning" to this insane rambling. Suppose that the index $i = 0, 3$ represents positions around a ring and that the map $\mathcal{M}_{i\to i+1}$ propagates a function from position i to position $i + 1$ defined modulo 4. Let us make contact with the one turn map analysis through an exercise.

Exercise 2-17

1. *Show that the one-turn map from position $i = 0$ back to $i = 0$ is given by the map*

$$\begin{aligned}\mathcal{M}_{0\to 4} &= \mathcal{D}_{\tan(\frac{\mu}{2})}\mathcal{Q}_{\sin(\mu)}\mathcal{O}_{k=-1}\mathcal{D}_{\tan(\frac{\mu}{2})}\\ &= \exp\left(:-\mu J:\right)\exp\left(:\frac{\left(x-\tan(\frac{\mu}{2})p\right)^4}{4}:\right).\end{aligned} \tag{2.72}$$

   Hint: Use the transfer maps $\vec{\zeta}_{i\to i+1} = \mathcal{M}_{i\to i+1}(x,p)$.

2. *Using a properly defined similarity transformation, show that the problem of stability at location i is always equivalent to stability at location $i = 0$.*

Here the "Ring" is defined by the theorist/experimentalist as the most efficient collection of maps necessary to answer a given set of physical questions. For example, suppose you are given the following problem: determine the area of stability of $Ring_4$ in phase space. It is clear that you can sit at position i = 0,1,2, or 3 and simulate the motion. The phase space volume of stable motion will be the same elsewhere because of the similarity transformation of Exercise 2-17. At this stage, the reader can elect to study the one-turn maps $\mathcal{M}_{0\to 4}$, $\mathcal{M}_{1\to 5}$, $\mathcal{M}_{2\to 6}$ or $\mathcal{M}_{3\to 7}$. Our convention is that $\mathcal{M}_{i\to i}$ is the identity map while $\mathcal{M}_{i\to i+\aleph}$ (or simply $\mathcal{M}_i$) is the one-turn map at position i. The Hebrew letter $\aleph$ denotes the number of maps in the ring. Therefore, given $Ring_4$, the reader is free to select any ring of cardinality $\aleph = 1$ to study the "dynamic aperture" or area of stability:

$$Ring_1 = \{\mathcal{M}_{i\to i+4} |\ \mathcal{M}_{i\to i+4} \text{ created with } Ring_4\}. \tag{2.73}$$

Of course, until now, all the "rings" we have studied in Part I have $\aleph = 1$. In other words, the search for the dynamic aperture requires, in theory, the study of a one-turn map. This one-turn map is constructed using the maps of the individual magnets. Mathematically we can think of the entire ring as being a single very large element.

There are a few important points to remember:

1. Stability is given by a ring with $\aleph = 1$. Therefore any map will do. This is why we emphasize the one-turn map so much.

2. If the ring needs to be observed at n locations, we need a ring with $\aleph = n$.

3. The maximal ring $Ring_\infty$ contains all the possible infinitesimal maps allowing us to propagate rays and functions between any two conceivable locations. Clearly, $Ring_\infty$ must be equivalent to the Hamiltonian.

4. The local Hamiltonian, i.e., $Ring_\infty$, has an existence of its own like the actual real machine. Rings which contain a finite number of maps are mental constructions of the theorist, the experimentalist, or the designer. It is the beauty of the Hamiltonian-free approach that it allows the "user" to choose what should be the ring on the basis of the things he is computing, measuring or designing.

We should make a comment here about $\aleph = 0$ and $\aleph = \infty$. The case of $\aleph = 0$ consists of a ring with no maps. Actually, when discussing non-dynamical issues we deal with $\aleph = 0$. For example, we may discuss the cost of a magnet, the color of the paint, etc. The ring as a physical object still exists, but as a dynamical object it has no existence (the author has used $\aleph = 0$ in a dynamical simulation: I forgot to activate the "print" statement in my tracking code and wasted computer time!). The case of $\aleph = \infty$ is the Hamiltonian. In this book, we will argue that dealing with the Hamiltonian is usually an overkill. Experiments have a finite number of knobs, monitors, and magnets. Simulation codes have a finite number of integration steps and most problems really require a finite ring from a conceptual point of view. Nevertheless, for simple models of the physical ring, we may find it useful to deal with $\aleph = \infty$. For any real ring, the reader should convince himself that it is impossible. We will get to $\aleph = \infty$ by taking the limit $\lim_{\aleph \to \infty} Ring_\aleph$. This is because we base our theory on the magnet metaphor as explained in the preface. And thus $Ring_\infty$ is a ring with an infinite number of tiny *related* magnets.

2.5.2 Phase Advance: the Floquet Ring

Let us take *Ring*$_4$ as an example. Presumably, for some reasons, we found it useful to examine these four locations. For example, $i = 0$ could be the collision point of a collider, $i = 1$ the injection point, etc. It is not really relevant: it suffices to assume that one has chosen this ring for some good reasons (the sarcastic reader is probably saying that the author concocted a ring in order to produce more exercises: he is right, but let's keep this a secret!).

First, we would like to create all the possible one-turn maps this ring has to offer. We now list them:

$$\begin{aligned} \mathcal{M}_{0\to4} &= \mathcal{DQOD} \\ \mathcal{M}_{1\to5} &= \mathcal{QODD} \\ \mathcal{M}_{2\to6} &= \mathcal{ODDQ} \\ \mathcal{M}_{3\to7} &= \mathcal{DDQO}\ . \end{aligned} \tag{2.74}$$

I dropped the subscripts since the two drifts are identical. Let us transform the linear part of these maps into a rotation using an algorithm which depends only on the map itself. We know from our theorem on freedom that there exists an infinite choice of such algorithms. We will choose a very famous one called the Courant-Snyder transformation:

$$\begin{gathered} \text{Select } \mathcal{A} \text{ such that } A = \begin{pmatrix} A_{11} & 0 \\ A_{21} & A_{22} \end{pmatrix} \\ \text{and } A_{11} > 0 \\ \Downarrow \ \text{according to (2.59)} \end{gathered}$$

$$A = \begin{pmatrix} \sqrt{\beta} & 0 \\ -\frac{\alpha}{\sqrt{\beta}} & \frac{1}{\sqrt{\beta}} \end{pmatrix}. \tag{2.75}$$

It is clear that the canonical transformation $\mathcal{A}$ can be viewed as a "functional" of the one-turn map. The particular functional we chose for the linear part is the Courant-Snyder functional denoted by $\mathcal{A}_{cs}[\mathcal{M}_i]$ where by $\mathcal{M}_i$ we denoted the one-turn map $\mathcal{M}_{i\to i+N}$.

Before we state our results, it is time to dig into the exercises.

Exercise 2-18

Show that in general a stable symplectic matrix M in one degree of freedom can be written as:

$$M = \begin{pmatrix} \cos(\mu) + \alpha\ \sin(\mu) & \beta \sin(\mu) \\ -\gamma \sin(\mu) & \cos(\mu) - \alpha\ \sin(\mu) \end{pmatrix}. \tag{2.76}$$

Hint: Use $M = ARA^{-1}$, Equations (2.75), *and* (2.59).

The next step is to evaluate the various Courant-Snyder matrices associated with the four one-turn maps of *Ring*$_4$.

Exercise 2-19
For the maps of Equation (2.74), *get all the Courant-Snyder transformations. The answers are:*

$$\begin{aligned}
A_{cs}[\mathcal{M}_0] &= \begin{pmatrix} 1 & 0 \\ 0 & 1 \end{pmatrix} \\
A_{cs}[\mathcal{M}_1] &= \begin{pmatrix} \frac{1}{\cos(\frac{\mu}{2})} & 0 \\ \sin(\frac{\mu}{2}) & \cos(\frac{\mu}{2}) \end{pmatrix} \\
A_{cs}[\mathcal{M}_2] &= \begin{pmatrix} \frac{1}{\cos(\frac{\mu}{2})} & 0 \\ -\sin(\frac{\mu}{2}) & \cos(\frac{\mu}{2}) \end{pmatrix} \\
A_{cs}[\mathcal{M}_3] &= \begin{pmatrix} \frac{1}{\cos(\frac{\mu}{2})} & 0 \\ -\sin(\frac{\mu}{2}) & \cos(\frac{\mu}{2}) \end{pmatrix}.
\end{aligned} \tag{2.77}$$

The linear Floquet[2.6] ring is defined using the set of $\aleph$ canonical transformations $\mathcal{A}[\mathcal{M}_i]$. To motivate how this comes about, let us look at the map $\mathcal{A}_0\mathcal{M}_0\mathcal{A}_0^{-1}$ (we dropped the reference to Courant-Snyder):

$$\begin{aligned}
\mathcal{A}_0\mathcal{M}_0\mathcal{A}_0^{-1} &= \mathcal{A}_0\mathcal{D}\mathcal{Q}\mathcal{O}\mathcal{D}\mathcal{A}_0^{-1} \\
&\Downarrow \\
&= \underbrace{\mathcal{A}_0\mathcal{D}\mathcal{A}_1^{-1}}_{\mathcal{R}_{0\to1}}\underbrace{\mathcal{A}_1\mathcal{Q}\mathcal{A}_2^{-1}}_{\mathcal{R}_{1\to2}}\underbrace{\mathcal{A}_2\mathcal{O}\mathcal{A}_3^{-1}}_{\mathcal{R}_{2\to3}\mathcal{O}^{\dagger}}\underbrace{\mathcal{A}_3\mathcal{D}\mathcal{A}_0^{-1}}_{\mathcal{R}_{3\to4}} \\
&= \mathcal{R}_{0\to1}\mathcal{R}_{1\to2}\underbrace{\mathcal{A}_2\mathcal{I}\mathcal{A}_3^{-1}}_{\mathcal{R}_{2\to3}}\underbrace{\mathcal{A}_3\mathcal{O}\mathcal{A}_3^{-1}}_{\mathcal{O}^{\dagger}}\mathcal{R}_{3\to4} \qquad (2.78) \\
&= \mathcal{R}_{0\to1}\mathcal{R}_{1\to2}\mathcal{R}_{2\to3}\underbrace{\exp\left(:\frac{x^4}{4\cos^4(\frac{\mu}{2})}:\right)}_{\mathcal{O}^{\dagger}}\mathcal{R}_{3\to4}.
\end{aligned}$$

[2.6] As explained in footnote 2.3, the term "linear Floquet ring" implies that the linear invariants are circles in this ring.

We denote by $\mathcal{R}_{i\to i+1}$ the linear part of $\mathcal{A}_i\mathcal{M}_{i\to i+1}\mathcal{A}_{i+1}^{-1}$. Since our canonical transformation was linear, if anything miraculous should happen, it can only be at the linear level. The miracle is that all the maps $\mathcal{R}_{i\to i+1}$ are rotations. They are the phase advances associated with the functional $\mathcal{A}_{cs}$ [ ].

Exercise 2-20

Compute all the matrices $R_{i\to i+1}$. Answer: $R_{1\to 2} = R_{2\to 3} = \mathrm{I}$ and $R_{0\to 1} = R_{3\to 4} = \begin{pmatrix} \cos(\frac{\mu}{2}) & \sin(\frac{\mu}{2}) \\ -\sin(\frac{\mu}{2}) & \cos(\frac{\mu}{2}) \end{pmatrix}$.

We are now in a position to construct the Floquet ring produced by the Courant-Snyder functional $\mathcal{A}_{cs}$ [ ]:

$$\mathcal{A}_{cs}\left[Ring_4\right] = \left\{\mathcal{R}(\frac{\mu}{2}), \mathcal{I}, \exp\left(: \frac{x^4}{4\cos^4(\frac{\mu}{2})} :\right), \mathcal{R}(\frac{\mu}{2})\right\}. \quad (2.79)$$

Of course, such a concept can now be extended to higher orders. Equation (2.53) provided us with an algorithm, a first order algorithm to be precise, for the computation of nonlinear transformations which depends only on the one turn map. Hence, it is a sensible question to ask what is the nonlinear Floquet ring associated with this definition of the nonlinear canonical transformation. Because the leading order nonlinearities are quartic, we start with a set of canonical transformations

$$\mathcal{B}_i = \exp\left(: F_i :\right), \quad (2.80)$$

where the functions F_i are homogeneous polynomials of degree four.

$$\begin{aligned} Ring_4^{1st} &= \mathcal{B}\left[\mathcal{A}_{cs}\left[Ring_4\right]\right] \\ &= \left\{e^{:F_0:}\mathcal{R}(\frac{\mu}{2})e^{-:F_1:}, e^{:F_1:}e^{-:F_2:}, \right. \quad (2.81) \\ &\quad \left. e^{:F_2:}\exp\left(: \frac{x^4}{4\cos^4(\frac{\mu}{2})} :\right)e^{-:F_3:}, e^{-:F_3:}\mathcal{R}(\frac{\mu}{2})e^{-:F_0:}\right\}. \end{aligned}$$

Presumably it is possible to select the F_i's in such away as to make the ring a quartic Floquet ring. This is the topic of the next exercise.

Exercise 2-21

The "1st" in $Ring^{1st}$ refers to first order in the octupole strength k_4. Without using the dreadful expression for F in Exercise 2-11.2, Equation (2.52), show that $Ring_4^{1st}$ is given by:

$$Ring_4^{1st} = \left\{\mathcal{R}(\frac{\mu}{2}), \mathcal{I}, \exp\left(: \frac{3}{8\cos^4(\frac{\mu}{2})}J^2 :\right), \mathcal{R}(\frac{\mu}{2})\right\}. \quad (2.82)$$

At this stage, it is time to pause and examine the results. First, it is kind of nice that the Floquet ring to octupole order, $Ring_4^{1st}$, shows a perturbation only at the location of the octupole. We can ask a few questions:

1. Why was the Courant-Snyder choice for the linear canonical transformation a good choice in the case of $Ring_4$?

2. The nonlinear phase, to leading order, advanced only at the nonlinear perturbation. Is this a lucky result?

3. Will this persist to all orders in the perturbation?

4. If the answer to 3 is no, then under what conditions will the extra phase be additive?

5. What does this imply?

Answer to question 1: we notice that the Lie polynomial of the nonlinear octupole perturbation depends only on the position function x. This is true in the body of multipole magnets away from the fringe field region. Therefore it makes sense in analytical calculations to choose a definition of $\mathcal{A}$, and hence of the phase advance, which does not complicate our expressions too much, i.e., $A_{12} = 0$. The answer to question 2 is no. If we define all our canonical transformations, linear as well as nonlinear, in terms of the one-turn maps only, then the new first order Floquet ring will always be a new ring of rotations with perturbation only at the location of each perturbation. The answer to question 3 is no. This is connected to the concept of matching. If we look at the one-turn maps before and after the octupoles, we see that they are different. An equivalent statement says that the normalizing transformations $\exp(:F_2:)$ and $\exp(:F_3:)$ are different. Because of this, the next Floquet ring $Ring_4^{2nd}$ will be a mess. This just answered question 4. Finally, the implications of these statements are profound. First, it says that the phase advance is probably not a invariant quantity, i.e., it depends on a gauge, in this case how we define the canonical transformation $\mathcal{A}$. It does say that the quantity is invariant between matched locations and hence measurable between such locations through an averaging process. This also has tremendous application in design theory. If you want to insert something into an existing ring or create a new ring, it pays to insert matched modules. Then their effects are localized and additive. The design theory with maps is based on the concept of matched[2.7] modules: phase trombones, $-I$ transforms, cells, sextupoles π apart, etc.

Exercise 2-22 *This exercise is devoted to the Courant-Snyder phase advance.*

1. *The matrix for the phase advance from a location k to a location j is given by $R_{k\to j} = A_j^{-1} M_{k\to j} A_k$. Show that the cosine and sine of this rotation are given by*

$$\cos\ \mu_{k\to j}^{CS} = \sqrt{\frac{\beta_k}{\beta_j}} M_{11}^{k\to j} - \frac{\alpha_k}{\sqrt{\beta_k \beta_j}} M_{12}^{k\to j}$$

and (2.83)

$$\sin\ \mu_{k\to j}^{CS} = \frac{M_{12}^{k\to j}}{\sqrt{\beta_k \beta_j}}.$$

2. *Show that the matrix $M_{k\to j}$, when expressed in terms of the invariants and the phase advance , is given by*

$$\begin{pmatrix} \sqrt{\frac{\beta_j}{\beta_k}}\left(\cos\mu_{k\to j}^{CS} + \alpha_k \sin\mu_{k\to j}^{CS}\right) & \sqrt{\beta_j\beta_k}\sin\mu_{k\to j}^{CS} \\ -\frac{1+\alpha_k\alpha_j}{\sqrt{\beta_j\beta_k}}\sin\mu_{k\to j}^{CS} + \frac{\alpha_k-\alpha_j}{\sqrt{\beta_j\beta_k}}\cos\mu_{k\to j}^{CS} & \sqrt{\frac{\beta_k}{\beta_j}}\left(\cos\mu_{k\to j}^{CS} - \alpha_j \sin\mu_{k\to j}^{CS}\right) \end{pmatrix}$$

(2.84)

2.6 PHASE ADVANCE AND HAMILTONIANS

The purpose of this section is to close the loop. We started with the one-turn map and showed that all the dynamics are contained in it. Conceptually and practically, it is the most important object of ring dynamics. Then we argued that the physical ring, the actual piece of hardware, has a physical existence of its own and is not a mathematical abstraction. In particular, we may be interested in observing its behavior at several locations simultaneously for all sorts of reasons. Therefore we introduced the concept of a *Ring*—a collection of maps judiciously chosen so as to minimize the amount of work one

[2.7] One says "matching modules" if the active point of view of the designer is taken.

has to perform conceptually and numerically. Finally, we argue that all of this must be mathematically equivalent to the usual Hamiltonian theory which we avoided and will keep avoiding when discussing global dynamical concepts. The reader may still want to see how this comes about. Therefore this last section is dedicated to the maximal ring $Ring_\infty$, i.e., the Hamiltonian. More precisely, how does one get all the nice analytical formulas of Hamiltonian theory by taking the limit $\aleph$ tending to infinity. We will also illustrate, by the example of Exercise 2-23.2, the dangerous over-kill and extra-complexity of the Hamiltonian approach.

2.6.1 The Hamiltonian

Let us go back to the case of $Ring_4$ as defined by Equation (2.70). We now assume that the Hamiltonian generating these maps is given by:

$$\begin{aligned} H = &\frac{H_{22}(s)}{2}p^2 + \frac{H_{11}(s)}{2}x^2 + H_{12}(s)xp \\ &-k_{oct}\delta_p(s - s_{oct})\frac{x^4}{4}. \end{aligned} \tag{2.85}$$

This Hamiltonian is assumed to have a period C and the function $\delta_p(s - s_{oct})$ is the periodic Dirac delta[2.8] function:

$$\delta_p(s - s_{oct}) = \sum_{n=-\infty}^{\infty} \delta(s - s_{oct} + nC). \tag{2.86}$$

For our example, we have assumed the most general form for the quadratic part of H in one degree of freedom. Usually, the functions H_{ij} are given by:

$$\begin{aligned} H_{11} &= k(s) \equiv \text{Focusing quadrupole strength} \\ H_{22} &= 1 \\ H_{12} &= 0\,. \end{aligned} \tag{2.87}$$

First of all, we will see that H can generate $Ring_4$ as it should since, in one degree of freedom, it has the most general form in its

[2.8] It is best to take the limit to a delta function in a Lie representation. This is actually the correct physical interpretation and it leads to the correct answer for the map across the delta function. In general if $H = -\delta_p(s)V_s$ then the map from 0^- to 0^+ is $\exp(:V_{s=0}:)$.

quadratic part. Conversely, however, it is not true that $Ring_4$ is generated only by Equation (2.85). In fact, two degree of freedom Hamiltonians, with coupled motion, can generate $Ring_4$. This is a very important fact, and proves once more the superiority of the map approach. The next exercise, which is very important, illustrates these points.

Exercise 2-23

1. *Find at least two possible forms for H as defined by Equation (2.85) which reproduce $Ring_4$.*

2. *Consider the following two-degree of freedom Hamiltonian:*

$$H = \frac{p_x^2 + p_y^2}{2} + k_{\text{dipole}}(s)\frac{x^2+y^2}{2} - k_{\text{skew}}(s)\; xy - k_{\text{oct}}\delta_p(s - s_{\text{oct}})\; \frac{x^4 + y^4 - 6x^2y^2}{4} \quad . \tag{2.88}$$

Notice that the focusing term k_{dipole} is identical in both planes. This is physically realisable if the focusing is inside a bend with a small radius and a quadrupole component. This was chosen to make the exercise easier. Using the skills acquired in Exercise 2-23.1 and any s-dependence of your choice (including Dirac delta's) for k_{dipole} and k_{skew} , construct a ring with $4 < \aleph < \infty$ which has at least one location where the motion is fully coupled between the planes $x - p_x$ and $y - p_y$. Choose your parameters such that $Ring_4$ is a sub-ring of your ring. This shows that in such a case, Hamiltonian theory can be a truly unwarranted horror if the location of interest sits outside the region of coupling. It is very important that the reader grasps this. A possible answer is:

$$Ring_{12} = \Big\{ \mathcal{D}_{\tan(\frac{\mu}{2})}, \mathcal{Q}_{\sin(\mu)}, e^{:xy:}, \mathcal{D}_{\tan(\frac{\pi}{4})}, \mathcal{Q}_{\sin(\frac{\pi}{2})}, \mathcal{D}_{\tan(\frac{\pi}{4})}, \mathcal{D}_{\tan(\frac{\pi}{4})}, \mathcal{Q}_{\sin(\frac{\pi}{2})}, \mathcal{D}_{\tan(\frac{\pi}{4})}, e^{-:xy:}, \mathcal{O}_{k=-1}, \mathcal{D}_{\tan(\frac{\mu}{2})} \Big\}$$

where

$$\mathcal{D}(L) = \exp\left(- : L\frac{p_x^2+p_y^2}{2} :\right) \text{ and } \mathcal{Q}(L) = \exp\left(- : L\frac{x^2+y^2}{2} :\right).$$

We go back to the Hamiltonian of Equation (2.85) and we ask what is $Ring_\infty$? To get this, we must investigate the effect of the transfer map from s to $s+\Delta s$ on an arbitrary function g. The answer is obtained with the help of the differential equation for the map discussed in Interlude I. If the Hamiltonian H generates the map $\mathcal{M}$ then:

$$\begin{aligned} \mathcal{M}_{s\to s+ds} &\overset{1^{st}\text{ order in }ds}{=} \mathcal{I} + ds\ :-H: \\ &\overset{1^{st}\text{ order in }ds}{=} \exp\left(ds\ :-H:\right)\ . \end{aligned} \tag{2.89}$$

2.6.2 The Evolution of the Invariant

We are ready to use the result of Equation (2.89) in the derivation of famous formulas of linear accelerator theory. First of all, we would like to see how the parameters α, β, and γ evolve under the action of the quadratic part of the Hamiltonian of Equation (2.85). We define the ring $Ring_\infty$ associated with the Hamiltonian:

$$\begin{aligned} Ring_\infty &= \left\{\mathcal{M}_{s\to s+ds} \mid \mathcal{M}_{s\to s+ds} \overset{\text{To order }ds}{=} \mathcal{I} + \ :ds - H:\right\} \\ &\overset{\text{or}}{=} \left\{\mathcal{M}_{s\to s+ds} \mid \mathcal{M}_{s\to s+ds} \overset{\text{To order }ds}{=} e^{ds:-H:}\right\}. \end{aligned} \tag{2.90}$$

The $Ring_\infty$ is just the Hamiltonian. It contains all possible maps of interest. Using the same algebra as before let us compute the one-turn map at $s+ds$ as a function of the one-turn map at s:

$$\begin{aligned} \mathcal{M}_{s+ds} &= \mathcal{M}^{-1}_{s\to s+ds}\mathcal{M}_s\mathcal{M}_{s\to s+ds} \\ &\overset{\text{Eq. (2.89)}}{=} \exp\left(ds: H:\right)\mathcal{M}_s \exp\left(ds: -H:\right) \\ &\overset{\text{Eqs. (2.43,2.44,2.45)}}{=} \exp\left(ds: H:\right)\exp\left(: -\frac{\mu}{2}\varepsilon_s :\right)\exp\left(ds: -H:\right) \\ &\overset{\text{See exercise 2-9}}{=} \exp\left(: -\frac{\mu}{2}e^{ds:H:}\varepsilon_s :\right) \\ &\Downarrow \\ \varepsilon_{s+ds} &= e^{ds:H_2:}\varepsilon_s = \varepsilon_s + ds\left[H_2, \varepsilon_s\right]\cdots \end{aligned} \tag{2.91}$$

The quantity H_2 is the quadratic part of H. We can now compute the commutator $[H_2, \varepsilon_s]$:

$$\begin{aligned} \frac{d\varepsilon_s}{ds} &= [H_2, \varepsilon_s] \\ &= 2(\alpha\ H_{11} - \gamma\ H_{12})x^2 + 2(\beta H_{11} - \gamma H_{22})xp \\ &\quad +2(\beta H_{12} - \alpha H_{22})p^2 \\ &\Rightarrow \begin{cases} \frac{d\gamma}{ds} = 2(\alpha\ H_{11} - \gamma\ H_{12}) \\ \frac{d\alpha}{ds} = \ \beta H_{11} - \gamma H_{22} \\ \frac{d\beta}{ds} = 2(\beta H_{12} - \alpha H_{22}) \end{cases}. \end{aligned} \tag{2.92}$$

The standard accelerator result is regained using the special case of Equation (2.87):

$$\frac{d\gamma}{ds} = 2\alpha\ k(s)$$
$$\frac{d\alpha}{ds} = \beta k(s) - \gamma$$
$$\frac{d\beta}{ds} = -2\alpha. \tag{2.93}$$

This formula is a simple consequence of the definition of $Ring_\infty$ and the knowledge that the one-turn map is stable. Needless to say that the nonlinear lattice functions, i.e., the functional form of the invariant in the general case, obey similar rules. One simply replaces the first line of Equation (2.92) by

$$\frac{dI_s}{ds} = [H, I_s]. \tag{2.94}$$

Here I_s is a nonlinear invariant. This formula is only valid for an invariant. We will derive it first with map manipulation and then using Hamiltonian theory.

With Maps: In general, consider a quantity I_s which is invariant under $\mathcal{M}_s$:

$$I_s = \mathcal{M}_s I_s$$
$$\mathcal{M}^{-1}_{s\to s+\Delta} I_s = \underbrace{\mathcal{M}^{-1}_{s\to s+\Delta}\mathcal{M}_s\mathcal{M}_{s\to s+\Delta}}_{\mathcal{M}_{s+ds}}\mathcal{M}^{-1}_{s\to s+\Delta} I_s$$
$$\Downarrow$$
$$I_{s+\Delta} = \mathcal{M}^{-1}_{s\to s+\Delta} I_s. \tag{2.95}$$

In Equation (2.95), the time shift Δ can be finite. The result of Equation (2.94) is gotten by replacing Δ by ds.

With the Hamiltonian: It is instructive to regain this through the Hamiltonian formalism. First, let us define **very precisely** what is meant by an invariant function in the context of Hamilton's equations. Suppose I_s is an invariant (like the Courant-Snyder invariant) and suppose furthermore that the map, from an arbitrary location s_0 to s, is denoted by $\vec{\zeta}_{s_0\to s}$

$$\forall \vec{z}_0 \in \mathbf{R}^2 \quad \frac{d}{ds} I_s((\vec{\zeta}_{s_0\to s}))(\vec{z}_0) = 0$$

or

$$\frac{d}{ds}\left(I_s \circ \vec{\zeta}_{s0\to s}\right) \quad = \quad \text{null function} \;= 0. \qquad (2.96)$$

The proof that follows will force us to leave the domain of rings by assuming that the initial invariant function I_{s_0} at s_0 is arbitrary. There is nothing too strange for readers who know about linacs and other linear systems where one propagates "invariants" whose initial values are dictated by outside considerations. Thus we phrase the problem as follows: given an initial function I_{s_0}, how should its functional form I_s evolved so as to remain invariant? We then apply the chain rule to Equation (2.96):

$$\begin{aligned}
& \frac{d}{ds}\left(I_s \circ \vec{\zeta}_{s0\to s}\right) = 0 && (2.97.a)\\
\overset{\text{Equation (2.12)}}{\Rightarrow}\; & \frac{d}{ds}\left(\mathcal{M}_{s0\to s} I_s\right) = 0 && (2.97.b)\\
\Rightarrow\; & \left(\frac{d}{ds}\mathcal{M}_{s0\to s}\right) I_s + \mathcal{M}_{s0\to s}\frac{d}{ds} I_s = 0 && (2.97.c)\\
\overset{\text{Equation (2.27)}}{\Rightarrow}\; & \mathcal{M}_{s0\to s} : -H : I_s + \mathcal{M}_{s0\to s}\frac{d}{ds} I_s = 0 && (2.97.d)\\
\Rightarrow\; & \frac{d}{ds} I_s =: H : I_s && (2.97.e)\\
\overset{\text{if } I_s = \mathcal{N} I_{s0}}{\Rightarrow}\; & \frac{d}{ds}\mathcal{N} =: H : \mathcal{N}. && (2.97.f)
\end{aligned}$$

The last step from (2.97.e) to (2.97.f) requires us, as pointed out above, to assume that the initial invariant is arbitrary. Of course, in a ring we will seek the periodic solution, i.e., the particular I_{s_0} which is periodic $I_{s_0+C} = I_{s_0}$. Thus one can argue that the periodic solution is not arbitrary and thus may not necessarily obey Equation (2.97.f). However the propagation of an invariant function is a local matter. The system has no way of knowing (by causality), if the function being propagated is a periodic invariant or not. And indeed, if it is, one can always change the system further down (i.e., at a latter time or s) to destroy the periodicity. Therefore the periodic solution is not different from any other invariant solution and must obey Equation (2.97.f).[2.9] Finally, we notice that the equation for $\mathcal{N}$ is the equation for the inverse map $\mathcal{M}^{-1}_{s0\to s}$.

[2.9] This entire issue, pointed out to me by Alex Dragt, is rather messy. In the map approach, the rules for the propagation of functions (i.e., plug-in or composition) are built in axiomatically from the start. In the Hamiltonian

Exercise 2-24
Show that $\mathcal{N}$ is $\mathcal{M}_{s0\to s}^{-1}$, thereby completing the proof of Equation (2.97). *Hint: Take the derivative of $\mathcal{M}_{s0\to s}\mathcal{M}_{s0\to s}^{-1}$.*

2.6.3 Computation of the Phase Advance

Let us go back to the quadratic Hamiltonian H_2 and the Courant-Snyder transformation defined in Equation (2.75).The new Floquet ring $\mathcal{A}_{cs}\left[Ring_\infty\right]$ is given by the expression:

$$\mathcal{A}_{cs}\left[Ring_\infty\right] = \left\{\mathcal{R}_{s\to s+ds} | \mathcal{R}_{s\to s+ds} = \mathcal{A}_s e^{ds:-H:}\mathcal{A}_{s+ds}^{-1}\right\} \tag{2.98}$$

Writing $\mathcal{R}_{s\to s+ds}$ as $\exp(-d\mu : J :)$, we then compute the map $\mathcal{A}_{s+ds}$ to first order in ds:

$$\begin{gathered}\exp(-d\mu : J :) = \mathcal{A}_s e^{ds:-H:}\mathcal{A}_{s+ds}^{-1}\\ \Downarrow \\ \mathcal{A}_{s+ds} = \mathcal{A}_s - ds\mathcal{A}_s : H : + d\mu : J : \mathcal{A}_s.\end{gathered} \tag{2.99}$$

Up to this point, we have not used the fact that we interested in the Courant-Snyder transformation. This transformation is characterized by the vanishing A_{12} entry of its associated transfer map. To get this entry, we act on the function "x" with $\mathcal{A}_{s+ds}$ and we require the coefficient of the function "p" to be zero:

$$\begin{aligned}\mathcal{A}_{s+ds}x &= \mathcal{A}_s x - ds\mathcal{A}_s[H,x] + d\mu : J : \mathcal{A}_s x\\ &= \sqrt{\beta_s}x - ds\mathcal{A}_s(-H_{22}p - H_{12}x)] + d\mu\sqrt{\beta_s}[J,x]\\ &= \sqrt{\beta_s}x + dsH_{22}\left(\frac{p}{\sqrt{\beta_s}} - \frac{\alpha_s x}{\sqrt{\beta_s}}\right) + dsH_{12}\sqrt{\beta_s}x - d\mu\sqrt{\beta_s}p\\ &= \left(\sqrt{\beta_s} + ds\frac{-H_{22}\,\alpha_s + H_{12}\beta_s}{\sqrt{\beta_s}}\right)x + \underbrace{\left(\frac{dsH_{22}}{\sqrt{\beta_s}} - d\mu\sqrt{\beta_s}\right)}_{\text{Must be zero for the Courant-Snyder Functional}}p\end{aligned}$$

$$\Downarrow$$

$$\frac{d\mu}{ds} = \frac{H_{22}}{\beta_s}. \tag{2.100}$$

Of course this result is also obtainable from the usual matrix for the phase advance as shown here in Exercise 2-25.

---

language, one keeps rediscovering them. Also the map approach does not assume any differentiable form for the function I_s and thus we can propagate functions with few (fitted invariants) or no derivatives (delta functions).

Exercise 2-25

Using the result of Equation (2.84) in Exercise 2-22, derive the famous formula for the phase advance (Equation (2.100)).

Exercise 2-26 (The Anti-Courant-Snyder Transformation)

Of course, the phase advance is not unique as we said earlier. The next exercise illustrates this point.

1. *Select $\mathcal{A}$ such that*

$$A = \begin{pmatrix} A_{11} & A_{12} \\ 0 & A_{22} \end{pmatrix}$$
$$A_{11} > 0. \tag{2.101}$$

Show that the matrix for A is given by:

$$A = \begin{pmatrix} \frac{1}{\sqrt{\gamma}} & -\frac{\alpha}{\sqrt{\gamma}} \\ 0 & \sqrt{\gamma} \end{pmatrix}. \tag{2.102}$$

2. *Show that the anti-Courant-Snyder phase advance is given by:*

$$\frac{d\mu}{ds} = \frac{H_{11}}{\gamma_s}. \tag{2.103}$$

3. *Relate the two phase advances to one another. The answer is:*

$$\int_{s_1}^{s_2} \frac{H_{22}}{\beta_s} ds = \int_{s_1}^{s_2} \frac{H_{11}}{\gamma_s} ds + \tan^{-1}(\alpha)\Big|_{\alpha_2}^{\alpha_1}. \tag{2.104}$$

For two particular definitions of the phase advance, we see from Exercise 2-26.3 that matching is a sufficient, but not a necessary condition for the equality of the phase advances. There is no contradiction with our previous statements because a phase advance is unique and measurable by averaging if and only if all possible definitions lead to the same answer. This only happens between fully matched locations.

2.6.4 Connection to Standard Floquet Theory

Let us consider the map $\vec{\zeta}_{0\to s}$ from location $s = 0$ to some value s. We have seen in our discussion of the Floquet ring that this map can be expressed in terms of a "phase advance"

$$\vec{\zeta}_{0\to s} = \vec{\alpha}_s \circ \vec{\rho}_{0\to s} \circ \vec{\alpha}_0^{-1} \tag{2.105}$$

whenever a total normalization is possible. Let us rewrite Equation (2.105) in terms of compositional maps and massage this expression using the knowledge gained in this chapter

$$\begin{aligned}
\mathcal{M}_{0\to s} &= \mathcal{A}_0^{-1} \underbrace{\mathcal{R}_{0\to s}}_{\substack{\exp\left(:\, H_{0\to s}(J)\, :\right) \\ \text{Phase Advance}}} \mathcal{A}_s \\
&= \mathcal{A}_0^{-1} \exp\left(:H_{0\to s}:\right) \mathcal{A}_s \\
&= \mathcal{A}_0^{-1} \exp\left(:H_{0\to s}:\right) \mathcal{A}_0 \mathcal{A}_0^{-1} \mathcal{A}_s \\
&= \exp\left(\mathcal{A}_0^{-1} : H_{0\to s} : \mathcal{A}_0\right) \mathcal{A}_0^{-1} \mathcal{A}_s \\
&= \exp\left(: H_{0\to s} \circ \vec{\alpha}_0^{-1} :\right) \mathcal{A}_0^{-1} \mathcal{A}_s .
\end{aligned} \tag{2.106}$$

This last equation is the usual Floquet theorem stated in a way appropriate to accelerator physics. Here, nothing is said about the variable s which could be a discontinuous variable (as it should be in rings) or a continuous parameter. To get the usual Floquet theorem we invent a continuous variable σ using the one turn map:

$$\begin{aligned}
\mathcal{M}_{0\to\sigma} = {} & \exp\left(: \frac{\sigma}{C} H_{0\to\aleph} \circ \vec{\alpha}_0^{-1} :\right) \exp\left(: -\frac{\sigma}{C} H_{0\to\aleph} \circ \vec{\alpha}_0^{-1} :\right) \\
& \times \exp\left(: H_{0\to s(\sigma)} \circ \vec{\alpha}_0^{-1} :\right) \mathcal{A}_0^{-1} \mathcal{A}_{s(\sigma)} .
\end{aligned} \tag{2.107}$$

It suffices that the variable σ, which is defined from $\sigma = 0$ to a pseudo-circumference of $\sigma = C$, reproduced the correct values of s for each of the $\aleph$ surfaces of section defined in $Ring_{\aleph}$. When defined in that manner, the phase advance is truly an arbitrary concept with little physical meaning. Nevertheless, this ties our treatment with the usual Floquet or Bloch theorem. Using Michelotti's notation(Section 2.1.2 of reference [11]), we rewrite Equation (2.107) as

$$\begin{aligned}
& \mathcal{M}_{0\to\sigma} = \exp\left(\sigma \widehat{C}\right) \mathcal{Q}(\sigma) \\
& \text{where} \quad \begin{cases} \widehat{C} = :\frac{1}{C} H_{0\to\aleph} \circ \vec{\alpha}_0^{-1} : \\ \mathcal{Q}(\sigma) = e^{-\sigma \widehat{C}} e^{:H_{0\to s(\sigma)} \circ \vec{\alpha}_0^{-1}:} \mathcal{A}_0^{-1} \mathcal{A}_{s(\sigma)} \end{cases} \\
& \text{and} \quad \mathcal{Q}(\sigma + C) = \mathcal{Q}(\sigma).
\end{aligned} \tag{2.108}$$

These maps are compositional maps—the associated transfer maps will be

$$\vec{\zeta}_{0\to\sigma} = \vec{Q}(\sigma) \circ \left\{ \exp\left(\sigma\hat{C}\right)\vec{I} \right\}. \tag{2.109}$$

For linear maps, there exists a matrix representation for each map in Equation (2.109), and therefore

$$M_{0\to\sigma} = Q(\sigma)\exp\left(\sigma C\right), \tag{2.110}$$

where M, C and Q are matrices.

As we have said in the introduction, standard accelerator theory boils down to a "Bloch wave" analysis. The Lie transforms and its companion the compositional map allow these concepts to be extended to nonlinear problems. To our knowledge, Equation (2.109) was first derived by Dragt[1] in the context of nonlinear compositional maps. This equation simply requires the introduction of a continuous pseudo-time like σ or the path length of a Frenet-Serrat curve; I do not like this because it violates our magnet metaphor and the theory does not require it anyway. More fundamentally, the introduction of an arbitrary σ is equivalent to allowing the map $\mathcal{A}_s$ to depend on things other than the one-turn map at s. Therefore, the phase advance between matched sections ceases to have any special meaning. This is something I find unacceptable in a clean and elegant theory. In his book, Michelotti points out repeatedly that Floquet theory applies to matrices of more than one dimension. By that statement he includes the 3 degrees of freedom of the full linear problem. But, it also means that the nonlinear map (if normalizable) can be put in Floquet form as indicated here. In Sect. 4.3, the matrix representation of the compositional map, restricted to a truncated power series of degree k (k-jets), is described in detail. This matrix representation ties together our Floquet theory of nonlinear maps, standard Hamiltonian perturbation theory and standard Floquet theory. As I have said before our methods extend naturally linear methods (Floquet) to nonlinear problems, and preserve the Hamiltonian tools (Lie methods).

Conclusion

In this chapter, we have seen how one starts with a stable symplectic map. We have seen that some problems dictate the observation of a ring at several places. Therefore we developed the concept of the *Ring*$_\aleph$. We have also shown that the usual Hamiltonian theory is the limiting case of $\aleph$ tending towards infinity. Our methods involved maps only and therefore are perfectly suited to systems which are described by a very discontinuous Hamiltonian for which the limit $\aleph \to \infty$ is impractical. Armed with that knowledge, we will now examine, in Chapter 3, local aspects of the dynamics.

CHAPTER 3
FROM THE HAMILTONIAN TO THE MAP

In Chapter 2 we have assumed the existence of a stable machine (or a marble gutter) and we proceeded to develop the formal tools necessary for the description of such a system. This approach is consistent with modern treatments of nonlinear dynamics. However, in many situations, the reverse point of view is necessary. One starts with the elemental building blocks of a ring and strings the beast together. This situation is encountered when one builds the actual ring or when one writes a simulation code capable of handling any situation. In the real world, there is no guarantee that the ring will be stable or that a closed orbit exists. Indeed, it is common to mispower, misalign, interchange electric leads, etc...As a result, the ring will not be stable when we first turn it on. There are other situations we may want to simulate which are linearly unstable.

It is clear that our code, like the machine, should rely on a purely local formalism which is independent of the final primary purpose of the ring namely *confinement*. It is the purpose of Chapter 3 to display the ideas of local integration for one dimensional examples.

The reader will see that local integration is pictorial even in theory. This fact casts some doubts on the necessity to have a "pictorial" version of Part III, but for completeness, let us go ahead. To avoid any reference to "non-dynamical" topics such as Maxwell's equations, we will examine the "Hot Wheel" racetrack problem.

This chapter will contain "advanced" topics which do not belong to rings exclusively: Lagrangians, change of independent variables, etc. all are standard topics which I put here so as to keep the flow of Part III uninterrupted.

3.1 HAMILTONIAN OF A SECTION OF GUTTER

In Sect. 3.1, we review through the gutter example some elements of Lagrangian and Hamiltonian dynamics. Our ultimate goal is to write a Hamiltonian for a section of gutter which uses distance along the gutter as the independent variable.

3.1.1 Lagrangian and Hamiltonian

Our starting point is the usual Lagrangian of Newtonian dynamics. The reader can glance at Interlude V for a theoretical review of the Lagrangian and Hamiltonian formalism if he wishes.

INTERLUDE V
FROM LAGRANGE TO HAMILTON

Lagrangian dynamics is derived from a least action principle. The action is defined as a functional in the space of path based on the Lagrangian:

$$A = \int_{t_0}^{t_1} L dt. \tag{3.1}$$

For the record, we remind the reader that the equations of motion are obtained by extremizing A while maintaining the end point values of the position constant.

$$\begin{aligned} \delta A &= \int_{t_0}^{t_1} \delta L \, dt \\ &= \int_{t_0}^{t_1} \left\{ L(\vec{x} + \delta\vec{x}, \dot{\vec{x}} + \delta\dot{\vec{x}}) - L \right\} dt \\ &= \int_{t_0}^{t_1} \left\{ \frac{\partial L}{\partial \vec{x}} \cdot \delta\vec{x} + \frac{\partial L}{\partial \dot{\vec{x}}} \cdot \delta\dot{\vec{x}} \right\} dt \\ &= \int_{t_0}^{t_1} \left\{ \frac{\partial L}{\partial \vec{x}} - \frac{d}{dt}\frac{\partial L}{\partial \dot{\vec{x}}} \right\} \cdot \delta\vec{x} \, dt + \underbrace{\left[\frac{\partial L}{\partial \dot{\vec{x}}} \cdot \delta\vec{x} \right]_{t_0}^{t_1}}_{=0}. \end{aligned} \tag{3.2}$$

It is easy to check that the integrand of Equation (3.2) vanishes for an arbitrary $\delta\vec{x}$ if and only if

$$\frac{\partial L}{\partial \vec{x}} - \frac{d}{dt}\frac{\partial L}{\partial \dot{\vec{x}}} = 0. \tag{3.3}$$

Of course, this is what we said earlier. In the Lagrangian view of dynamics, the equations of motion are obtained from an extremal principle in configuration space. To get the Hamiltonian, we define a Legendre transformation on L as

$$\begin{aligned} H(\vec{x},\vec{p};t) &= \vec{p}\cdot\dot{\vec{x}} - L(\vec{x},\dot{\vec{x}};t) \\ \text{where } \vec{p} &= \frac{\partial L}{\partial \dot{\vec{x}}} \,. \end{aligned} \tag{3.4}$$

Exercise 3-1

By computing the total differential dH of H in two different ways, show the usual result of Hamiltonian dynamics:

$$\frac{d\vec{x}}{dt} = \frac{\partial H}{\partial \vec{p}}, \quad \frac{d\vec{p}}{dt} = -\frac{\partial H}{\partial \vec{q}}, \quad \textit{and} \quad \frac{dH}{dt} = \frac{\partial H}{\partial t} = -\frac{\partial L}{\partial t}. \tag{3.5}$$

END OF INTERLUDE

Consider a nonrelativistic particle in a constant gravitational field. The Lagrangian is given by

$$L = \frac{1}{2}(\dot{x}^2 + \dot{y}^2 + \dot{z}^2) - y. \tag{3.6}$$

The equations of motion are given by the usual rules:

$$\begin{aligned} \frac{d}{dt}\frac{\partial L}{\partial \dot{x}} &= \frac{\partial L}{\partial x} \longrightarrow \ddot{x} = 0 \\ \frac{d}{dt}\frac{\partial L}{\partial \dot{y}} &= \frac{\partial L}{\partial y} \longrightarrow \ddot{y} = -1 \\ \frac{d}{dt}\frac{\partial L}{\partial \dot{z}} &= \frac{\partial L}{\partial z} \longrightarrow \ddot{z} = 0. \end{aligned} \tag{3.7}$$

A Hamiltonian can be obtained by the usual Legendre transformation on the Lagrangian:

$$\begin{aligned} \text{given } p_i = \frac{\partial L}{\partial \dot{x}_i} \;\rightarrow\; & H(\vec{p}) = \vec{p}\cdot\dot{\vec{x}}(\vec{p}) - L(\vec{x},\vec{p}) \\ & \Downarrow \\ & H = \frac{1}{2}(p_x^2 + p_y^2 + p_z^2) + y. \end{aligned} \tag{3.8}$$

These results should be well known to the reader. Let us proceed by defining a new variable u by the relation

$$y = u + f(x, z). \tag{3.9}$$

At this stage the function $f(x, z)$ need not have any physical meaning. We now express the Lagrangian in terms of (x, u, z):

$$L(x, \dot{x}, u, \dot{u}, z, \dot{z}) = \frac{1}{2}[\dot{x}^2 + (\dot{u} + f_x\dot{x} + f_z\dot{z})^2 + \dot{z}^2] - u - f,$$
$$\text{where } f_q = \frac{\partial f}{\partial q}. \tag{3.10}$$

Lagrangian dynamics is based on the minimization of a functional, the action $\int_{t_0}^{t_1} Ldt$ (see Interlude V), and therefore, in the new variables (x, u, z) the equations of motion are obtained by identical rules.

To proceed further, we identify the function $f(x, z)$ as the gutter function providing the holonomic constraint $y = f$. This constraint is simply $u = 0$. Following Arnold[12], this system is governed by the Lagrangian

$$\mathcal{L} \equiv L|_{u=\dot{u}=0} \quad . \tag{3.11}$$

This can be seen by adding a "constraining potential" $\frac{1}{2}\Omega u^2$ to the original Lagrangian L and letting the stiffness Ω tend towards infinity. Therefore our "gutter" Lagrangian is just:

$$\mathcal{L}(x, \dot{x}, z, \dot{z}) = \frac{1}{2}[\dot{x}^2 + (f_x\dot{x} + f_z\dot{z})^2 + \dot{z}^2] - f. \tag{3.12}$$

We can write a Hamiltonian for this constrained Lagrangian using the usual Legendre transformation:

$$H = \frac{1}{2}\frac{p_x^2(1 + f_z^2) + p_z^2(1 + f_x^2) - 2f_xf_zp_xp_z}{1 + f_x^2 + f_z^2} + f. \tag{3.13}$$

Exercise 3-2

The usage of x, y and z is common in a Cartesian geometry. This is not necessarily convenient if the gutter is "donut" shaped in order to bend the marble trajectory. Instead consider the cylindrical frame of reference (r, y, s) defined as

$$x = r\ \cos(\frac{s}{r_0})$$
$$z = r\ \sin(\frac{s}{r_0}). \tag{3.14}$$

We assume that the gutter is along the s direction centered around some value of r denoted by r_0. The constraint equation is then:

$$\begin{aligned} u &= 0 \\ \textit{where } y &= u + f(r, s). \end{aligned} \tag{3.15}$$

1. *Show that the Lagrangian of the constrained motion is*

$$L(r, \dot{r}, s, \dot{s}) = \frac{1}{2}[\dot{r}^2 + (f_r\dot{r} + f_s\dot{s})^2 + \frac{r^2}{r_0^2}\dot{s}^2] - f. \tag{3.16}$$

2. *Show that the Hamiltonian is given by*

$$H = \frac{1}{2}\frac{p_r^2(1 + \frac{r_0^2}{r^2}f_s^2) + \frac{r_0^2}{r^2}p_s^2(1 + f_r^2) - 2\frac{r_0^2}{r^2}f_r f_s p_r p_s}{(1 + f_r^2) + \frac{r_0^2}{r^2}f_s^2} + f. \tag{3.17}$$

3.1.2 A Coordinate as the Independent Variable

Surely most readers must have played with or seen train sets or race car sets at least once in their lives. These sets are characterized by a collection of tracks which fit snugly into one another. This can be achieved by assuming that the constraining equation at both ends of a piece of gutter is uniquely determined for all pieces of gutter, let us say by $y = f_\infty(x)$. Hence a gutter of length D ($0 < z < D$) can be represented as follows:

$$f(x, z) = f_0(x) + \sum_{n=1}^{\infty} f_n(x) \cos(2\pi n\frac{z}{D})$$

subject to

$$f_\infty(x) = f_0(x) + \sum_{n=1}^{\infty} f_n(x). \tag{3.18}$$

In Equation (3.18) and in the following discussion, the variables (x, z) can be replaced by (r, s) if we study a circular gutter. The absence of sine functions in the expansion insures that the gutter's edges will fit into one another and, in particular, that there is no kink at the edges, i.e., we have a zero slope at $z = 0, D$. Great![3.1] Going back to

[3.1] This condition is not necessary; it is analogous to a drift tube in an accelerator. Accelerators themselves are surrounded by the earth's magnetic field which, at low energy, could be non negligible.

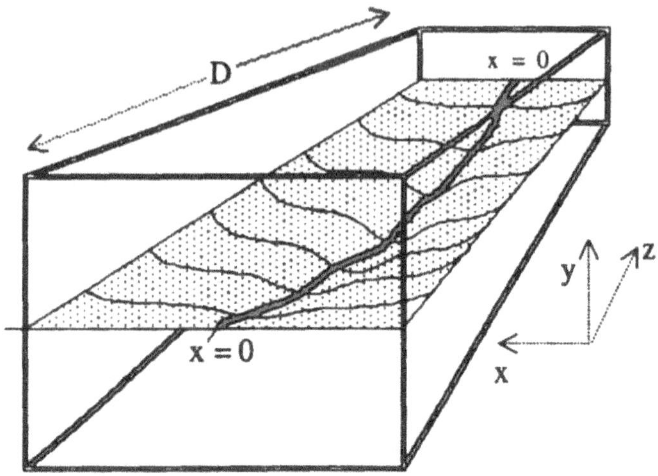

Figure 3.1: Geometry of a Gutter.

our intuition, it is clear that visually we know if a marble is within a given section of the gutter if we see it in it! Translated mathematically this indicates to us that a Lagrangian or Hamiltonian parameterized by the time t is not very suitable when integrating a system made of pieces of gutter which are best described by a local equation such as Equation (3.18). The same is true of magnets. We know when a particle enters the field of an electro-magnetic element by knowing its position relative to some entrance and exit plane. For example, the boundaries of a quadrupole are two planes perpendicular to its axis of symmetry. The equivalent of the gutter boundary condition $y = f_{\infty}(x)$ is simply the existence of a field-free region at both ends of the magnet (we remind the reader that the magnet is a metaphor whose exact translation into a dynamical theory is possible if and only if its fringe fields are localized; see the preface and Part III.) As we said earlier, the electro-magnetic case may be simpler mathematically than the "pictorial" gutter analogy. We are now poised to ask a fundamental question: can we use the variable z (or s) to parameterize the gutter and is this new motion described by some Hamiltonian? Let us proceed first with exercises followed by a theoretical interlude.

Exercise 3-3

This first set of exercises adopts the Lagrangian point of view. The reader may want to use the examples of Equations (1.3) *and* (1.5) *as a guide (see Chapter* 1).

1. *Going back to the results of Interlude* V, *it is clear that one must minimize the action (Equation* (3.1)). *Let us assume that in some region of state space* $(x, \dot{x}, y, \dot{y}, z, \dot{z}, t)$ *the function* $z(t)$ *is monotonic. This is clearly the case of a marble wandering along the gutter or that of a particle passing through a magnet. What is the z-parameterized Lagrangian* $\overline{L}$? *Answer, denoting by a prime the derivative with respect to* z:

$$\overline{L}(x, x', y, y', t, t', z) = L(x, \frac{x'}{t'}, y, \frac{y'}{t'}, z, \frac{1}{t'}, t)\; t'. \quad (3.19)$$

2. *Show that the new canonical momenta are given by*

$$\begin{aligned} \overline{p}_x &= \frac{\partial \overline{L}}{\partial x'} = \frac{\partial L}{\partial \dot{x}} = p_x \\ \overline{p}_y &= \frac{\partial \overline{L}}{\partial y'} = \frac{\partial L}{\partial \dot{y}} = p_y \\ \overline{p}_t &= \frac{\partial \overline{L}}{\partial t'} = L - \frac{1}{t'}x'\frac{\partial L}{\partial \dot{x}} - \frac{1}{t'}y'\frac{\partial L}{\partial \dot{y}} - \frac{1}{t'}\frac{\partial L}{\partial \dot{z}} \\ &= L - \vec{p}\cdot\dot{\vec{x}} = -H. \end{aligned} \quad (3.20)$$

3. *Show that it follows that the new Hamiltonian* $\overline{H}$ *is given by*

$$\overline{H} = \vec{\overline{p}}\cdot\vec{x}' - \overline{L} = -p_z. \quad (3.21)$$

In Exercise 3-3 the Lagrangian is used to provide a direct proof. This proof has a certain elegance[9]. In the next exercise, a pure brute force "Hamiltonian" proof is suggested. The most elegant proof will follow in Interlude VI.

Exercise 3-4 (Brute Force on the Hamiltonian)

Consider a Hamiltonian $H(\vec{x}, \vec{p}; t)$. *Consider the following equation as defining* p_t:

$$\begin{aligned} p_t &= -H(x_1, \vec{x}^{>}, p_1, \vec{p}^{>}; t) \\ \text{where } \vec{x}^{>} &= (x_2, \cdots, x_N) \\ \text{and } \vec{p}^{>} &= (p_2, \cdots, p_N). \end{aligned} \quad (3.22)$$

Consider a new set of variables $(t, \vec{x}^{>}, p_t, \vec{p}^{>})$ replacing the original phase space.

1. *Show that the equation of motion for the new variables in terms of x_1 are:*

$$\begin{aligned}
\frac{dt}{dx_1} &= \left(\frac{\partial H}{\partial p_1}\right)^{-1} \\
\frac{dp_t}{dx_1} &= -\left(\frac{\partial H}{\partial t}\right)\left(\frac{\partial H}{\partial p_1}\right)^{-1} \\
\frac{dx_i}{dx_1} &= \left(\frac{\partial H}{\partial p_i}\right)\left(\frac{\partial H}{\partial p_1}\right)^{-1} \\
\frac{dp_i}{dx_1} &= -\left(\frac{\partial H}{\partial x_i}\right)\left(\frac{\partial H}{\partial p_1}\right)^{-1} .
\end{aligned} \tag{3.23}$$

2. *What happens if $\frac{\partial H}{\partial p_1}$ vanishes? If x_1 is the length along the gutter, what is the physical significance of this? Partial answer: The marble must go from $z = 0$ to $z = D$ without reversing direction. In other words it is not trapped in the gutter. In the case of a magnet, the magnetic field is weak. The particle cannot reverse direction.*

3. *Define implicitly using Equation (3.22) a new function K by the relation $K = -p_1(t, \vec{x}^{>}, p_t, \vec{p}^{>}; x_1)$. Show by brute force that K is the Hamiltonian for the motion of $(t, \vec{x}^{>}, p_t, \vec{p}^{>})$ as explicitly given in Equation (3.23). Hint: Compute the total differential dK.*

Using the results of Exercise 3-3 we can derive the z-dependent Hamiltonians corresponding to Equations (3.13) and (3.17). They are given by the equations

$$\begin{aligned}
K_x &= \frac{-\sqrt{-2(1+f_x^2)p_t - 2f(1+f_x^2) - p_x^2}\sqrt{1+f_x^2+f_z^2} - f_x f_z p_x}{1+f_x^2} \\
K_r &= \frac{-\frac{r}{r_0}\sqrt{-2(1+f_r^2)p_t - 2f(1+f_r^2) - p_r^2}\sqrt{1+f_r^2+\frac{r_0^2}{r^2}f_s^2} - f_r f_s p_r}{1+f_r^2}
\end{aligned} \tag{3.24}$$

These Hamiltonians are quite ugly and, as we pointed out, they turn out to be simpler in the magnetic case. The reader will notice that we regain the Cartesian gutter if the ratio $\frac{r}{r_0}$ is nearly one. This corresponds to gutters which bend the marble by a very small angle. More precisely, the radius of curvature of the gutter is large compared to the oscillations (Δr) within the gutter. In accelerators, this is the case of large machines. We explore this large gutter limit in the next exercise.

Exercise 3-5

1. *Suppose that a Hamiltonian K has a trajectory $\vec{z}_0(t) = (\vec{x}_0, \vec{p}_0)$. Furthermore, suppose that we are interested in deviations from this trajectory and that we define a deviation variable $\vec{\zeta} = \vec{z} - \vec{z}_0(t)$. Show that the Hamiltonian for the motion of $\vec{\zeta}$ is given by $K^d(\vec{\zeta}) = K(\vec{\zeta} + \vec{z}_0(t)) - \vec{\zeta} \cdot \vec{\nabla} K_{\vec{z}_0}$. In other words, the Hamiltonian in deviation variables is the old one expressed in terms of the new variable minus its linear part. This ensures that the origin of phase space is sent into itself, as one expects.*

2. *Neglecting boundary conditions ($f_s = 0$) show that, for a pure racetrack bending gutter ($f = f'r$), a design particle of kinetic energy K_0 is at equilibrium at $r = r_0$ if*

$$K_0 = \frac{1}{3} H_0 = \frac{1}{2} r_0 f'. \tag{3.25}$$

3. *At equilibrium, neglecting the time of flight information (i.e., set $p_t = -H_0$), show that the linear motion around the equilibrium orbit is given by the Hamiltonian*

$$\begin{aligned} K_2 &= \frac{1}{2(1+f'^2)\sqrt{2K_0}} p_r^2 + \frac{9f'^2}{8\sqrt[3]{3K_0}} \Delta r^2 \\ \Delta r &= r - r_0. \end{aligned} \tag{3.26}$$

The reader will notice that Equation (3.26) contains a small amount of focusing proportional to the square of the gutter curvature. This result is easily explained as follows: if a marble starts above the design trajectory $r = r_0$ with $p_r = 0$ it will fall towards the design orbit because it lacks kinetic energy. The centripetal acceleration is less that the gravitational pull. Conversely, if it starts below, it will move upwards because the gravitational pull is insufficient. Amazingly, this type of focusing is found in sector bends in circular rings. There, as we will see in Part III, the issue is somewhat more confusing because this effect vanishes in parallel face bends to be replaced, in the vertical plane, by an effect of equal magnitude (to first order) due to Maxwellian fringe field effects.

We conclude Sect. 3.1 with a theoretical interlude which rephrases the results of this section in a more geometrical way. In particular, we get a very elegant demonstration of the formula for the position-parameterized Hamiltonian. As we will see, it follows from the Lagrangian techniques of Exercise 3-3 by generalizing the action principle to the full state space—positions, momenta and time are varied independently of one another.

INTERLUDE VI
CHANGING THE INDEPENDENT VARIABLE

Following Arnold[12], we generalize the action principle as follows:

$$
\begin{aligned}
\delta A = \delta \int_{t_0}^{t_1} L dt &= \delta \int_{t_0}^{t_1} \left(\vec{p}\cdot\dot{\vec{x}} - H\right)\, dt \Big|_{\delta p = \delta \frac{\partial L}{\partial \dot{x}}} \\
?\Downarrow ? & \qquad\qquad\qquad\qquad (3.27) \\
&= \delta \int_{t_0}^{t_1} \left(\vec{p}\cdot\dot{\vec{x}} - H\right)\, dt = \delta \int_{\vec{x}_0,t_0}^{\vec{x}_1,t_1} \vec{p}\cdot d\vec{x} - H dt \Big|_{\delta x,\, \delta p,\, \delta t \text{ arbitrary}}.
\end{aligned}
$$

The first line of Equation (3.27) is the least action principle in configuration space. The last line is a generalized least action principle in extended phase space. This principle, if true, will provide an immediate derivation of the position-dependent Hamiltonian needed for the gutter (Equation (3.24)) or the magnetic ring. The proof of (3.27) is simple:

$$
\begin{aligned}
\delta A &= \delta \int_{\vec{x}_0,t_0}^{\vec{x}_1,t_1} \vec{p}\cdot d\vec{x} - H dt \\
&= \int \delta\vec{p}\cdot d\vec{x} + \vec{p}\cdot d\delta\vec{x} - \delta H dt - H d\delta t \\
&= \vec{p}\cdot\delta\vec{x} - H\delta t \Big|_{\vec{x}_0,t_0}^{\vec{x}_1,t_1} \\
&\quad + \int \left(\delta\vec{p}\cdot d\vec{x} - d\vec{p}\cdot\delta\vec{x} - \tfrac{\partial H}{\partial \vec{x}}\delta\vec{x}dt - \tfrac{\partial H}{\partial \vec{p}}\delta\vec{p}dt - \tfrac{\partial H}{\partial t}\delta t dt + \delta t dH\right)dt \\
&= \int \Big(\underbrace{\left\{d\vec{x} - \tfrac{\partial H}{\partial \vec{p}}dt\right\}}_{=0}\cdot\delta\vec{p} - \underbrace{\left\{d\vec{p} + \tfrac{\partial H}{\partial \vec{x}}dt\right\}}_{=0}\cdot\delta\vec{x} + \underbrace{\left\{dH - \tfrac{\partial H}{\partial t}dt\right\}}_{=0}\delta t\Big)dt \\
&= 0. \qquad\qquad (3.28)
\end{aligned}
$$

To reparameterize the motion in terms of a new variable, let us say x_1, we suppose that we have solved the equation of motion within a region of phase space such that $\frac{\partial H}{\partial p_1}$ is always of the same sign. Then according to the implicit function theorem, we can solve for p_1 is in terms of the other variables $(t, H, \vec{x}^{>}, \vec{p}^{>})$ as defined in Equation (3.22). Within the region of extended phase space where the inversion is possible, we can minimize the action while viewing p_1 as a function of the other coordinates $(t, H, \vec{x}^{>}, \vec{p}^{>}; x_1)$. Clearly this amounts to a relabeling and the new Hamiltonian for the x_1-dependent motion is therefore $K = -p_1$.
END OF INTERLUDE

3.2 APPROXIMATIONS OF INDIVIDUAL SECTIONS

3.2.1 Types of Elements: Gutter and Magnets

Of course, each type of magnet or gutter will have a primary purpose. For example, the circular gutters will bend the marble, the straight gutter will propagate it and focus it transversely. In the design stage of a racetrack one uses this approximate description of the motion through an element to piece together a tentative design. Quite often this analysis is restricted to the linear (paraxial) properties of the gutter segments (or magnets). In the case of modern rings, here are a few types of magnets encountered:

1. Bends: bend the particles so as to close the ring.

2. Quadrupoles: provide transverse focusing and defocusing. If paired with alternating signs, they provide overall focusing.

3. Drifts: unavoidable.

4. Radio-Frequency (RF) Cavities: to provide longitudinal focusing and acceleration if needed.

5. Sextupoles: provide energy dependent focusing (chromaticity) to cancel the energy dependence of the quadrupole focusing. Strong nonlinearities are introduced in the process.

6. Solenoid: used in colliders near the particle detectors. They usually couple the horizontal and vertical planes.

7. Skew Quadrupoles: decouple the motion due to solenoids and error tilts in other elements.

8. Wigglers and Undulators are strings of bends with alternating signs which wiggle the beams usually to produce light.

9. Higher Order Multipoles are stuff which we don't want but is normally present because of construction errors.

10. Fringe Field Effect: You can't avoid Maxwell's equations. This is important in small machines and modern colliders such as B-factories.

11. Etc...

Of course, the real purpose of these magnets belongs to a treatise on the physics and design of accelerators. We invite the reader to the appropriate references. To understand the basic concepts of machine design the first five items in the list suffice. In fact, for transverse stability one can concentrate on the first three items as is done in many introductory books. Of course, our gutter analogy has a similar set of elemental pieces. Suppose we concentrate on gutters with the boundary condition $f_\infty = 0$. Then our track set will contain the following classes of elements:

1. Bend: a gutter which bends the marble by an angle $\frac{D}{r_0}$.

2. Convex gutter: this is displayed in Figure 3.1.

3. Concave gutter: for those who like to play with fire.

4. Drift: a flat board, $f(x,z) = f_\infty = 0$.

The magnet designer as well as the racetrack designer will need a simple model to understand approximately the purpose of his building blocks. (We will neglect the time variable and assume a monoenergetic marble gutter. The reader can rederive the full two degrees of freedom problem.) Assuming large gutters, we already have an approximate result for the pure bend. Namely, it is a drift:

$$K_2 = \frac{1}{2(1+f'^2)\sqrt{2K_0}}p_r^2 + \frac{9f'^2}{8\sqrt[3]{3K_0}}\Delta r^2 \cong \frac{1}{2\sqrt{2K_0}}p_r^2. \quad (3.29)$$

The relationship between the incoming and the outgoing ray is given by:

$$\begin{pmatrix} r \\ p_r \end{pmatrix}^{fin} = \begin{pmatrix} 1 & \frac{D}{\sqrt{2K_0}} \\ 0 & 1 \end{pmatrix} \begin{pmatrix} r \\ p_r \end{pmatrix}^{in}. \quad (3.30)$$

In the case of the straight focusing gutter, we expand the Cartesian Hamiltonian in Equation (3.24) to leading order in f:

$$K_x \cong \frac{p_x^2}{2\sqrt{2K_0}} + \frac{f}{\sqrt{2K_0}} + \cdots \tag{3.31}$$

Equation (3.31) is a large gutter approximation. Again, we will have a similar approximation in the case of large circular rings. To get an idea of the motion produced by this Hamiltonian, we can average the function f over z. Such a thing is often done for wigglers and undulators in rings.

$$\begin{aligned}\langle K_x \rangle_z &= \frac{p_x^2}{2\sqrt{2K_0}} + \frac{f_0}{\sqrt{2K_0}} \\ &= \frac{p_x^2}{2\sqrt{2K_0}} + \frac{1}{\sqrt{2K_0}} \sum_{n=2}^{\infty} \frac{f_0^{[n]}}{n!} x^n.\end{aligned} \tag{3.32}$$

We see that the averaged Hamiltonian of the straight gutter contains many terms. Clearly if $f_0^{[2]}$ is positive definite it will provide focusing. The matrix for this element is given by

$$\begin{pmatrix} r^{fin} \\ p_r^{fin} \end{pmatrix} = \begin{pmatrix} \cos(\frac{D\sqrt{f_0^{[2]}}}{\sqrt{2K_0}}) & \frac{\sin(\frac{D\sqrt{f_0^{[2]}}}{\sqrt{2K_0}})}{\sqrt{f_0^{[2]}}} \\ -\sqrt{f_0^{[2]}}\sin(\frac{D\sqrt{f_0^{[2]}}}{\sqrt{2K_0}}) & \cos(\frac{D\sqrt{f_0^{[2]}}}{\sqrt{2K_0}}) \end{pmatrix} \begin{pmatrix} r^{in} \\ p_r^{in} \end{pmatrix}.$$

Using the matrices we just derived the "designer" can start putting the gutter together. Of course, gutters are different from magnets. It is possible to design a gutter with positive definite curvature (convex everywhere). Such gutters would be stable under perturbation even if the paraxial motion is on the integer tune. Magnets are more interesting since quadrupoles are forced by nature to be focusing in one plane and defocusing in the other plane. We will write down without proof an ultrarelativistic Hamiltonian for large machines.

$$H = \frac{p_x^2 + p_y^2}{2(1+\delta)} + \frac{x^2}{2r_0^2} - \frac{x\delta}{r_0} + k_2\frac{x^2 - y^2}{2} + V(x,y). \tag{3.33}$$

In Equation (3.33), the quantity r_0 is the design bending radius of a bend and k_2 is the quadrupole component of the element. We lump all other multipoles in the potential $V(x,y)$. Notice that it is not possible for a bend to provide focussing in both transverse planes in a very large machine. This magnet Hamiltonian contains the variable δ which is the total momentum deviation. In the ultra-relativistic case, this is the same as the energy deviation and therefore this variable is conjugate to time which is just path length in this case. The reader will notice that the origin of phase space is sent onto itself.

Using the results we derived for the gutter, we can write down an equivalent averaged large gutter Hamiltonian for the bend-convex-concave gutter:

$$H = \frac{1}{2(1+f'^2)\sqrt{2K_0}} p_r^2 + \frac{9f'^2}{8\sqrt[3]{3K_0}} \Delta r^2 + \sum_{n=2}^{\infty} \frac{f_0^{[n]}}{n!} x^n. \tag{3.34}$$

Exercise 3-6

The correct form of the approximate Hamiltonian of Equation (3.33) *in three degrees of freedom is*

$$\begin{aligned} H \;=\; & \frac{p_x^2 + p_y^2}{2(1 - \frac{2p_t}{\beta_0} + p_t^2)} - \frac{x p_t}{\beta_0 r_0} + \frac{x^2}{2r_0^2} + \frac{(1-\beta_0^2)p_t^2}{2\beta_0^2} \\ & + k_2 \frac{x^2 - y^2}{2} + V(x, y). \end{aligned} \tag{3.35}$$

Where did the $\frac{(1-\beta_0^2)p_t^2}{2\beta_0^2}$ *come from? Hint: Send a nonrelativistic particle down the design axis of a drift tube, what can be said about the dependence of the path length on the energy? What can be said about the time of flight? What about the ultra-relativistic case?*

While the gutter problem is essentially a "joke," the approximate circular ring Hamiltonian of Equation (3.35) is not. In fact it is at the root of serious simulation codes found in accelerators. It will be discussed in Chapter 14.

3.2.2 Connecting Gutters or Magnets: LEGO Blocks

Before describing the techniques needed to approximate or compute the motion through a segment of gutter, we need to describe in more detail how the various segments are put together. This will allow us to treat the misplacement or misalignment of a segment or magnet. This is essential in writing a realistic simulation code.

A local description of a ring forces us to invent a theoretical framework where all the magnets are objects which can be moved around independently of one another. As we said previously, there are two

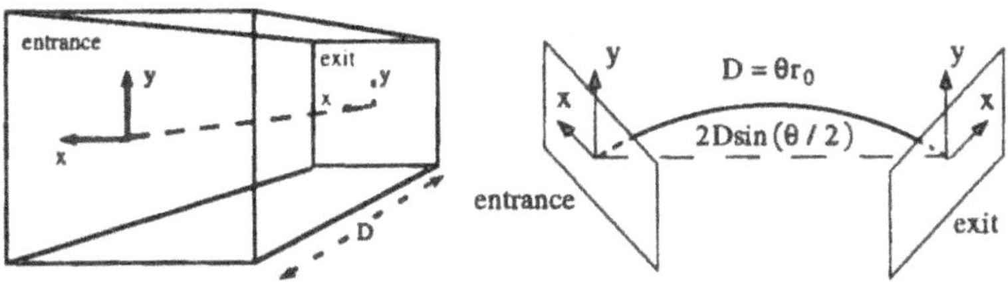

Figures 3.2 and 3.3: Straight and Cylindrical LEGO Blocks.

types of gutters: the bends and straight sections. Ideally, to leading order, the bend curves the trajectory and the straight section does nothing to it. In reference [13], Hirata and I name these ideal elements "LEGO" blocks by analogy to the famous toy. The LEGO block is defined by two surfaces in phase space and their relative position in configuration space. For example, a straight element in the magnet case is a block of length D connecting two planes parallel to each other. For a gutter with boundary condition $f_\infty = 0$, a straight element connects two parallel lines separated by a distance D. These planes are called the entrance and exit plane. They are both equipped with identical frames of reference. This is depicted in Figures 3.2 and 3.3 for magnets. The person doing the layout needs only to assume that a model (even a wrong model) exists which can carry a particle from the entrance face and the exit face.

To connect two blocks it suffices that their frame of reference be matched. The origin and the $x - y$ axes must coincide. Again the reader should realize that the actual map which propagates phase space from the entrance to the exit of a block can be anything. By adopting such a point of view, we disconnect completely modeling, positioning in the lattice, and analysis. If one thinks of the actual experimental conditions then one realizes that the particle will travel through a magnet according to its relative position within this magnet (local modeling) independent of its position in the lattice. Particle transport is a local business as far as a particle is concerned and therefore it should be local in any comprehensive simulation code.

The blocks must be put together to form a lattice. This is done with the theoretical layout of the ring. The layout is a map (in the cartographic sense) of the position of each ideal segment. In other words, it is a collection of planes connected by arcs of circles and straight lines. In the simple case of a planar ring (or gutter), the bend and straight blocks can be put together without the help of torsion. This should be self-evident.

3.2.3 Misplacing Segments: the Euclidean Group

N.B. *Before starting this discussion, we warn the reader of a coming confusion. We define all our translations and rotations as acting "actively" on the magnets. Therefore a translation of* d *in the* x *direction is* $x \longmapsto x - d$. *A rotation in the* $x - y$ *plane of* 90 *degrees moves a positive* x*-coordinate into a negative* y*-coordinate. Thus the* x*-axis moves into the old* y*-axis.*

Even in the planar world of gutters, we must allow for the misalignment, tilt or translation of a segment. If we did not use a length as our time-like variable, this would be easy. Indeed, we are simply saying that one should be able to let the full planar Euclidean group (1 rotation and 2 translations) act on the Hamiltonian. The action of this group on the usual phase space (x, p_x, y, p_y, z, p_z) is linear. If $\vec{\alpha}$ is a linear transformation belonging to the Euclidean group, then the new Hamiltonian is $H \circ \vec{\alpha}$. If $\mathcal{M}$ were the compositional map for a time T, then the new map would be $\mathcal{AMA}^{-1}$ where is $\mathcal{A}$ is associated with $\vec{\alpha}$ (or $\vec{\zeta}^{new} = \vec{\alpha}^{-1} \circ \vec{\zeta} \circ \vec{\alpha}$).

In reality, we cannot use the usual Hamiltonian as pointed out in Sect. 3.1.2. For example, in a straight segment, the time-like variable is normally the z-coordinate measuring length along the propagation axis. This means that the full Euclidean group must involve transformations which couple the time-like variable and phase space. Hence the new map for a rotated "LEGO" block cannot in general be obtained by a simple similarity transformation such as $\mathcal{AMA}^{-1}$. For simplicity, we first examine the transformations which do not couple the propagation and transverse directions. It is clear that this kind of transformation can only exist for the straight LEGO block. There are the two translations: the x and the z translations. The x translation happens in a plane completely perpendicular to the direction of propagation and is therefore the trivial shift in coordinate one would expect. This is depicted in Figure 3.4. The new map is simply

$$\mathcal{M}^{new} = \exp\left(: d_x p_x :\right) \mathcal{M} \exp\left(: -d_x p_x :\right). \tag{3.36}$$

The map of the translated element is always expressed in terms of the entrance and exit plane of the ideal element. These planes are

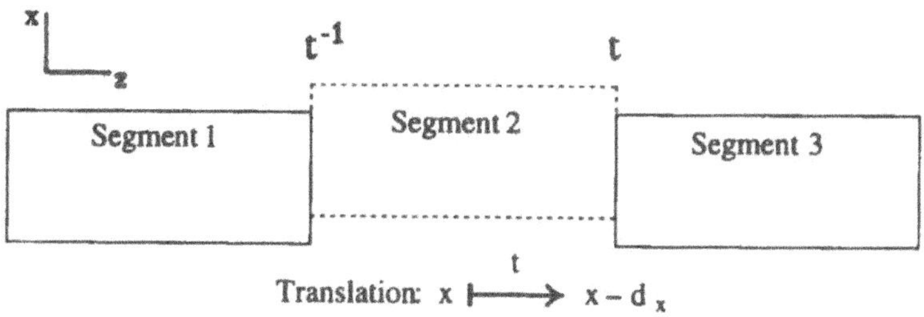

Figure 3.4: Transverse Translation of a Straight Block.

defined once and for all in the layout of the gutter or ring. We have defined the translation "actively" which means that the segment is moved along the positive x-axis for a positive d_x.

The longitudinal translation is given by propagation through a flat gutter. It is depicted in Figure 3.5. Mathematically, the new map representing the straight LEGO block is

$$\begin{aligned} \mathcal{M}^{new} &= \exp\left(: -d_z K_z^{f=0} :\right) \mathcal{M} \exp\left(: d_z K_z^{f=0} :\right) \\ &= \exp\left(: d_z p_z :\right) \mathcal{M} \exp\left(: -d_z p_z :\right). \end{aligned} \tag{3.37}$$

There is a certain pleasing symmetry between Equations (3.36) and (3.37). By definition, motion along the direction of propagation is given by the Hamiltonian which happens to be the momentum of the original Hamiltonian parameterized by the usual time. Hence our usual view of the momentum operator as a translation operator holds in both directions. Again the translated element fits snuggly into the layout. Of course, an arbitrary translation of the segment can be expressed as

$$\mathcal{M}^{new} = T_{\vec{d}} \mathcal{M} T_{\vec{d}}^{-1}$$

where

$$\begin{aligned} T_{\vec{d}} &= \exp\left(: d_x p_x :\right) \exp\left(: d_z p_z :\right) \\ &= \exp\left(: d_x p_x :\right) \exp\left(: -d_z K_z^{f=0} :\right). \end{aligned} \tag{3.38}$$

Notice that the Poisson bracket of the Hamiltonian of a drift and p_x vanishes. Hence the two translations commute.

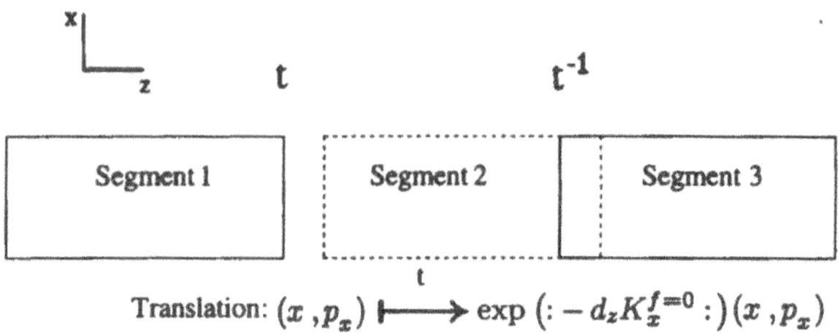

Figure 3.5: Longitudinal Translation of a Straight Block.

The planar Euclidean group also contains the $x - z$ rotation. This operation is problematic. In Figure 3.6 we picture the rotation of a block around the origin[3.2] O_{12} of its entrance plane. By definition the map $\mathcal{M}$ which is being rotated connects the two planes attached to a segment. The planes have origins O and O′. Let us denote the rotation around O_{12} by $\mathcal{Y}(\theta)$. It is not important for the moment to know the exact analytical form of this operator (a drift in polar coordinates!). The questions we need to answer are:

1. Why isn't the map for the rotated segment just $\mathcal{Y}(\theta)\mathcal{M}\mathcal{Y}^{-1}(\theta)$?
2. Can we fix this without any gruesome geometrical computation?

The answer to question 1 can be found by examining Figure 3.7. In this figure the points O and O′ are attached to the segment. The points O_{12} and O_{23} are points permanently attached to the layout; they are theoretical monuments against which everything is measured. The map $\mathcal{Y}(\theta)\mathcal{M}\mathcal{Y}^{-1}(\theta)$ carries a ray in phase space from the $(O, \vec{e}_x, \vec{e}_z)$ frame to the $(O', \vec{e}_x, \vec{e}_z)$, where $\vec{e}_x$ and $\vec{e}_z$ are unit vectors attached to the segment. Even though we only mention configuration space, it is clear that we must ensure that the total phase space is smoothly joined. This is ensured by the boundary condition $f = 0$ for the gutter. For magnets, we assume that the magnet is

[3.2]The author must apologize for conflicting notations. In Chapter 10 the subscripts 1 and 2 as in O_{12} refer to the planes rather than the magnets as they do here. The reader will see in Part III that we give first emphasize a layout of reference planes before ever defining the map of a magnet.

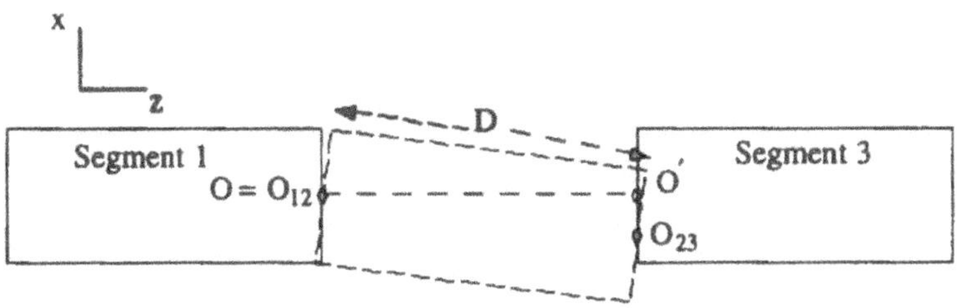

Figure 3.6: $x - z$ Rotation of a Straight Block at the Entrance.

embedded in a field free region.[3.3] In the case of this pure $x - z$ rotation around O, we can easily workout geometrically what needs to be added to $\mathcal{Y}(\theta)\mathcal{M}\mathcal{Y}^{-1}(\theta)$. First the ray must be translated transversely from the $(O', \vec{e}_x, \vec{e}_z)$ frame to the $(O'', \vec{e}_x, \vec{e}_z)$ frame. This is given by the operator $\exp{(: d_1 p_x :)}$ where d_1 is $D\sin{(\theta)}$ by elementary geometry. Next, we propagate from the frame $(O'', \vec{e}_x, \vec{e}_z)$ to the layout frame $(O_{23}, \vec{e}_x, \vec{e}_z)$. This is done by the forward propagator $\exp{\left(: -d_2 K_z^{f=0} :\right)}$. And once more d_2 is easily found to be $D\left(1 - \cos(\theta)\right)$. Therefore, the total map for this operation is given by

$$\mathcal{M}^{new} = \mathcal{Y}(\theta)\mathcal{M}\mathcal{Y}^{-1}(\theta)e^{:D\sin(\theta)p_x:}e^{:-D(1-\cos(\theta))K_z^{f=0}:}. \quad (3.39)$$

In general, especially in the case of the full Euclidean group required for magnets, it would be nice if we could find a way to avoid any geometrical arguments. The answer to question 2 relies on a trivial but profound observation: all the geometrical corrections depend on the length D of the magnet. Therefore, if the magnet had length zero we would expect the formula $\mathcal{M}^{new} = \mathcal{Y}(\theta)\mathcal{M}\mathcal{Y}^{-1}(\theta)$ to hold. At first glance, this looks like a pious wish with little practical application. However, keeping in mind the example of Figures 3.6 and 3.7, consider

[3.3] Moving a magnet in a region of nonzero field is more tricky. For example, in the interaction region of colliders, quadrupoles and bends can be inside a solenoid. It remains true that a z-translation is obtained by forward and backward propagation, however it is not a drift anymore. The theory is messier.

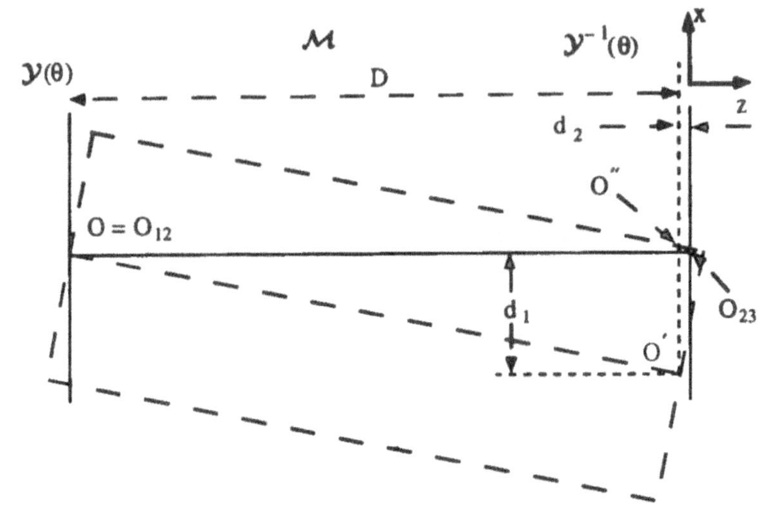

Figure 3.7: $x - z$ Rotation of a Straight Block at the Entrance: Details.

the following mathematical gymnastic on the map $\mathcal{M}$:

$$\mathcal{M} = \underbrace{\mathcal{M}\, e^{:DK_s^{f=0}:}}_{\substack{\text{Segment of length}\\ \text{zero}}} \underbrace{e^{:-DK_s^{f=0}:}}_{\substack{\text{Drift Segment}\\ \text{of length } D}}$$

$$= \mathcal{M}_{\text{thin}}^{z=0} \underbrace{e^{:-DK_s^{f=0}:}}. \tag{3.40}$$

The map $\mathcal{M}_{\text{thin}}^{z=0}$ is effectively a zero length segment located at the entrance plane (i.e., in the $(O_{12}, \vec{e}_x, \vec{e}_z)$ plane). Conceptually, we have modified the layout of the ring as indicated in Figure 3.8. We have decomposed the second segment from $(O_{12}, \vec{e}_x, \vec{e}_z)$ to $(O_{23}, \vec{e}_x, \vec{e}_z)$ into two segments: one thin segment from $(O_{12}, \vec{e}_x, \vec{e}_z)$ to $(O_{12+}, \vec{e}_x, \vec{e}_z)$ (which is artificially thick for drawing purposes) and a drift segment from $(O_{12+}, \vec{e}_x, \vec{e}_z)$ to $(O_{23}, \vec{e}_x, \vec{e}_z)$. We know that prior to rotation, the product of the thin map and the drift is equal to the original map. This is true by construction as shown in Equation (3.40). In Figure 3.8, we rotate the ray by an angle θ into the $(O_{12}, \vec{e}_{x'}, \vec{e}_{z'})$ frame, we then propagate through the thin sandwich. Finally, we rotate back

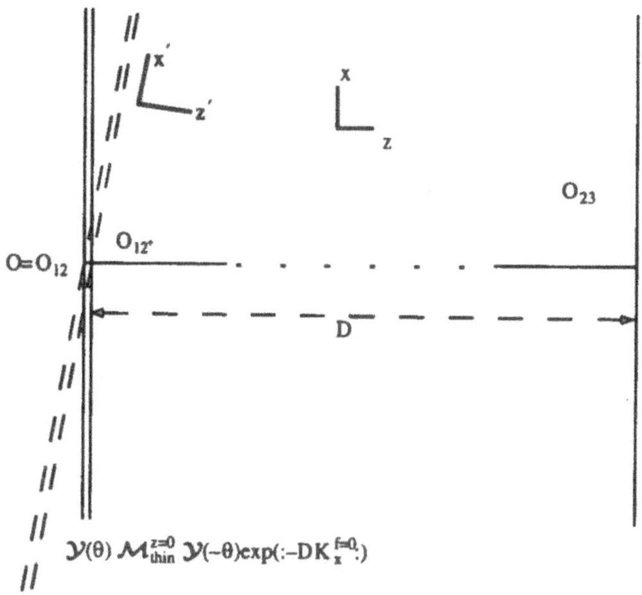

Figure 3.8: $x - z$ Rotation of a Zero Length Block.

into $(O_{12+}, \vec{e}_x, \vec{e}_z)$. But, of course, O_{12+} and O_{12} are the same points. Therefore, to get to the face of the next segment, i.e., $(O_{23}, \vec{e}_x, \vec{e}_z)$, we simply need to drift a length D. So, in general using the trickery of negative drifts, we can create thin LEGO blocks which are exactly equivalent and easily rotated. Hence, the relation

$$\mathcal{M} = \mathcal{Y}(\theta)\mathcal{M}_{thin}^{z=0}\mathcal{Y}^{-1}(\theta)\exp\ \left(: -DK_z^{f=0} :\right) \tag{3.41}$$

must be equivalent to that of Equation (3.39). The equivalence of the geometric construction and the thin lens is now the topic of an exercise.

Exercise 3-7

In this exercise we derive explicitly the map for $\mathcal{Y}(\theta)$ and the connection between the geometrical and thin lens construction.

1. *Consider the map $\mathcal{Y}(\theta)$. Convince yourself that this can be viewed as a drift in polar coordinates: the particle drifts in free space but changes its coordinate frame as it travels. This is depicted in Figure 3.9 where a particle moves on a straight line from frame $(\mathrm{O}, \vec{e}_x, \vec{e}_z)$ to frame $(\mathrm{O}, \vec{e}_{x(\theta)}, \vec{e}_{z(\theta)})$.*

2. *Exercise 3-7.1 implies that one can derive this transformation either from pure geometry or from Hamilton's equation. This canonical transformation $\mathcal{Y}(\theta)\,(x, p_x)$ is*

$$x(\theta) = \frac{x}{\cos(\theta)\left(1 - \frac{p_x \tan(\theta)}{\sqrt{1-p_x^2}}\right)}$$
$$p_x(\theta) = p_x \cos(\theta) + \sin(\theta)\sqrt{1-p_x^2}. \quad (3.42)$$

   In Equation (3.42), we did not include the time of flight and we chose the kinetic energy to be 1/2, or equivalently, we say that we measure the momenta in units of $\sqrt{2K_0}$.

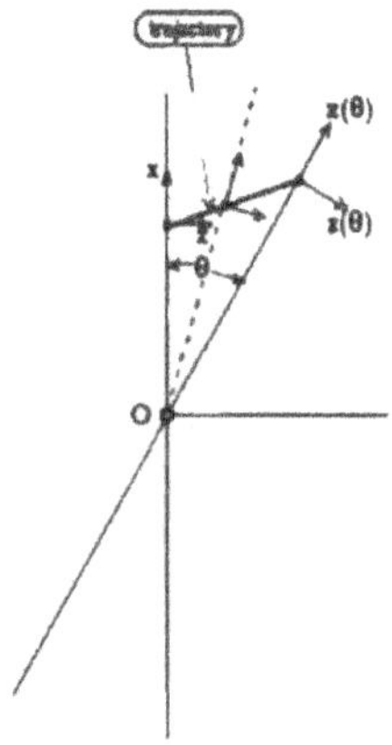

Figure 3.9: The $x - z$ Rotation or Coordinate Transformation $\mathcal{Y}(\theta)$.

   Using the polar Hamiltonian $K_r^{f=0}$ of Equation (3.24), argue that the correct expression for the function map $\mathcal{Y}(\theta)$ is just

$$\mathcal{Y}(\theta) = \exp\left(: \theta x\sqrt{1-p_x^2} :\right). \quad (3.43)$$

3. *It follows that the canonical transformation of Equation (3.42) must obey*

$$\frac{dx}{d\theta} = \frac{\partial\left(-x\sqrt{1-p_x^2}\right)}{\partial p_x}$$
$$\frac{dp_x}{d\theta} = \frac{\partial\left(x\sqrt{1-p_x^2}\right)}{\partial x}. \quad (3.44)$$

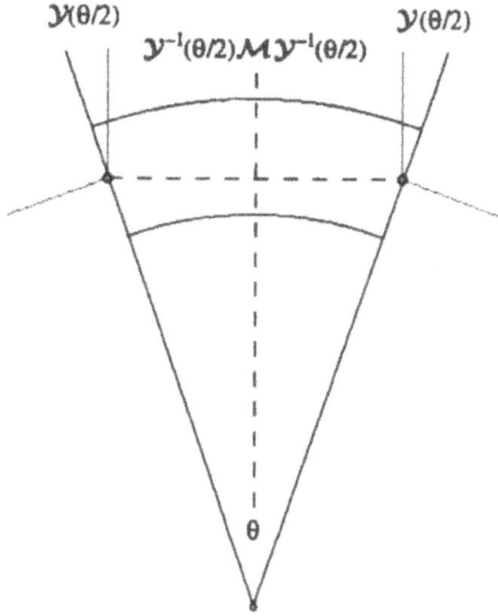

Figure 3.10: Making a Bend Block Straight.

Check this result.

4. *Compare the Lie exponents of the maps in Equations* (3.39) *and* (3.41). *Show for this example that the methods are equivalent.*

A full analysis of the Euclidean group and, in particular, the relationship between a drift in polar coordinates and the $x - z$ rotation will be explored in Part III. Before closing this section, we solve the problem of the Euclidean group acting on a cylindrical LEGO block of angle θ as displayed in Figure 3.3. Amazingly, we have all the tools necessary to solve this problem. The reader will recall that, when we needed a zero length segment, we crunched the segment with negative drifts. It amounted to a change of the layout with a complete mathematical equivalence. Hence, let us follow the same path: is there any way to transform the circular segment into the straight one? The answer is yes and it is pictorially displayed in Figure 3.10. It suffices to rewrite the map $\mathcal{M}$ of the bend element as

$$\mathcal{M} = \mathcal{Y}(\frac{\theta}{2})\underbrace{\mathcal{Y}^{-1}(\frac{\theta}{2})\mathcal{M}\mathcal{Y}^{-1}(\frac{\theta}{2})}_{\text{straight block}}\mathcal{Y}(\frac{\theta}{2}). \tag{3.45}$$

Then we make the straight block into a thin lens at $z = d$:

$$\mathcal{M} = \mathcal{Y}(\tfrac{\theta}{2})\mathcal{D}(d)\underbrace{\mathcal{D}(-d)\mathcal{Y}^{-1}(\tfrac{\theta}{2})\mathcal{M}\mathcal{Y}^{-1}(\tfrac{\theta}{2})\mathcal{D}(-D+d)}_{\substack{\text{thin straight block}\\ \text{at } z = d}}\mathcal{D}(D-d)\mathcal{Y}(\tfrac{\theta}{2}). \tag{3.46}$$

The elements of the Euclidean group act on the thin straight element of Equation (3.46) directly through a similarity transformation.

This concludes our discussion of the Euclidean group and its dynamical representation. The full group will be examined in Part III in great detail.

3.3 SYMPLECTIC INTEGRATION

In our treatment we make a big "fuss" about the dichotomy between the global analysis and the local analysis. Nevertheless the two aspects are linked. In particular, it should be clear that these aspects must be linked in the design phase of an accelerator. Various modules are patched together so as to produce a bigger module which has the desired global properties. Designers find it useful to use matched modules as their building blocks. As we saw in Sect. 2.5.2, properties of a given chunk of ring, such as the phase advance, are additive when matched modules are connected.

Simulation is a simpler task at least in theory. If the designer provides the theorist with the full Hamiltonian, then it is just a matter of solving the equations. Unfortunately, when nonlinearities are present, we have no way to produce the exact solutions. Furthermore, we have no global theorem to make grandiose statements on long-term stability. In the end, we must use numerical integration. There are two philosophies concerning numerical integration and we will now describe them for the sake of completeness.

3.3.1 Integration or Modeling?

Read this section diagonally if you have elected to skip initially parts of this book treating the global properties of a ring. Section 3.3.1 *shows that the local simulation of a ring requires a knowledge of global properties if efficiency is important.*

Let us assume for the sake of argument that the Hamiltonian given to us is absolutely exact. It describes exactly the machine or gutter, including errors and nonlinearities. And, why not, let us assume that there are no other physical phenomena which may disturb this knowledge. In other words, we are God! Great, but[3.4] unlikely! Anyway, let us be arrogant fools for awhile. Under these assumptions, it is clear that an exact solution of the local equations of motion would be fantastic since by assumption it would correspond to the solution of the true problem of interest. The reader may laugh again but this is almost the situation found in astronomy. After all, the masses of the planets, the laws of gravitation, the invariance of the total energy and of the total angular momentum are well-known facts. Somehow it seems dangerous to replace the solar system by a model which violates these laws or modifies the God-given planetary parameters. Let me say that astronomers have replaced the solar system by bizarre models for the sake of speed, but not without a certain apprehension. There is a lot of debate concerning long term integration of the solar system when approximate equations are used.

Therefore, armed with this divine knowledge, we feel compelled to perform an exact integration and thus use a very stable and accurate method. Typically, high order predictor-corrector methods are used. With these techniques it is possible to reach machine precision with a relatively small number of steps, thus minimizing truncation error. Of course, the various invariants, such as the energy, are preserved with an accuracy consistent with the accuracy of the integrator. This is what we would call "true integration." Since we are looking for an exact solution to the exact problem, integration involves purely local quantities. The reader may be surprised or even "horrified" to learn that this type of simulation is never performed in circular accelerators. Why? The answer to this question is central to all simulations done in ring dynamics.

Let us assume that we do not know precisely the Hamiltonian, that there are unavoidable construction or position errors and that the Hamiltonian has a horrible s or z dependence. This is clearly the case of an accelerator. Even relatively small accelerators have hundreds of elements whose magnetic fields are not exactly known. Also, the s-dependence of the Hamiltonian is a horrible function simply because various functions such as the quadrupole focusing strength k_2

[3.4]Actually we are more a local deity because God Himself would know the global topology and history of the system and thus would not need any integration scheme!

are localized. They vanish outside the physical quadrupoles. More important, the same is true for the nonlinear multipoles such as the omnipresent chromatic sextupoles. It is unlikely that such a system is integrable or possesses any true invariant besides the obvious ones.[3.5] Given these considerations, we are tempted to replace the real problem by a new problem which is simpler and faster to integrate. This simpler problem should have the same global properties as the old ones as far as we can gauge. In other words, we are hoping that the fundamental stability question can be answered on the approximate system. It is hoped that simulation done on this system will reflect the behavior of the true system whatever it may be. Even though our simulation is local in character, the final model we will simulate is determined by a partial understanding of its global properties. In fact, because the design goals of a ring are defined partially in terms of global properties, it is certainly important to have these properties preserved by our model. We have one more reason to talk about global properties before local ones: not only does designing a lattice require an understanding of global properties, but even efficient simulation can only be done by a lucid analyst.

This is rather abstract. To illustrate this point, we consider the following gutter Hamiltonian:

$$H = \frac{p_x^2}{2} + f_s^{[2]}\frac{x^2}{2} + f_s^{[3]}\frac{x^3}{6},$$

$$\text{where } f_s = 0 \text{ for } 0 < s < \frac{1}{2}$$

$$\text{and} \quad \begin{cases} f_s^{[2]} = f_0^{[2]} = 4 \\ f_s^{[3]} = f_0^{[3]} = 6 \end{cases} \text{for } \frac{1}{2} < s < 1. \tag{3.47}$$

This Hamiltonian is assumed to have a period equal to one. Therefore the one-turn map is obtained by integrating the equation of motion from $s = 0$ to $s = 1$. Of course, this Hamiltonian has two segments, first a drift and then a convex gutter. The phase space of this system at $s = 3/4$ is shown in Figure 3.11.

In Figure 3.12, the "exact Hamiltonian" of Equation (3.47) is replaced by the impulse approximation. The effect of the potential is lumped at $s = 3/4$. The Hamiltonian for this system is

$$H = \frac{p_x^2}{2} + \tilde{f}_s^{[2]}\frac{x^2}{2} + \tilde{f}_s^{[3]}\frac{x^3}{6},$$

[3.5]The energy p_t is conserved in the absence of an RF cavity. In a position-dependent Hamiltonian, it appears as a conserved momentum which is trivially preserved by any reasonable integrator.

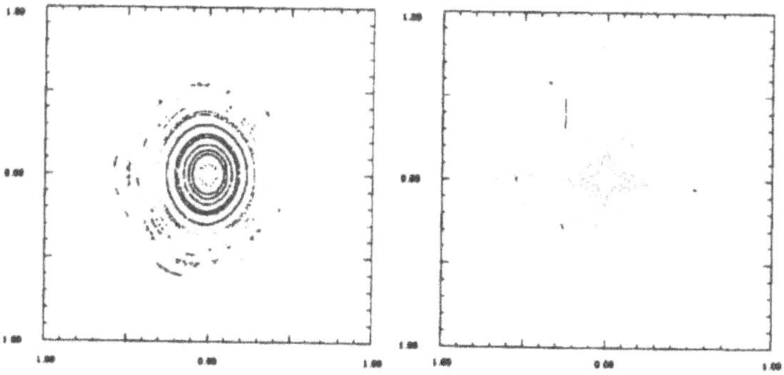

Figures 3.11 and 3.12: Exact Integration and the one-kick approximation.

$$\text{where} \quad \begin{cases} \widetilde{f}_s^{[2]} = \frac{f_0^{[2]}}{2}\delta_p(s-\frac{3}{4}) = 2\delta_p(s-\frac{3}{4}) \\ \widetilde{f}_s^{[3]} = \frac{f_0^{[3]}}{2}\delta_p(s-\frac{3}{4}) = 3\delta_p(s-\frac{3}{4}) \end{cases}, \tag{3.48}$$

where $\delta_p(s)$ is the periodic Dirac delta function. The reader will notice that this system is exactly and trivially solvable. The one-turn map at $s = 3/4$ is given by

$$(x^{in}, p_x^{in}) \longmapsto (x^{fin}, p_x^{fin})$$

$$\begin{aligned} \overline{x} &= x^{in} + \frac{3}{4}p_x^{in} \\ p_x^{fin} &= p_x^{in} - \frac{f_0^{[2]}}{2}\overline{x} - \frac{f_0^{[3]}}{4}\overline{x}^2 \\ x^{fin} &= \overline{x} + \frac{1}{4}p_x^{fin}, \end{aligned}$$

$$\text{with } f_0^{[2]} = 4 \text{ and } f_0^{[3]} = 6. \tag{3.49}$$

There is a striking difference between the exact solution and the approximate impulsive map as Figures 3.11 and 3.12 illustrate.

Exercise 3-8

In Exercise 2-2 of Sect. 2.2 we have seen how maps are critically sensitive to the tune.

1. *Compute the matrix and the tune of the one-turn map produced by the Hamiltonian of Equation* (3.47) *at* $s = 3/4$. *Answer: The*

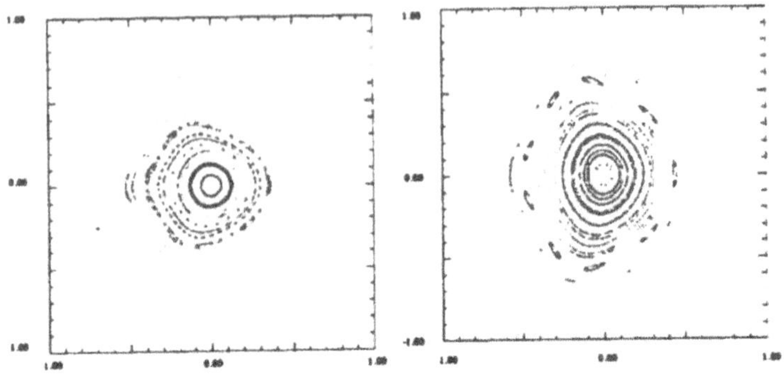

Figures 3.13 and 3.14: Adjusting the tune and also adjusting the Floquet Ring.

matrix is

$$\begin{pmatrix} x \\ p_x \end{pmatrix}^{fin} \approx \begin{pmatrix} 0.11957 & 0.80581 \\ -1.2232 & 0.11957 \end{pmatrix} \begin{pmatrix} x \\ p_x \end{pmatrix}^{in} . \quad (3.50)$$

and the tune is $\nu = 0.2309$.

2. *Show that the tune of the impulsive system is* $\nu = \frac{1}{4}$. *Does this explain the distortion of Figure* 3.12*?*

3. *Modify* $f_0^{[2]}$ *to achieve a tune of* 0.2309. *Plot the phase space. Answer:* $f_0^{[2]} = 3.522$ *and the plot is on Figure* 3.13.

4. *We can ask how the impulsive system should be slightly altered in order to reproduce qualitatively the results of the exact solution. For that task, compute the matrix* M_{inp} *of the impulsive system at the center of the impulse. Allow the total length to vary by some quantity* ΔC. *Answer:*

$$M_{inp} = \begin{pmatrix} 1 - \frac{1+\Delta C}{4} f_0^{[2]} & 1 + \Delta C \\ -\frac{f_0^{[2]}}{4}\left(2 - \frac{1+\Delta C}{4} f_0^{[2]}\right) & 1 - \frac{1+\Delta C}{4} f_0^{[2]} \end{pmatrix} . \quad (3.51)$$

5. *Consider the following ring for the impulsive system (as defined in Sect.* 2.5.1*) with maps starting at the center of the impulse:*

$$Ring_2 = \left\{ \exp\left(: -\tfrac{1}{4}x^3 :\right), \mathcal{M}(f_0^2, \Delta C), \exp\left(: -\tfrac{1}{4}x^3 :\right) \right\} . \quad (3.52)$$

where $\mathcal{M}$ is the compositional map corresponding to the linear matrix of Equation (3.51). *This ring represents the impulsive system at the center of the nonlinear impulse (or kick). Using the Courant-Snyder transformation defined in Equation* (2.75), *show that a possible Floquet ring is:*

$$Ring_2^F = \left\{\exp\left(:-\tfrac{1}{4}\beta^{3/2}x^3:\right), \exp(:-\mu J:), \exp\left(:-\tfrac{1}{4}\beta^{3/2}x^3:\right)\right\},$$
$$\text{where} \quad \begin{cases} \cos\ \mu = 1 - \frac{1+\Delta C}{4} f_0^{[2]} \\ \qquad \beta = \frac{1+\Delta C}{\sin\ \mu} \end{cases} . \tag{3.53}$$

6. *Choose the two parameters* $\left(f_0^{[2]}, \Delta C\right)$ *so as to make the Floquet rings of the exact and impulsive systems identical. Plot the results at* $s = 3/4$ $(Ring_2)$. *Answer: the new parameters are* $(f_0^{[2]}, \Delta C) = (4.3696, -0.1942)$ *and the plot is in Figure* 3.14.

A comparison of Figures 3.11 and 3.14 suggests that it is possible to reproduce the main aspects of a system by a pretty coarse integration provided that one is willing to restore some global properties by refitting the available parameters. One notices that the agreement is not perfect and certainly the presence of the chain of nine islands is suspicious. (This could be improved by adding a second impulse i.e., another kick.) In any event, the uncertainties associated with our system, the lack of invariants, and the questions we ask, compel us to use simulations of the kind depicted in Figure 3.14. We can say that accelerator physicists do symplectic simulations rather than symplectic integration: the integrator is itself part of the model. Therefore, efficient simulations will require an understanding of the global properties of a ring. Of course, on an infinitely fast and accurate computer, local integration is totally local. Even in symplectic modeling, once the number of integration steps has been selected through global gymnastics, the implementation of the integrator should be once again local.

3.3.2 Explicit Symplectic Integration

Looking back at Sect. 3.3.1, we can view symplectic modeling as the replacement of an insoluble Hamiltonian (Equation (3.47)) by a soluble one (Equation (3.48)). Accelerator physicists have in the past viewed all their "kick codes" in that manner. It turns out that explicit symplectic integration is a generalization of this point of view.

Consider a Hamiltonian H which can be locally written as the sum of two s-independent parts H_1 and H_2 where each term is exactly solvable when considered as an individual Hamiltonian. To describe explicit symplectic integration it is best to use the formalism of Sect. 2.3. We start with the differential equation for the map as given by Equation (2.27) in Interlude I. This equation can be solved immediately because we assumed that our Hamiltonian is s-independent.

$$\frac{d}{ds}\mathcal{M}(s) = \mathcal{M}(s) : -H :$$

$$\Downarrow \; s\text{-inpendence}$$

$$\mathcal{M}(s) = \exp\,(: -sH :)\,. \tag{3.54}$$

Similarly, the same equation can be solved for the Hamiltonians H_1 and H_2:

$$\mathcal{M}_k(s) = \exp\,(: -sH_k :) \;\text{ where } k = 1, 2. \tag{3.55}$$

By assumption the compositional maps are solvable: if we generate a regular canonical transformation by letting $\mathcal{M}_k$ act on the identity transfer map (see Equation (2.18) in Sect. 2.3), the resulting expressions are all in terms of "well-known" functions which can be evaluated to machine precision on a computer. Given these facts, the person with a "symplectic integration" point of view asks himself: can one use the maps $\mathcal{M}_k$ to approximate the exact map $\mathcal{M}(s) = \exp\,(: -sH :)$? It is useful to introduce a theorem which will be proven in Part II Sect. 6.1.2. It is the Baker-Campbell-Hausdorff theorem (or BCH theorem). Consider two formal exponential operators $\exp(\varepsilon A)$ and $\exp(\varepsilon B)$. Their product $\exp(\varepsilon A)\ \exp(\varepsilon B)$ can be formally expressed as follows:

$$\begin{aligned}\exp(\varepsilon C) &= \exp(\varepsilon A)\exp(\varepsilon B)\\ C &= A + B + \varepsilon\, \ell\\ \ell &= \frac{1}{2}\{A, B\} + \cdots \in \text{ Lie Algebra of commutators of } A \text{ and } B.\end{aligned} \tag{3.56}$$

We can use this theorem to construct immediately an approximation of $\exp\,(: -sH :)$:

$$\exp\,(: -sH :) = \underbrace{e^{:-sH_1:}}_{\mathcal{M}_1(s)}\underbrace{e^{:-sH_2:}}_{\mathcal{M}_2(s)}\, e^{\frac{s^2}{2}\{:H_1:, :H_2:\}+\cdots}\,. \tag{3.57}$$

Here, we also wrote down explicitly the correction term of order s^2 in terms of the commutator denoted by curly brackets. One can check that the following symmetrized expression is an even better integrator:

$$\exp(:-sH:) = \underbrace{e^{:-\frac{s}{2}H_1:}}_{\mathcal{M}_1(s/2)}\underbrace{e^{:-sH_2:}}_{\mathcal{M}_2(s)}\underbrace{e^{:-\frac{s}{2}H_1:}}_{\mathcal{M}_1(s/2)}e^{s^3\ell+\cdots} \quad . \tag{3.58}$$

We can apply our new knowledge to the Hamiltonian in Equation (3.47). We will notice immediately that the functions $H_1 = \frac{p_x^2}{2}$ and $H_2 = f_s^{[2]}\frac{x^2}{2} + f_s^{[3]}\frac{x^3}{6}$ are exactly solvable in the problematic second region $\frac{1}{2} < s < 1$. They induce the canonical transformations

$$\begin{aligned}
\mathcal{M}_1(d) &= \exp\left(-d:\frac{p_x^2}{2}:\right)\\
\mathcal{M}_2(d) &= \exp\left(-d: f_0^{[2]}\frac{x^2}{2} + f_0^{[3]}\frac{x^3}{6}:\right)\\
&\Downarrow\\
\vec{\zeta}_1 &= \mathcal{M}_1(d)(x,p_x) = (x + dp_x, p_x)\\
\vec{\zeta}_2 &= \mathcal{M}_2(d)(x,p_x)\\
&= (x, p_x - d\, f_0^{[2]}\, x - \frac{1}{2}d\, f_0^{[3]}\, x^2).
\end{aligned} \tag{3.59}$$

Given the fact that our total compositional map from $s = 0$ to $s = 1$ is just

$$\mathcal{M} = \underbrace{\mathcal{M}_1(\frac{1}{2})}_{\text{first segment}}\underbrace{\exp(-:\frac{1}{2}H_{s=\frac{3}{4}}:)}_{\text{second segment}}, \tag{3.60}$$

we can approximate the second segment by the second-order integrator of Equation (3.58)

$$\begin{aligned}
\mathcal{M} &\approx \underbrace{\mathcal{M}_1(\frac{1}{2})}_{\text{first segment}}\underbrace{\mathcal{M}_1(\frac{1}{4})\mathcal{M}_2(\frac{1}{2})\mathcal{M}_1(\frac{1}{4})}_{\text{second segment}}\\
&= \mathcal{M}_1(\frac{3}{4})\mathcal{M}_2(\frac{1}{2})\mathcal{M}_1(\frac{1}{4}).
\end{aligned} \tag{3.61}$$

This map $\mathcal{M}$ is the compositional map associated with the total transfer map $\vec{\zeta}_1(1/4)\circ\vec{\zeta}_2(1/2)\circ\vec{\zeta}_1(3/4)$ which is given precisely by Equation (3.49). Hence, one can view the impulsive approximation of Sect. 3.3.1 as the application of one step of a second-order explicit symplectic integrator.

Exercise 3-9
The reader well-versed in special functions might already be complaining. The second segment admits an exact solution in terms of elliptic functions. We will limit ourselves to ordinary trigonometric functions and consider anything else as "non-elementary." Nevertheless, there is another split of the Hamiltonian into two pieces which is amenable to implicit symplectic integration. What is it? Hint: s-independent quadratic Hamiltonians are always solvable in terms of elementary functions.

Explicit symplectic integration is actually connected to integration in Lie groups. The formulas we used for first and second order integration are based on the formal existence of the operator $\widehat{C}$ as insured by the BCH formula (Equation (3.56)); the symplectic nature of the flow was not relevant. We will see that it is possible to extend this formalism to higher order and to a larger number of splits of the Hamiltonian in Part III. Finally, we have applied explicit symplectic integration to the approximate large gutter Hamiltonian neglecting boundary effects which are intrinsically s-dependent. It is obvious that the large ring Hamiltonian of Equation (3.35) is also amenable to explicit symplectic integration. What about the real gutter Hamiltonian and the small (correct) ring Hamiltonian? The reader can already convince himself that Equation (3.24), which describes a gutter, is hopelessly hard. In fact, it is only possible to integrate it using *implicit* symplectic integration. Remarkably, the body of an ideal magnet, despite nasty square-roots, is amenable to explicit symplectic integration! Only fringe fields present a problem to accelerator physicists. Again it seems that the gutter problem is mathematically more complicated than the real magnet problem.

3.4 MAP EXTRACTION

We have described the tools necessary in a local simulation code. Such a code allows the propagation of a ray around the ring or the gutter: mathematically speaking it produces maps (i.e., tracking data) for segments of the ring. Its representation consists of a sequence of commands describing propagation through the layout. Each block contains elements of the Euclidean group sandwiching a few steps of symplectic integration. Even for a small ring, this represents a few hundred function evaluations on a computer. A mathematician would correctly state that we have an analytic representation of our ring. In practice, one can stare at the local representation (the integration routines) for hours or days without gaining any understanding of the global properties of the one-turn map it represents. This is not surprising since we emphasized locality in designing a simulation code.

Of course, I am being facetious: the purpose of a simulation code is to simulate. Mathematically, it means that we are capable of evaluating some map $\vec{\zeta}$ of a given beam line for an arbitrary $\vec{z}_0$, i.e., to compute $\vec{\zeta}(\vec{z}_0)$. The entire discussion about modeling in Sect. 3.3.1 implies that we view this evaluation as an "exact" evaluation. Because our understanding of the map will come from a finite number of evaluations, we say loosely, but realistically, that we do not know the one-turn map; complex analysis of tracking data and/or the production of approximate maps is necessary to gain insight into the global dynamics.

In Chapter 2 (as well as in Sect. 3.3.1), we have done calculations on simple representations of the map. In particular, we have expanded maps around their stable fixed point and looked at the linear (and nonlinear) parts. Some very important (global) knowledge can be extracted from this approximate representation such as the tune, the linear invariants, etc... In our case, the Taylor series representation was available because a simple map was studied. In general, thanks to automatic differentiation, it is possible to computerize Taylor series extraction for the messiest and biggest simulation code. The extraction of $\vec{\zeta}(\vec{z}_0)$, the usual tracking data, can then be viewed as a special case as we will see. We will conclude Part I, with a few comments on these topics.

3.4.1 Tracking

The reader must now assume that he has a tracking code representing a complex problem. So, pick a big ring and a big code, and start tracking! What will happen? In circular rings, one sees phase space plots of the types shown in the figures encountered in this chapter. Near the center of the ring, ellipses will be drawn with various eccentricities and tilts. Further away, the phase space will look like a distorted ellipse. Depending on the tune it may take the shape of a polygon with those nasty islands appearing. At larger amplitude, the plots will seem chaotic. Finally, rays at very large amplitude will simply take off to infinity (beam pipe).

The author (and the reader) can concoct all sorts of scenarios which combine, mix, exclude or suppress some of the effects we just mentioned. In the end, our simulations must and will display these effects. So one must ask questions about the pictures appearing on the screen, starting with the topology near the fixed point. One will see that the motion, in one degree of freedom, will be confined to ellipses and that these ellipses are invariants of a linear approximation of the map. Furthermore, because they are quasi-invariants of the map, they are intimately connected to a Lie representation of the map. It is clear that we will have closed the intellectual loop.

1. If we seek confinement in a periodic system we are led towards the one-period map and its Lie representation.

2. If we are given a confined periodic system to analyze, either through experiment or simulation, we will see the quasi-invariants of the one-period map drawn before our own eyes. Therefore, we will be compelled to introduce them in the theory.

The latter was the story of my life from 1985 to 1991. Circular rings are complete systems from a mathematical point of view. Personally, I started thinking about the field of circular rings by "running" simulation codes. The end results were ellipse-like curves being drawn. Therefore a global theory emerged independently of the inner (local) entrails of the simulation code. This is the central message of this book. If you do experiments, real or numerical, you are led to a map theory as the only intuitive theory which matches your everyday experience.

The results of tracking, which can be simple phase plots or can assume more sophisticated forms when combined with global analysis, [3.6] are incomplete representations of the one-turn map. In other words, we cannot plot the entire phase space. Nevertheless, the pictorial nature of a plot is very powerful: human beings can process visual data at a phenomenal rate when compared to present day computers. Nevertheless, can a tracking code produce routinely a more "analytic" approximation of the one-turn map? Yes, this is possible through automatic differentiation.

3.4.2 Automatic Differentiation: Taylor Series Maps

The mathematician views our tracking code as an explicit representation of the transfer map. If we denote by $\vec{\Delta}$ the set of parameters controlling our system, then the tracking code is a $\vec{\Delta}$ dependent map-

[3.6] Fourier transforms, phase space plots in some Floquet variables, fitted invariants, mode analysis as proposed by Laskar[14], etc.

ping of phase space:

$$\begin{aligned}
&\mathbf{R}^2 \xrightarrow{\vec{\zeta}} \mathbf{R}^2 \\
&\vec{z} = (x, p; \vec{\Delta}) \longmapsto \vec{\zeta}(\vec{z}; \vec{\Delta}) \\
&\vec{\Delta} \in \mathbf{R}^P \text{ where } P = \text{ Dimension of parameter space.}
\end{aligned} \tag{3.62}$$

The dimensionality of the parameter space P is at most the total number of inputs in the code. For many problems it is simply zero. For example, in Exercise 3-8, we allowed the convexity and the length of the gutter to change. This means that we considered the case $P = 2$ in Equation (3.51). If we assume that $\vec{\zeta}(\vec{z}; \vec{\Delta})$ is differentiable in the parameter $\vec{\Delta}$, we can imagine a Taylor series expansion of a function around some initial condition $\vec{z}_0$ and some design value of the parameter $\vec{\Delta}_0$:

$$\vec{z}_0 + \vec{z} \quad \longmapsto \quad \vec{\zeta}(\vec{z}_0 + \vec{z}; \vec{\Delta}_0 + \vec{\Delta}) \tag{3.63}$$

$$\vec{\zeta}(\vec{z}_0 + \vec{z}; \vec{\Delta}_0 + \vec{\Delta}) \overset{\text{To order } k}{=} \vec{\zeta}(\vec{z}_0; \vec{\Delta}_0) + d\vec{\zeta}^{[1]} + d\vec{\zeta}^{[2]} + \cdots + d\vec{\zeta}^{[k]}.$$

In Equation (3.63), the functions $d\vec{\zeta}^{[j]}$ are j^{th}-degree homogeneous polynomials in the deviation variables $(\vec{\zeta}, \vec{\Delta})$. This particular expansion agrees with the original map of the simulation code to degree k in $(\vec{\zeta}, \vec{\Delta})$.

Generally a simulation code will provide the zeroth order component of Equation (3.63) and no more. However, due to automatic differentiation, it is possible to generate this Taylor series expansion automatically on a computer. In fact, with the help of advanced languages such as C++, one can easily obtain the routines for the Taylor series extraction from the original coding, by redefining the meaning of the various functions and arithmetical symbols at compile time. In languages such as FORTRAN, standard function calls and operations are replaced by subroutine calls. Our goal is not to describe the topic of automatic differentiation, but to assure the reader that it is possible to extract *routinely* the polynomials of Equation (3.63). It should be pointed out that automatic differentiation is not a "small difference" numerical procedure. It is an exact method. From now on, we will always assume that a simulation code is capable of producing the exact Taylor series expansion consistent with its model, and if, for some reason it is not the case, we will explicitly state so.

In circular rings, there exists a central fixed point even in the presence of errors around which the particles orbit. It is in the neighborhood of this point that the machine is designed. Therefore let us assume that we have found (numerically) the exact location of this closed orbit. If we expand around this orbit, we obtain a new Taylor series

$$\vec{\zeta}(\vec{z}_0+\vec{z};\vec{\Delta}_0+\vec{\Delta}) \overset{\text{to order } k}{=} \underbrace{\vec{z}_0}_{\vec{\zeta}(\vec{z}_0;\vec{\Delta}_0)} + \underbrace{d\vec{\zeta}^{[1]} + d\vec{\zeta}^{[2]} + \cdots + d\vec{\zeta}^{[k]}}_{=\vec{\zeta}^{[k]}}$$
$$\Downarrow$$
$$\vec{z} \longmapsto \vec{\zeta}^{[k]}. \tag{3.64}$$

The Taylor series around this point has very special properties since a self-consistent perturbation theory can be built around it. Indeed if we iterate the exact map $\vec{\zeta}$ for n turns

$$\vec{z} \longmapsto \vec{\zeta}^{n} = \underbrace{\vec{\zeta} \circ \cdots \circ \vec{\zeta}}_{\text{n times}}(\vec{z}) \tag{3.65}$$

then the expansion of nth iterate $\vec{\zeta}^{n}$ around the fixed point $\vec{z}_0$, is given by $\left(\vec{\zeta}^{[k]}\right)^{n}$ truncated at the kth degree.

The self-consistency of a truncated map around its fixed point is central to perturbation theory. We will close this chapter with this topic.

3.4.3 Algebra of k-jets: Differential and Lie Algebras

The extraction of maps by automatic differentiation is based on the algebra of truncated power series which incidently is not a differential algebra even though it is isomorphismic to one. For maps around a fixed point, the concept of k-jets and differential algebra becomes very useful in formal manipulations. Consider a differentiable function of phase space f (we drop the parameter space for ease of notation) such that

$$f(\vec{z}_0) = 0 \tag{3.66}$$

where $\vec{z}_0$ is a fixed point of the one-turn map $\vec{\zeta}$. We can define the Taylor series expansion of f around $\vec{z}_0$ to order k:

$$f(\vec{z}_0+\vec{z}) = f^{[k]} + O(\|\vec{z}\|^{k+1}) + \cdots . \tag{3.67}$$

Let us now say that two functions f and g are k-equivalent around $\vec{z}_0$ (denoted by $f \rightsquigarrow g$) if and only if their Taylor series to degree

k around $\vec{z}_0$ are the same. This forms an equivalence relation and k-equivalent functions form a class. The k-jets are the equivalence classes of the relation "$\rightsquigarrow$."

Consider two functions f and g belonging to the class of $f^{[k]}$ and $g^{[k]}$ respectively. Let us denote the set of infinitely differentiable functions at $\vec{z}_0$ by C_∞ and the class of $f^{[k]}$ by $\left\{f^{[k]}\right\}$. Then it is true that

$$\begin{aligned} \text{if } a, b \in \left\{f^{[k]}\right\} &\text{ and } c, d \in \left\{g^{[k]}\right\} \\ &\Downarrow \\ a + c \rightsquigarrow b + d &\rightsquigarrow f^{[k]} + g^{[k]} \\ ac \rightsquigarrow bd &\rightsquigarrow f^{[k]} g^{[k]}. \end{aligned} \tag{3.68}$$

Functions that have only terms of degree higher than k in their Taylor series expansion are equivalent to the null function $\mathbf{0}$. The function 0 maps all phase points to the real number zero. We can define the addition and multiplication of equivalence classes (k-jets) by using any element of the class. In particular, we can use the defining k^{th}-degree Taylor expansions $f^{[k]}$ and $g^{[k]}$. The new algebra made out of equivalent classes is the factor or quotient algebra usually denoted by $C_\infty/\{0\}$.

All of this is pretty obvious and is a mathematical restatement of the self-consistency of the algebra of truncated power series under addition and multiplication. The extension to transfer maps is trivial because maps are a collection of $2N$ functions where $2N$ is the dimensionality of phase space. Two maps are equivalent if all their components are equivalent:

$$\vec{\zeta} \rightsquigarrow \vec{\xi} \Leftrightarrow \forall i \; \zeta_i \rightsquigarrow \xi_i \; . \tag{3.69}$$

Again, we can label (and confuse) the equivalence class with the canonical representative $\vec{\zeta}^{[k]}$. The set of maps is the Cartesian product $C_\infty^{2N} = C_\infty \times \cdots \times C_\infty$ and the set of k-jet maps is the product of the quotient spaces $C_\infty^{2N}/\{\vec{0}\} = C_\infty/\{0\} \times \cdots \times C_\infty/\{0\}$.

Actually, the absence of constant terms in the Taylor series (Equation (3.66)) has not yet been used. It comes in because our ultimate goal is to represent the composition of maps as k-jets in a self-consistent theory. We are in a position to use Equation (3.66) in a critical way. The absence of constant terms in the functions defining transfer maps implies

$$\text{if } \vec{\xi}, \vec{\zeta} \in \overline{C}_\infty^{2N} \text{ then } \vec{\xi} \circ \vec{\zeta} \rightsquigarrow \vec{\xi}^{[k]} \circ \vec{\zeta}^{[k]}, \tag{3.70}$$

where $\overline{C}_\infty$ is the set of infinitely differentiable functions with no constant terms. We can compose maps or jets self-consistently. In addition, let us define a differential operator $\vec{F}\cdot\vec{\nabla}$, where $\vec{F}\in\overline{C}_\infty^{2N}$. One can easily show that

$$\left.\begin{array}{c} f \rightsquigarrow g \\ \vec{F} \rightsquigarrow \vec{G} \end{array}\right\} \Rightarrow \vec{F}\cdot\vec{\nabla}f \rightsquigarrow \vec{G}\cdot\vec{\nabla}g. \tag{3.71}$$

The operator $\vec{F}\cdot\vec{\nabla}$ is a derivation (it obeys Leibnitz's rule). We can commute, multiply and add two differential operators. We can also multiply them by scalars. Because the operators are derivations, we call this algebra a "differential algebra." It follows from Equation (3.71) that the operators trivially extended to the k-jets also form a differential algebra. We recall the important property of Lie operators discussed in Interlude II,

$$e^{\vec{F}\cdot\vec{\nabla}}f = f\circ\left(e^{\vec{F}\cdot\vec{\nabla}}\vec{I}\right), \tag{3.72}$$

which is due to Leibnitz' rule, and which can be trivially extended to k-jets if the derivation properties are preserved:

$$e^{\vec{F}\cdot\vec{\nabla}}f \rightsquigarrow f\circ\left(e^{\vec{F}\cdot\vec{\nabla}}\vec{I}\right). \tag{3.73}$$

As usual, when dealing with jets, we can perform any calculation with any representative of the class,

$$\text{if } g = e^{\vec{F}\cdot\vec{\nabla}}f \text{ then } g^{[k]} \rightsquigarrow f^{[k]}\circ\left(e^{\vec{F}^{[k]}\cdot\vec{\nabla}}\vec{I}\right) \rightsquigarrow e^{\vec{F}^{[k]}\cdot\vec{\nabla}}f^{[k]}, \tag{3.74}$$

and can confuse the canonical representative of a class with the equivalence class itself and write

$$g^{[k]} = f^{[k]}\circ\left(e^{\vec{F}^{[k]}\cdot\vec{\nabla}}\vec{I}\right). \tag{3.75}$$

In summary, the benefits of automatic differentiation are

1. We can get a Taylor series expansion of the map around any orbit through the implementation of the truncated power series algebra.

2. The algebra of truncated power series is equivalent to the algebra of k-jets.

3. It is natural to expand around the central fixed point (or sometimes around other fixed points) because in that case composition of maps is closed and well defined for k-jets.

4. It also follows that the Lie operators needed to represent maps around fixed points form a differential algebra even when restricted to k-jets. Therefore all the theorems we derived for compositional Lie maps apply to k-jets Lie maps acting on k-jets. The entire theory is self-consistent around fixed points.

5. Even though we focussed on phase space, the definition of a k-jet can include a vector $\vec{\Delta}$ of parameter space.

Exercise 3-10
In this exercise, we revisit in spirit Exercise 2-6 in the context of k-jets.

1. *Consider the Lie operator* $:f:$ *associated with the function* $f = -L\frac{(p+\varepsilon)^2}{2}$. *Write down the Lie operator in the* $\vec{F}\cdot\vec{\nabla}$ *notation.*

2. *Repeat Exercise* 2-6.1 *by letting the map act on the function* sin (x). *In particular, do not cheat in evaluating* exp $(:f:)$ sin $(x$ *derive a differential equation for this function and solve it. Compare the result with direct substitution.*

3. *Show that the various ways to compute this function as a 2-jet in the variables* (x,p) *are not compatible unless* $\varepsilon = 0$.

The self-consistency of map production and analysis is very important in circular rings in part because we are doing modeling. However, it is not sacrosanct. In single pass systems, where absolute accuracy is more important, there are many instances which force us to sacrifice self-consistency.

This concludes the pictorial chapter. We are now ready to enter a more complete discussion of the mathematical tools needed for understanding and describing the physics of circular rings.

PART II
Global Description: Purpose

CHAPTER 4
CLASSIFICATION OF ONE-TURN MAPS

In this chapter, we intend to classify as fully as we can the main types of one-turn maps we encounter in rings. This classification is based on perturbation theory, i.e., maps viewed as k-jets. We will examine the motion around the fixed point and make some statements about the full map based on normalization methods. As such it is incomplete in its ability to predict the behavior of a ring: chaos is absent from normalized maps.

So why do we care? Our understanding of the behavior of a ring is usually based on comparisons between some exactly solvable models and the actual complex model represented in the simulation code or the ring itself. For example, we examine departure from linearity simply because in the region where the system is linear all its properties are understood. We can concentrate on the nonlinear effects. So, it can be said that a great part of accelerator theory is an attempt to map the actual system into a solvable one and study deviations from this solvable model. A simple and elegant theory will thus be capable of providing a systematic method to execute this task.

Sometimes, in the presence of chaos, one even tries to map the actual system into a simpler one which is not exactly solvable. For example, in the study of the beam-beam force in the weak-strong approximation, it has been found useful to map the real system into a simpler two-resonance model. The heuristic Chirikov criterion is then applied leading to useful results.

In this book, the emphasis is on the known rather than the unknown. Therefore, we will try to restrict ourselves to rigorously solvable problems. We will find that normalization theory gives a unifying treatment and technique for the study of exactly solvable models derived from the real system. There are two complementary ways to proceed.

1. First, the techniques based on a Taylor series expansion of the

map around a fixed point. These techniques will make use of the Lie formalism originally developed by Hori in two seminal papers[15, 16]. These techniques are not restricted to Taylor series representations, but do require the ability to manipulate an analytic representation of the map or the Hamiltonian. We will find that all of standard accelerator theory boils down to a normalization of a map (usually the one-turn map).

2. Secondly, there are some techniques where the map is computed by fitting tracking data. These techniques allow for the computation of quasi-invariants valid at large amplitude. The map and the canonical transformation used to compute the invariants are expressed as Fourier series in the angles, whose coefficients are interpolated functions of the actions. More recently Laskar has proposed other numerical techniques to study departure from integrability, they are also connected to Fourier analysis[14].

In this book, the first technique is emphasized because it has a more day-to-day applicability. Normalization with fitted maps is still an area of active research.

4.1 THE QUEST FOR INVARIANTS

In Figures 2.1, 2.2, 2.3, and 2.4, we displayed the phase space of a simple map of $\mathbf{R}^2$. Then in Chapter 2, we argued that a large portion of accelerator theory is an attempt to find and parameterize the quasi-invariant surfaces (curves in 2-d) which one sees on a phase plot. Therefore, given a map $\vec{\zeta}$, the central theoretical quest of accelerator theory is to understand the nature of the functional equation

$$\text{find } f \text{ such that } f \circ \vec{\zeta} = f, \tag{4.1}$$

and, given an (approximate) solution for f, understand numerically and analytically how it depends on the control parameters. Of course, in realistic systems, the presence of chaos is an indication that Equation (4.1) has no solutions. However, to the extent that chaotic behavior is minimal (by design), it is still interesting to find approximate solutions of Equation (4.1). The procedure for finding an approximate solution will depend on the purpose of the calculation. The usefulness of the approximation is often checked by tracking the approximate invariant with a simulation code. The reader may look at page 200 for Poincaré's views on this issue.

We would like to study how features of the phase plot depend on the invariant(s) on which a particle sits. This is a daunting task, but this is what ring theory is all about. We are trying to understand and control the one-turn map on the basis of simple objects such as invariants.

We know that the Hamiltonian and canonical transformations on it are useful tools in the computation of approximate invariants. Unfortunately, the core of this book is based on the realization that in rings, small or big, the Hamiltonian is an extremely complex object. Thus I declare without reservation that a theory based on finite maps is superior both conceptually and computationally. Consequently, we could try to solve Equation (4.1) by brute force. In fact, this was suggested in Exercise 2-2.3. But, how more pleasing and powerful it would be, if the machinery of canonical transformations was applied to maps. In other words, is there a way to carry over the Hamiltonian formalism to maps?

In Sect. 2.3.2, we defined a "plug-in" or compositional map. In this more abstract setting, Equation (4.1) takes the form

$$\mathcal{M}f = f. \tag{4.2}$$

Finally, and of critical importance for the connection to Hamiltonian methods, it is found that maps of the form

$$\mathcal{M} = \exp\left(: g :\right) \tag{4.3}$$

are "plug-in" maps. In fact, their action on an arbitrary function f is equivalent with the composition of f with $\vec{\zeta} = \exp(: g :)\vec{I}$, i.e., $\exp(: g :)f = f \circ \vec{\zeta}$. The transformation $\vec{\zeta}$ is the action of the map $\exp(: g :)$ on the identity transformation $\vec{I}$.

Conversely, if all our transfer maps are symplectic maps, then their associated compositional maps can be represented by products of Lie transformations, i.e., products of $\exp(: g :)$-type maps.

In the presence of a deterministic non-Hamiltonian force (dissipation), such as classical radiation, we can ask similar questions. For example, Equation (4.1) can be modified for a dissipative system

$$f \circ \vec{\zeta} = e^{\alpha(f)} f. \tag{4.4}$$

The function α can be viewed as an f-dependent growth coefficient. Limit cycles are easily found to be given by the equation $\alpha(f) = 0$.

As in the case of symplectic maps, the compositional map can be expressed with the help of Lie transformations of the form $\exp(\vec{F}\cdot\vec{\nabla})$.

Now we will see how perturbation theory is a search for solutions of Equation (4.1) and other related functional equations.

4.2 EXISTENCE OF A CENTRAL CLOSED ORBIT

In order to solve Equation (4.2), we would like to expand around "the" closed orbit. Well, does it exist? If we jump to Part III, we can prove the existence of the closed orbit by construction. We remember that the designer uses ideal elements which bend the ideal particle on arcs of circles. Provided that the matrix around this ideal machine does not have unit eigenvalues, we know by the implicit function theorem that the existence of a fixed point is guaranteed for machines in the neighborhood of the ideal one. Because it is a fact that designers construct their machines around an ideal model, we could stop here and assume the existence of a closed orbit. However, we can justify its existence from a more abstract (and global) point of view.

Let us list the properties of an ideal ring:

►A. At a surface of section, which we place at location $i = 1$, we would like an open ball in $\mathbf{R}^{2N}$ to stay globally invariant under the repeated effect of the one-turn map $\vec{\zeta}_1$

$$\forall n > 0 \; \vec{\zeta}_1^{\,n}(B_1) = \underbrace{\left(\vec{\zeta}_1 \circ \cdots \circ \vec{\zeta}_1\right)}_{n \text{ times}}(B_1) \subset B_1. \tag{4.5}$$

►B. We then call the maximal ball $\widehat{B}_1$ satisfying ►A the "dynamic aperture."

►C. If the transfer map from location 1 to location k is given by $\vec{\zeta}_{1\to k}$, then the dynamic aperture B_k at k is obtained from B_1 by the formula

$$B_k = \vec{\zeta}_{1\to k}(B_1). \tag{4.6}$$

►D. Then, one can also show (Brower fixed point theorem in topology) that there must exist a fixed point for the one-turn map $\vec{\zeta}_1$:

$$\exists \vec{f}_1 \in B_1 \;\rightarrow\; \vec{\zeta}_1(\vec{f}_1) = \vec{f}_1. \tag{4.7}$$

Of course, at location k, the fixed point is

$$\vec{f}_k = \vec{\zeta}_{1\to k}(\vec{f}_1). \tag{4.8}$$

The initial statement ▶A is a dream. It is dubious that one can ever produce a model of the single particle dynamics of an accelerator which is realistic and which obeys ▶A except in the case of a classically radiating electron. Hamiltonian maps in two degrees of freedom or more will always undergo Arnold diffusion which strictly speaking turns the dynamic aperture into a set of measure zero (i.e., a collection of fixed points). But it is certainly fine to have dreams! And, based on this dream, it is clear why the design process of an accelerator should start with an ideal machine which, by construction, has a closed orbit. In fact, we will see later that this ideal orbit is important (but not essential) in defining the layout of the ring in the local treatment of Part III as it was shown in the gutter example of Chapter 3. Thus the accelerator physicist starts with an ideal orbit which is then modified during the modeling and simulation processes. Because we always stay away from integer tunes for reasons of stability (see Chapter 2), the implicit function theorem guarantees that this fixed point survives.

4.3 THE *K*-JETS AROUND THE CLOSED ORBIT

So now that we know that the dream ring should have a fixed point, let us assume that the real one does. Therefore, our first inclination is to expand the motion around this closed orbit as we described in Sect. 3.4. Given a one-turn map $\vec{\zeta}$, we expand it around its fixed point $\vec{z}_0$ to order k. Such maps, denoted by $\vec{\zeta}^{[k]}$, equipped with truncation rules at degree k, belong to the (differential) algebra of k-jets as explained in Sect. 3.4.3.

Let us consider a single k-jet, $f^{[k]}$, which can be viewed as the kth degree expansion of an arbitrary function f. Because we are around the fixed point we will assume that all our functions have no constant[4.1] terms, i.e., $f(\vec{0}) = 0$. Therefore, using the homomorphism

[4.1] The constant term of a function may be important (tunes for example), however when we manipulate k-jets we set them to zero. It is when we reinterpret the k-jet as an actual function that we put then back in. There is a temptation to redefine k-jets as a triplet: the initial value of the function, its final value, and the actual k-jet. The constant terms are left out of any jet calculation so as to preserve the differential algebra but their existence

of Equation (3.70), we can rewrite the k-jet equivalent of Equation (4.2):

$$\mathcal{M}f = f \circ \vec{\zeta} = f$$
$$\Downarrow$$
$$f^{[k]} \circ \vec{\zeta}^{[k]} = f^{[k]}. \tag{4.9}$$

To proceed further, we have two choices—either brute force or the use of Lie transforms $\vec{F} \cdot \vec{\nabla}$ and their associated Lie transformations $\exp(\vec{F} \cdot \vec{\nabla})$. Let us first start with brute force.

To solve Equation (4.9), we expand all the jets involved in the monomial basis. Any jet $f^{[k]}$ can be expanded in terms of monomials:

$$f^{[k]} = \sum_{|\vec{m}|=1}^{k} A_{\vec{m}} \, P_{\vec{m}}^{[k]}$$
where

$$\mathbf{R}^{2N} \xrightarrow{P_{\vec{m}}} \mathbf{R}^{2N}$$
$$\vec{z} \longmapsto P_{\vec{m}}(\vec{z}) = z_1^{m_1} \cdots z_{2N}^{m_{2N}}. \tag{4.10}$$

Of course, the k-jet map $\vec{\zeta}^{[k]}$ can be expanded in a similar manner:

$$\vec{\zeta}^{[k]} = \sum_{|\vec{m}|=1}^{k} \vec{Z}_{\vec{m}} \, P_{\vec{m}}^{[k]}. \tag{4.11}$$

Using the monomial representation, we can express Equation (4.9) in matrix form:

$$\begin{aligned} f^{[k]} \circ \vec{\zeta}^{[k]} &= \sum_{|\vec{m}|=1}^{k} A_{\vec{m}} \left\{ P_{\vec{m}}^{[k]} \circ \vec{\zeta}^{[k]} \right\} \\ &= \sum_{|\vec{m}|=1}^{k} \sum_{|\vec{n}|=1}^{k} A_{\vec{m}} \left\{ \mathbf{M}_{\vec{m}\vec{n}} P_{\vec{n}}^{[k]} \right\}. \end{aligned} \tag{4.12}$$

---

allows us to restore the k-jet to the status of a function. I prefer to keep the constant term implicit but invisible so as to keep the notation clean. In Interlude XI one can see a typical algorithm (Newton search) which entails successive metamorphoses between jets and functions. Of course this is true for any fitting algorithm.

Symbolically, we can write the effect of the compositional map on an arbitrary function as

$$\mathcal{M}f = f' \Rightarrow (\mathcal{M}f)^{[k]} = f'^{[k]} \Leftrightarrow \mathbf{M}^{\mathsf{T}}\vec{\mathbf{A}} = \vec{\mathbf{A}}'. \tag{4.13}$$

The component vector $\vec{\mathbf{A}}$ has $\frac{(k+2N)!}{k!(2N)!}-1$ entries corresponding to all the possible monomials from degree 1 to k. The matrix $\mathbf{M}^{\mathsf{T}}$ represents the action of the compositional map $\mathcal{M}$ acting on the components $\vec{\mathbf{A}}$. The search for invariants reduces to a linear algebra problem of finite, but often large, dimension. Thus a jet is invariant if and only if

$$\mathbf{M}^{\mathsf{T}}\vec{\mathbf{A}} = \vec{\mathbf{A}}. \tag{4.14}$$

In other words, we are searching for the unit eigenvectors of the matrix $\mathbf{M}^{\mathsf{T}}$.

Exercise 4-1

In this exercise, we revisit Exercise 2-2.3. The computation of the invariant around the fixed point using "brute force."

1. *Find the matrix* $\mathbf{M}^{\mathsf{T}}$ *for 2-jets in the case of* $d = 0$. *Answer:*

$$\mathbf{M}^{\mathsf{T}} = \begin{pmatrix} 1 & -\frac{1}{2} & 0 & 0 & 0 \\ \frac{1}{2} & \frac{3}{4} & 0 & 0 & 0 \\ 0 & 1 & 1 & -\frac{1}{2} & \frac{1}{4} \\ 0 & 1 & 1 & \frac{1}{2} & -\frac{3}{4} \\ 0 & \frac{1}{4} & \frac{1}{4} & \frac{3}{8} & \frac{9}{16} \end{pmatrix}. \tag{4.15}$$

Here we chose the basis $\{x, p, x^2, xp, p^2\}$ *for the 2-jets.*

2. *Find the eigenvalues of* $\mathbf{M}^{\mathsf{T}}$. *Answer:*

$$\lambda = 1, \frac{7 \pm i\sqrt{15}}{8}, \frac{17 \pm i7\sqrt{15}}{32}. \tag{4.16}$$

One should recognize $\exp(\pm i\mu)$ *and* $\exp(\pm i2\mu)$.

3. *Show that the quadratic invariant is* $x^2 + \frac{1}{2}xp + p^2$.

Direct solution of the functional equation is not very illuminating. Nevertheless, on some occasions, when the map is presented to us in a nonanalytic form, such as tracking data, there are some compelling reasons to solve the functional equation directly with methods not involving Lie transforms.

4.4 PERTURBATION THEORY: LIE METHODS

Instead of brute force, we want to make use of the relationship between the compositional map and Hamiltonian techniques. As we mentioned in Interludes I and II, the Lie transforms allow us to exploit the Hamiltonian and differential algebraic nature of the compositional maps. In the case of k-jets around a closed orbit, there is a clearly defined homomorphism between functions and k-jets. This is a property which we will exploit in the rest of Part II. It ties together standard perturbation theory and the map-based approach of this book.

The basic idea behind Lie methods applied to one-turn maps emerges from the realization that the one-turn compositional map can be expressed in terms of Lie transforms. In general, the k-jet compositional map $\mathcal{M}$ associated with $\vec{\zeta}$ can be expressed as follows:

$$\mathcal{M} = \underbrace{\mathcal{L}}_{\substack{\text{Linear}\\ \text{part}}} \underbrace{\exp\left(\vec{F}^{[2]} \cdot \vec{\nabla}\right) \ldots \exp\left(\vec{F}^{[k]} \cdot \vec{\nabla}\right)}_{\text{nonlinear jets}}$$

$$\text{with} \quad \forall \vec{x} \rightarrow F_i^{[j]}(\lambda \vec{x}) = \lambda^j F_i^{[j]}(\vec{x}). \tag{4.17}$$

This is the standard Dragt-Finn factorization. If the map $\vec{\zeta}$ is well behaved, one can show that the Lie transforms $\vec{F}^{[n]} \cdot \vec{\nabla}$ are themselves well behaved. Therefore, this is a perfectly reasonable way to represent a compositional jet-map as well as a real map. In the case of a Hamiltonian system, this factorization becomes:

$$\mathcal{M} = \underbrace{\mathcal{L}}_{\substack{\text{Linear}\\ \text{part}}} \underbrace{\exp\left(: f_3 :\right) \ldots \exp\left(: f_{k+1} :\right)}_{\text{nonlinear jets}},$$

$$\text{where} \quad : f_{n+1} : \equiv -\mathrm{J}\vec{\nabla} f_{n+1} \cdot \vec{\nabla}. \tag{4.18}$$

The Dragt-Finn factorization takes advantage of the quasi-triangular structure of $\mathbf{M}^{\mathrm{T}}$. If we are interested in k-jets only, then it is necessary to compute the Dragt-Finn map only up to the factor $\exp(\vec{F}^{[k]} \cdot \vec{\nabla})$.

And finally, as we will show later, the usage of the Lie operators $\vec{F}^{[k]} \cdot \vec{\nabla}$ will keep the maps, in the presence of radiation, near the symplectic group. This is essential in an elegant theory which must reduce to the usual Hamiltonian theory as the operators become symplectic.

4.4.1 The Linear Part of the Map

The topic of this section is covered in Sect. 2.4 in a less abstract setting.

Linear accelerator theory is based on the analysis of the linear compositional map $\mathcal{L}$. Let us assume that the ring is truly linear and represented by $\mathcal{L}$. This map will transform a homogeneous polynomial f_n of degree n into a new homogeneous polynomial of degree n. The proof is trivial:

$$
\begin{aligned}
\forall\lambda \in \mathbb{R} \text{ and } \forall\vec{z} \in \mathbb{R}^{2N} \quad (\mathcal{L}f_n)(\lambda\vec{z}) &= f_n(L\lambda\vec{z}) \\
\because\ L \text{ is a matrix (linear map)} \qquad & \\
&= f_n(\lambda L\vec{z}) \\
\because\ f_n \text{ is homogeneous} \qquad & \\
&= \lambda^n f_n(L\vec{z}). \qquad (4.19)
\end{aligned}
$$

It follows that the dynamics of k-jets and polynomials of degree k is rigorously equivalent. However, more importantly, if we can fully normalize (diagonalize) the action of $\mathcal{L}$ on linear monomials, we will have normalized the full map $\mathcal{L}$ or equivalently the matrix $\mathbf{L}^{\mathsf{T}}$ associated with $\mathcal{L}$. Again the reasons for this follow from the definition of $\mathcal{L}$:

$$
\begin{aligned}
&\text{if } \exists v_a\ a = 1, 2N \qquad v_a(\vec{z}) = \sum_{i=1,2N} V_a^i z_i \text{ and } \mathcal{L}v_a = \lambda_a v_a \\
&\qquad\text{then} \\
&\mathcal{L}v_1^{m_1} \cdots v_{2N}^{m_{2N}} = (\mathcal{L}v^{m_1}) \cdots (\mathcal{L}v^{m_{2N}}) \\
&\qquad = \lambda_1^{m_1} \cdots \lambda_{2N}^{m_{2N}} v_1^{m_1} \cdots v_{2N}^{m_{2N}}. \qquad (4.20)
\end{aligned}
$$

From Equation (4.20) we conclude that a monomial $v_1^{m_1} \cdots v_{2N}^{m_{2N}}$ is invariant under $\mathcal{L}$ if and only if $\lambda_1^{m_1} \cdots \lambda_{2N}^{m_{2N}}$ is equal to one.

Therefore, we create the matrix $\mathbf{L}^{\mathsf{T}}$ for 1-jets in the usual projection basis:

$$\text{Basis} = \{I_1, \cdots, I_{2N}\}$$

$$\begin{aligned} &\mathbf{R}^{2N} \xrightarrow{I_i} \mathbf{R}^{2N} \\ &\vec{z} \longmapsto I_i(\vec{z}) = z_i. \end{aligned} \tag{4.21}$$

Quite often, as we did in Part I, the projection functions are denoted as $\{z_1, \cdots, z_{2N}\}$ or even as $\{x_1, p_1, \cdots x_N, p_N\}$. The reader must exercise caution when notation "overloading" takes place. The matrix $\mathbb{L}^\intercal$ for 1-jets is easily recognized to be $L^\intercal$ (see Sect. 2.4 if in doubt).

The next step is to put the matrix $L^\intercal$ in real canonical Jordan normal form. For example, if we have a four-dimensional stable map with distinct tunes on the unit circle, then it can be normalized as follows:

$$\begin{aligned} &A^\intercal L^\intercal A^{\intercal-1} = R^\intercal \\ &\text{where } R^\intercal = \begin{pmatrix} R_1^\intercal & 0 \\ 0 & R_2^\intercal \end{pmatrix} \\ &\text{and } R_i^\intercal = \begin{pmatrix} \cos(\mu_i) & -\sin(\mu_i) \\ \sin(\mu_i) & \cos(\mu_i) \end{pmatrix}. \end{aligned} \tag{4.22}$$

We can go further and diagonalize the matrix $R^\intercal$ using a complex transformation (see Sect. 2.4.5):

$$\begin{aligned} &S^\intercal = C^\intercal R^\intercal C^{\intercal-1} = \begin{pmatrix} e^{-i\mu_1} & 0 & 0 & 0 \\ 0 & e^{i\mu_1} & 0 & 0 \\ 0 & 0 & e^{-i\mu_2} & 0 \\ 0 & 0 & 0 & e^{i\mu_2} \end{pmatrix} \\ &C^\intercal = \frac{1}{2}\begin{pmatrix} 1 & -i & 0 & 0 \\ 1 & i & 0 & 0 \\ 0 & 0 & 1 & -i \\ 0 & 0 & 1 & i \end{pmatrix}. \end{aligned} \tag{4.23}$$

With the help of $C^{\intercal-1}$, the eigenvectors of $R^\intercal$ are found to be

$$\vec{R}_\pm^k = \left(0, 0, \cdots, \underbrace{1}_{R_{2k-1}}, \underbrace{\pm i}_{R_{2k}}, \cdots, 0, 0\right) \quad k = 1, N \quad , \tag{4.24}$$

from which we deduce the eigenfunctions of the compositional map $\mathcal{R}$ whose matrix is $R^\intercal$

$$\begin{aligned} &r_{k\pm}(\vec{z}) = z_{2k-1} \pm i z_{2k} \quad k = 1, N \\ &\mathcal{R} r_{k\pm} = \exp(\mp i\,\mu_k)\, r_{k\pm}. \end{aligned} \tag{4.25}$$

It is easy to check that the functions $J_k = r_{k+}r_{k-}$ are invariants of $\mathcal{R}$. Also, one can check that the map

$$\exp\left(: - \sum_{k=1}^{N} \mu_k J_k : \right)$$

also leaves the functions J_k invariant. Since this map must be a compositional map according to the results of Interlude II, it suffices to check its effects on linear functions (i.e., 1-jets) to prove that it is the map $\mathcal{R}$. It then follows that the map $\mathcal{L}$ is given by

$$\begin{aligned} \mathcal{L} &= \mathcal{A}^{-1} \exp\left(: - \sum_{k=1}^{N} \mu_k J_k : \right) \mathcal{A} \\ &= \exp\left(: - \sum_{k=1}^{N} \mu_k \mathcal{A}^{-1} J_k : \right) \end{aligned} \tag{4.26}$$

where $\varepsilon_k(\vec{z}) = (2\mathcal{A}^{-1} J_k)(\vec{z}) = (A^{-1}_{2k-1\ j}\ z_j)^2 + (A^{-1}_{2k\ j}\ z_j)^2$.

In the case of a truly linear ring, the study of L^{T} restricted to 1-jets provides all the invariants of the full map $\mathcal{L}$. The reader will recognize the functions ε_k as the Courant-Snyder invariants of a fully coupled system. It is worthwhile to point out that the eigenfunctions of the full map $\mathcal{L}$ are just

$$\begin{aligned} h^k_{\pm} &= \mathcal{A}^{-1} r^k_{\pm} \\ &= A^{-1}_{2k-1\ j}\ x_j \pm i\ A^{-1}_{2k\ j}\ x_j. \end{aligned} \tag{4.27}$$

A large section of accelerator theory is based on a linear analysis of the motion—Courant-Snyder theory, coupled Courant-Snyder theory, computation of damping in the presence of radiation, i.e., synchrotron integrals, linear lattice design, etc...—these fields can be unified by a map treatment based on the idea of normalization.

4.4.2 The Nonlinear Part

Ultimately we are interested in the full $\left\{\frac{(k+2N)!}{k!(2N)!} - 1\right\} \times \left\{\frac{(k+2N)!}{k!(2N)!} - 1\right\}$ matrix $\mathbf{M}^{\mathsf{T}}$. Here, it is best to think in terms of the Dragt-Finn representation.

Using the linear map $\mathcal{A}$, which normalizes $\mathcal{L}$ exactly, we transform $\mathcal{M}$ into:

$$
\begin{aligned}
\mathcal{N}_1 &= \mathcal{A}\mathcal{M}\mathcal{A}^{-1} \\
&= \underbrace{\mathcal{A}\mathcal{L}\mathcal{A}^{-1}}_{\text{Linear part}} \underbrace{\mathcal{A}\exp\left(\vec{F}^{[2]}\cdot\vec{\nabla}\right)\mathcal{A}^{-1}\ldots\mathcal{A}\exp\left(\vec{F}^{[k]}\cdot\vec{\nabla}\right)\mathcal{A}^{-1}}_{\text{nonlinear jets}} \\
&= \underbrace{\mathcal{R}}_{\text{Linear part}} \underbrace{\exp\left(\mathcal{A}\vec{F}^{[2]}\cdot\vec{\nabla}\mathcal{A}^{-1}\right)\ldots\exp\left(\mathcal{A}\vec{F}^{[k]}\cdot\vec{\nabla}\mathcal{A}^{-1}\right)}_{\text{nonlinear jets}} \\
&= \underbrace{\mathcal{R}}_{\text{Linear part}} \underbrace{\exp\left(\overrightarrow{\mathcal{A}\vec{F}^{[2]}}\cdot\vec{\nabla}\right)\ldots\exp\left(\overrightarrow{\mathcal{A}\vec{F}^{[k]}}\cdot\vec{\nabla}\right)}_{\text{nonlinear jets}}. \qquad (4.28)
\end{aligned}
$$

The clumsy notation $\overrightarrow{\mathcal{A}\vec{F}}$ emphasizes that the transformational properties of the vector field are not those of a function, i.e., $\overrightarrow{\mathcal{A}\vec{F}} \neq \vec{F}\circ\vec{\alpha}$ where $\vec{\alpha}(\vec{x}) = A\vec{x}$. In fact, they are given by

$$\left.\overrightarrow{\mathcal{A}\vec{F}}\right|_k = \left(\vec{F}\cdot\vec{\nabla}\alpha_k^{-1}\right)\circ\vec{\alpha}, \qquad (4.29)$$

which, in the symplectic case, reduces to

$$
\begin{aligned}
&\text{if } \exists f \rightarrow \vec{F}\cdot\vec{\nabla} =: f : \\
&\Rightarrow \mathcal{A} : f : \mathcal{A}^{-1} =: \mathcal{A}f :=: f\circ\vec{\alpha} : . \qquad (4.30)
\end{aligned}
$$

Exercise 4-2
Prove Equation (4.29).

The goal of perturbation theory is to simplify $\mathcal{N}_1$ using a canonical transformation which will not affect the linear part. This can be done by a succession of transformations

$$
\begin{aligned}
\mathcal{N}_1 &= \mathcal{A}\mathcal{M}\mathcal{A}^{-1} \\
\mathcal{N}_2 &= \exp\left(\vec{A}^{[2]}\cdot\vec{\nabla}\right)\mathcal{N}_1\exp\left(-\vec{A}^{[2]}\cdot\vec{\nabla}\right) \\
&\vdots \\
\mathcal{N}_k &= \exp\left(\vec{A}^{[k]}\cdot\vec{\nabla}\right)\mathcal{N}_{k-1}\exp\left(-\vec{A}^{[k]}\cdot\vec{\nabla}\right) \\
&\Downarrow
\end{aligned}
$$

$$\mathcal{M} = \mathcal{A}^{-1} \exp\left(-\vec{A}^{[2]} \cdot \vec{\nabla}\right) \cdots \exp\left(-\vec{A}^{[k]} \cdot \vec{\nabla}\right)$$
$$\times \mathcal{N}_k \exp\left(\vec{A}^{[2]} \cdot \vec{\nabla}\right) \cdots \exp\left(\vec{A}^{[k]} \cdot \vec{\nabla}\right) \mathcal{A}. \qquad (4.31)$$

At the jth step of the normalization, the map is left unchanged at orders lower than j. Suppose we factorize the map $\mathcal{N}_{j-1}$

$$\mathcal{N}_{j-1} = \mathcal{R} \exp\left(\vec{N}^{[2]} \cdot \vec{\nabla}\right) \cdots \exp\left(\vec{N}^{[j-1]} \cdot \vec{\nabla}\right)$$
$$\times \exp\left(\vec{F}^{[j]}_{(j-1)} \cdot \vec{\nabla}\right) \cdots \exp\left(\vec{F}^{[k]}_{(j-1)} \cdot \vec{\nabla}\right), \qquad (4.32)$$

where the operator $\vec{F}^{[n]}_{(j-1)} \cdot \vec{\nabla}$ is the Dragt-Finn Lie transform of order n at the $(j-1)^{th}$ iteration. The operators $\vec{N}^{[m]} \cdot \vec{\nabla}$, $m = 1, j-1$, are the normalized operators and thus do not change during the jth iteration. Because of the nature of the Dragt-Finn factorization, we can derive an equation for $\vec{N}^{[j]} \cdot \vec{\nabla}$:

$$\vec{N}^{[j]} \cdot \vec{\nabla} = \mathcal{R}^{-1} \vec{A}^{[j]} \cdot \vec{\nabla} \mathcal{R} - \vec{A}^{[j]} \cdot \vec{\nabla} + \vec{F}^{[j]}_{(j-1)} \cdot \vec{\nabla}$$
$$= \left\{\exp\{-\mathrm{ad}(\vec{N}^{[1]} \cdot \vec{\nabla})\} - \mathcal{I}\right\} \vec{A}^{[j]} \cdot \vec{\nabla} + \vec{F}^{[j]}_{(j-1)} \cdot \vec{\nabla}$$
$$\text{given } \mathcal{R} = \exp\left(\vec{N}^{[1]} \cdot \vec{\nabla}\right)$$
$$\text{and } (\mathrm{ad}\widehat{O}_1)\widehat{O}_2 = \{\widehat{O}_1, \widehat{O}_2\} = \widehat{O}_1\widehat{O}_2 - \widehat{O}_2\widehat{O}_1. \qquad (4.33)$$

In the rest of this section, we will restrict the discussion to symplectic maps for simplicity. In Sect. 6.4.3 we will reproduce the results of this section for a general map. In the case of a symplectic map, Equation (4.33), after using several homomorphisms (see Sect. 6.1.1), reduces to

$$n^{[j]} = \left\{\exp\left(:-n^{[1]}:\right) - \mathcal{I}\right\} a^{[j]} + f^{[j]}_{(j-1)}$$
$$\text{given } \mathcal{R} = \exp\left(:n^{[1]}:\right), \qquad (4.34)$$

where $:g: = \vec{G} \cdot \vec{\nabla}$. The reader might have noticed that nothing needs to be said about the nature of the normalized map $\mathcal{R}$. Nevertheless, it is instructive to go back to our rotation example. In this case, Equation (4.34) becomes

$$n^{[j]} = \left\{\exp\left(:\vec{\mu} \cdot \vec{J}:\right) - \mathcal{I}\right\} a^{[j]} + f^{[j]}_{(j-1)}, \qquad (4.35)$$

which can be simplified if we use the eigenfunctions of $\mathcal{R}$:

$$\text{if} \quad \begin{cases} n^{[j]} = \displaystyle\sum_{|\vec{m}^+|+|\vec{m}^-|=j} n_{\vec{m}}\, r_{1+}{}^{m_1^+} r_{1-}{}^{m_1^-} \cdots r_{N-}{}^{m_N^-} \\ \text{where } \vec{m} = (\vec{m}^+, \vec{m}^-) \text{ similarily for } a^{[j]} \text{ and } f^{[j]}_{(j-1)} \end{cases}$$

$$\Rightarrow n_{\vec{m}} = \left\{\exp\{i\vec{\mu}\cdot(\vec{m}^+ - \vec{m}^-)\} - 1\right\} a_{\vec{m}} + f_{\vec{m}}. \tag{4.36}$$

Because $\mathcal{R}$ is a rotation, we can immediately "purify" the map using Equation (4.36) into

$$a_{\vec{m}} = \frac{f_{\vec{m}}}{1 - \exp\{i\vec{\mu}\cdot(\vec{m}^+ - \vec{m}^-)\}} \text{ if } \frac{1}{2\pi}\vec{\mu}\cdot(\vec{m}^+ - \vec{m}^-) \notin \mathbf{Z},$$
$$a_{\vec{m}} = 0 \text{ if } \frac{1}{2\pi}\vec{\mu}\cdot(\vec{m}^+ - \vec{m}^-) \in \mathbf{Z}. \tag{4.37}$$

This equation is central to perturbation theory and it exhibits the famous small denominators. In the case for which the tunes $\nu_i = \mu_i/2\pi$ are prime amongst each other (i.e., $\vec{\nu}\cdot(\vec{m}^+ - \vec{m}^-) \in \mathbf{Z} \Rightarrow \vec{m}^+ - \vec{m}^- = \vec{0}$) then it follows that the purified Lie operator $:n^{[j]}:$ is the generator of a rotation

$$n^{[j]} = \sum_{|\vec{m}^+| = \frac{j+1}{2}} n_{\vec{m}}\, (r_{1+}r_{1-})^{m_1^+} \cdots (r_{N+}r_{N-})^{m_N^+}$$
$$= \sum_{|\vec{m}^+| = \frac{j+1}{2}} n_{\vec{m}}\, 2^{\frac{j+1}{2}} J_1{}^{m_1^+} \cdots J_N{}^{m_N^-}. \tag{4.38}$$

The final map $\mathcal{N}_k$ is also a rotation and can be written exactly with one Lie exponent:

$$\mathcal{N}_k = \mathcal{R}\exp\left(:n^{[3]}:\right)\exp\left(:n^{[5]}:\right)\cdots\exp\left(:n^{[k]}:\right)$$
$$= \exp\left(:\sum_{j=0}^{(k-1)/2} n^{[2j+1]}\right). \tag{4.39}$$

The Lie exponents of the Dragt-Finn representation commute with each other because they only depend on the actions J_i's. The full compositional map $\mathcal{N}_k$ leaves the actions invariant. To get an expression for the normalized map in phase space we expressed it in action-angle variables using Equation (2.54). In the sloppy notation for derivatives, the effect of $\mathcal{N}_k$ on some vector $\vec{z}$ is given by

$$z_{2j-1}^{final} = x_j^{final} = \sqrt{2J_j^{final}}\cos\left(\phi_j^{final}\right)$$

$$= \sqrt{2J_j}\cos(\phi_j + \Delta_j)$$

$$\text{where } \Delta_j = -\frac{\partial}{\partial J_j}\sum_{m=0}^{(k-1)/2} n^{[2m+1]}. \tag{4.40}$$

A similar expression follows for the momenta. Hence the map in phase space is simply:

$$\begin{aligned} z_{2j-1}^{final} &= \cos(\Delta_j)\, z_{2j-1} + \sin(\Delta_j)\, z_{2j} \\ z_{2j}^{final} &= \cos(\Delta_j)\, z_{2j} - \sin(\Delta_j)\, z_{2j-1}. \end{aligned} \tag{4.41}$$

Exercise 4-3
Making use of the exact definition for the action-angle variables, revisit Exercise 2-13 and explain the following formulas.

$$\mathcal{N} = \mathcal{T}_{\varpi}\mathcal{N}_k\mathcal{T}_{\varpi}^{-1}$$

$$\Rightarrow \mathcal{N}\begin{pmatrix} I_{2j-1} \\ I_{2j} \end{pmatrix} = \begin{pmatrix} I_{2j-1} - \partial_{2j}\sum_{m=0}^{(k-1)/2} n^{[2m+1]}\circ\varpi \\ I_{2k} \end{pmatrix}$$

$$\Delta_j = \left\{ -\partial_{2j}\sum_{m=0}^{(k-1)/2} n^{[2m+1]}\circ\varpi \right\}\circ\varpi^{-1}. \tag{4.42}$$

Lie methods allow us to use Hamiltonian perturbation theory on a representation of the nonlinear one-turn map. It gives us a way to extend to the nonlinear problem, at least formally, the normalization done on the linear part $\mathcal{L}$. Equations (4.33) or (4.34), which must be solved at every order in the perturbation, have a structure which depends critically on the linear map $\mathcal{L}$. Therefore, in a book devoted to accelerator theory, it is important to classify the linear problems which we encounter most often. This is the topic of Sect. 4.5. However, we must now reflect on what perturbation theory achieves or fails to achieve.

4.4.3 Information in k-jets: Zero...What?!

It is the contention of this book that the study of the one-turn map $\vec{\zeta}$ or its compositional companion $\mathcal{M}$ is the heart of ring theory. To be more precise we are interested in $\vec{\zeta}^n$, the nth iterate of the map, for large values of n. We have seen in Chapter 2 how the structure exhibited by the map $\vec{\zeta}^n$ can be very complex: islands, stable and unstable manifolds, chaos, etc. This fact found in simulations cannot be ignored. Needless to say, powers of the compositional map must lead to even more complex behavior: after all, the map $\mathcal{M}^n$ contains $\vec{\zeta}^n$ since $\vec{\zeta}^n = \mathcal{M}^n \vec{I}$. It is therefore natural to ask what information, if any is to be found in the k-jet restriction of $\mathcal{M}^n$. The answer to this is: *none*! It is actually very easily seen. Consider this sequence of equations where all the maps are k-jets:

$$\begin{aligned} &\mathcal{M}^n = \mathcal{B}^{-1}\mathcal{R}^n\mathcal{B} \\ \overset{k\text{-jets}}{\Longleftrightarrow}\ &\mathbf{M}^{\top n} = \mathbb{B}^{\top -1}\mathbb{R}^{\top n}\mathbf{B}^{\top}. \end{aligned} \tag{4.43}$$

Here $\mathcal{B}$ is the total normalization transformation

$$\mathcal{B} = \exp\left(: a^{[k]} :\right) \cdots \exp\left(: a^{[2]} :\right) \mathcal{A}. \tag{4.44}$$

We know that for the case of tunes which are prime amongst each other, the k-jet compositional map $\mathcal{R}$ is a rotation for all possible values of k. Hence the matrix $\mathbf{M}^{\top n}$ is gotten from Equations (4.12) and (4.41) by expanding the cosine and the sine of $n\Delta_k$. In all cases the matrix for $\mathbf{M}^{\top n}$ contains secular terms which grow with n. The effect of the canonical transformation $\mathcal{B}$ is negligible for large n. This result applies to all maps with the exception of resonant maps which form a set of measure zero in parameter space. We conclude that maps are rotations when viewed as k-jets no matter how large k is. This is obviously nonsense. As a corollary, one should not raise to a power a truncated Taylor series map such as $\vec{\zeta}^{[k]}$. It is bound to lose all meaning.

In contrast, when dealing with a perturbed Schrödinger equation, we do expand the perturbed wave function in terms of an infinite set of eigenfunctions of the unperturbed Hamiltonian. There, as for k-jet maps, we truncate the series. This leads to shifted eigenenergies whose accuracy can be improved by including more terms in the relevant expansion. As is shown in Sect. 9.1.3, the mechanics of classical and quantum perturbation theory are so similar that one would naïvely expect similar results. What is the problem with classical perturbation theory? It can be shown that usually what one gets from k-jets (i.e., the matrices $\mathbb{R}^{\top}$ and $\mathbb{B}^{\top}$) is usually a divergent series. Hence it is meaningless unless it is re-interpreted properly.

The purpose of perturbation theory, as we stated in the beginning of this chapter, is to produce exactly solvable models which are to be compared with actual simulations or experiments. Therefore, we are not interested in the trivial properties of k-jets, but in using k-jets to produce solvable models. By definition, the model is not a k-jet, but a new map or Hamiltonian which is studied as exactly as possible. For example, if our tracking data seem to lie on invariant tori (near the origin) and it is our purpose to describe this analytically, we are tempted to approximate the real map as $\mathcal{M} \approx \mathcal{B}^{-1}\mathcal{R}\mathcal{B}$ where $\mathcal{R}$ is a rotation. We then check numerically if the invariants of $\mathcal{B}^{-1}\mathcal{R}\mathcal{B}$ (i.e., $\mathcal{B}^{-1}J_i$) are good approximate invariants of $\mathcal{M}$. Another good example is resonant extraction: the k-jet map is not turned into a rotation but into a one-resonance k-jet which is then studied as a real map. What usually emerges is not a rotation, but an integrable map containing separatrices and islands.

In conclusion, nonlinear perturbation theory gives us a "pair of glasses" which we can use to filter data obtained from a more reliable source. This is to be contrasted with linear perturbation theory for which exact results and complete classification exist. This leads us to our next topic: the classification of linear maps. It is important for two reasons. First and foremost, many of the design criteria for a ring are defined in terms of the "paraxial optics." Even in extremely nonlinear machines, the beam occupies a small region of phase space during operation. Therefore, the interaction of the beam with users is defined primarily in terms of the linear map. For example, luminosity in a collider or brilliance in a light source may involve beta functions, etas, etc. Secondly, as we have just seen, perturbation theory is an attempt to extend to nonlinear jets the behavior observed in the linear régime. A system executing oscillations is transformed paraxially into a rotation, and, using k-jets, into an amplitude-dependent rotation. For all these reasons, we will now explore systematically the linear maps which one usually encounters in ring dynamics.

4.5 LINEAR PROPERTIES AROUND A FIXED POINT

The topology of phase space will be greatly influenced by the fixed points of the map as well as by the fixed point of the iterates of the map. Thus it is of tremendous theoretical importance to understand the motion around a fixed point and of enormous practical importance to understand the linear motion around the central fixed point.

To get a quick idea of the topology, it is best to display the eigenvalues of the linear map. For clarity, we pictured the case of a map in $\mathbf{R}^4$. Generally, in the symplectic case, the reader can imagine that additional planes are provided with eigenvalues that are distinct and on the unit circle.

4.5.1 The Symplectic Linear Map

The most important case of all, which motivated the original Courant-Snyder theory, is displayed in Figure 4.1. This is the case which we used extensively in Chapter 2 as well as in Sect. 4.4.1. In this case, the matrix L can be brought into a rotation by a linear symplectic transformation A as explained in Equation (4.22). Physically, this matrix represents transverse motion in a ring in the absence of an RF-cavity. In the presence of a cavity, the full six-dimensional phase space cannot be ignored. That is to say the energy of a particle and its time of flight (i.e., path length for ultra-relativistic particles) are interacting with the transverse variables. The matrix L can be brought into a rotation with three angles.

Figure 4.2 represents a rare case which we do not normally encounter in a machine, at least not for the "design" values of the control parameters. If we assume that the eigenvalues start as in Figure 4.1, then they can collide and leave the unit circle as we change some control parameters. This collision is called a Krein collision. It is also displayed in Figure 4.6.

In Figure 4.4, we display the classic integer resonance. One of the tune approaches a multiple of 2π and is then allowed to leave the unit circle. Of course, there is also the case of the half-integer tune as shown in Figure 4.5. The local topology of all these unstable situations has application in the study of resonant extraction and in the understanding of nonlinear resonances.

For all these symplectic cases we display the normalized Lie transformation $\mathcal{R}$. For the rotation case of Figure 4.1, the Lie operator for $\mathcal{R}$ is given by:

$$
\begin{aligned}
:h: &= -\sum_{i=1}^{N} \mu_i : \frac{z_{2i-1}^2 + z_{2i}^2}{2} : \\
&= - : \vec{\mu} \cdot \vec{J} : . \qquad (4.45)
\end{aligned}
$$

Therefore, we claim that there exists a linear symplectic transformation $\mathcal{A}$ such that

$$\mathcal{R} = \mathcal{A}^{-1} \mathcal{L} \mathcal{A} = \exp\left(: h :\right), \qquad (4.46)$$

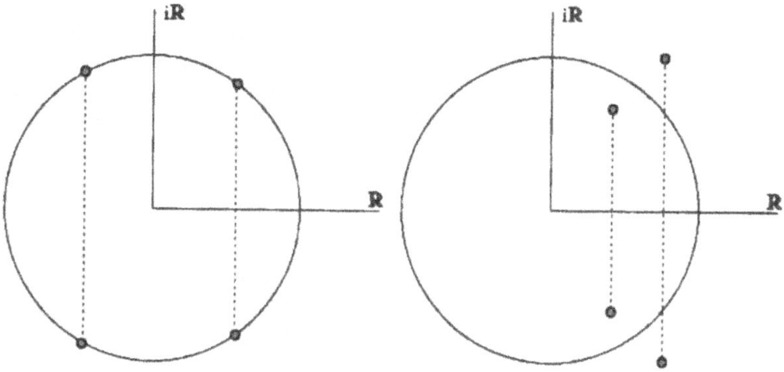

Figures 4.1 and 4.2: Rotation and Hyperbolic Spiral.

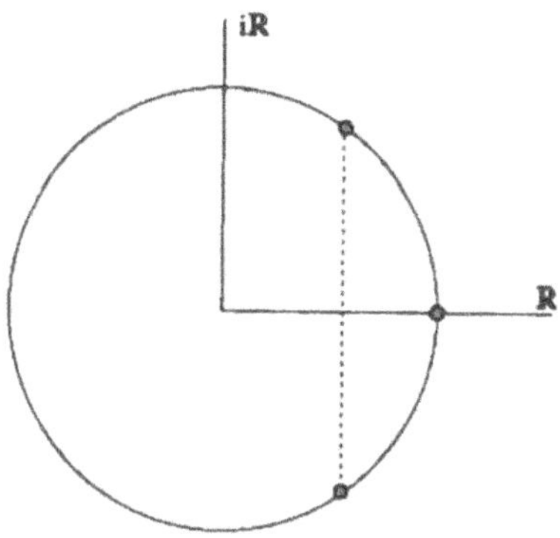

Figure 4.3: Coasting Beam.

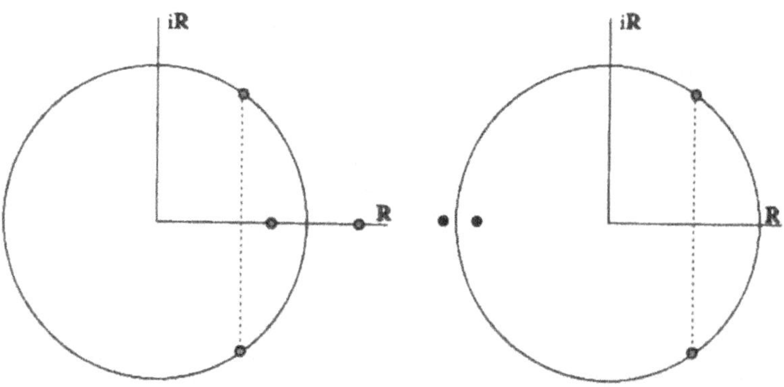

Figures 4.4 and 4.5: Integer and Half-Integer Cases.

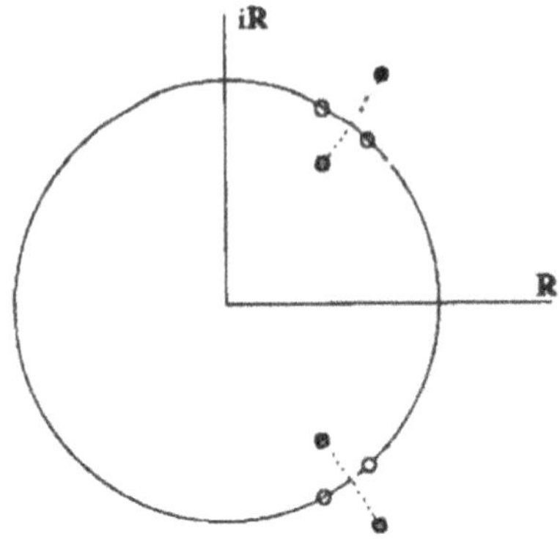

Figure 4.6: On a linear sum resonance $\mu_1 + \mu_2 = k2\pi$, the eigenvalues can leave the unit circle.

and that this transformation is constructed according to Equation (2.48). In fact, the procedure was described in detail in Sect. 2.4.2. Of course, if we indulge in the use of the resonance basis

$$(z_{2j-1}, z_{2j}) \overset{\vec{\mathcal{C}}^{-1}}{\longmapsto} \Big(\underbrace{z_{2j-1} + iz_{2j}}_{r_{j+}(\vec{z})} \, , \underbrace{z_{2j-1} - iz_{2j}}_{r_{j-}(\vec{z})} \Big) \tag{4.47}$$

as we did in Equation (4.23), then the map $\mathcal{R}$ becomes $\mathcal{S}$:

$$\begin{aligned} \mathcal{S} &= \exp\,(\widehat{s}) = \mathcal{C}\mathcal{R}\mathcal{C}^{-1} \\ \text{where } \widehat{s} &= \sum_{n=1}^{N} -i\,\mu_n (z_{2n-1}\partial_{2n-1} - z_{2n}\partial_{2n})\,. \end{aligned} \tag{4.48}$$

Notice that all homogeneous monomials are eigenfunctions of $\mathcal{S}$.

Next, let us examine the case of the coasting beam. We assume that one of the canonical variables, the energy, is a constant. For example, let it be the $(2N-1)^{th}$ variable, as is customary in accelerator dynamics [4.2]

$$:h_c:=:\left\{-\sum_{i=1}^{N-1}\mu_i\frac{z_{2i-1}^2+z_{2i}^2}{2}\right\}+\frac{1}{2}C\alpha_c z_{2N-1}^2: . \quad (4.49)$$

In the usual case $N = 3$, the transverse planes undergo oscillations and the "longitudinal" or temporal plane is drifting. The quantity α_c is proportional to what is known as the momentum compaction in accelerator jargon.[4.3] Indeed, the 6th temporal variable advances by a constant amount $z_6^{final} = z_6 + C\alpha_c z_5$, which depends on the constant energy variable z_5. In resonance basis, which affects only the transverse plane, the linear map is given by

$$\mathcal{S} = \exp(\widehat{s}) = \mathcal{CRC}^{-1}$$

where

$$\widehat{s}=\sum_{n=1}^{N-1}\{-i\,\mu_n(z_{2n-1}\partial_{2n-1}-z_{2n}\partial_{2n})$$
$$+C\alpha_c z_{2N-1}\partial_{2N}\}. \quad (4.50)$$

The case of the integer resonance of Figure 4.4 is similar to a rotation. The regular trigonometric functions are replaced by hyperbolic functions. In this case, the operator $:h:$ is

$$:h:=-\sum_{i=1}^{N}\mu_i:\frac{z_{2i-1}^2-\epsilon_i^2 z_{2i}^2}{2}:$$

$$\epsilon_i=\begin{cases}\sqrt{-1} & \because\ i^{th}\text{ plane on the unit circle}\\ 1 & \because\ i^{th}\text{ plane on the real axis}\end{cases} . \quad (4.51)$$

[4.2] We prefer to use $\vec{z} = (x, p_x, y, p_y, \delta, \ell)$ instead of $\vec{z} = (x, p_x, y, p_y, \ell, -\delta)$. The reasons are related to computer programming. When the energy (i.e., δ) is a constant, one can drop the time of flight (i.e., ℓ). In this case, the transfer map is in $\mathbf{R}^4$ and δ is a control parameter. The time of flight is ignored. The only advantage of the other notation is letting δ remain a "momentum-like" variable. Mathematically, it makes no difference and thus we opted for the practical choice.

[4.3] This quantity is scaled by the circumference of the ring C. That circumference C is not sacred and is a left over of the Frenet-Serrat type of Hamiltonian. It should not bother the reader and can be put to one. It is simply a choice of scale used to achieve a dimensionless definition.

The reader can check that the matrix for the transfer map associated with the hyperbolic case is

$$\begin{pmatrix} \cosh(\mu) & -\sinh(\mu) \\ -\sinh(\mu) & \cosh(\mu) \end{pmatrix}. \tag{4.52}$$

We can generalize the resonance basis, i.e., linear eigenfunctions of $\mathcal{R}$, to include hyperbolic points

$$\begin{aligned} r_{k\pm}(\vec{z}) &= z_{2k-1} \pm \epsilon_k z_{2k} \quad k = 1, N\ , \\ \mathcal{R} r_{k\pm} &= \exp\left(\mp\, \epsilon_k\ \mu_k\right) r_{k\pm}. \end{aligned} \tag{4.53}$$

In this basis, the map $\mathcal{S}$ is

$$\begin{aligned} \mathcal{S} &= \exp\left(\widehat{s}\right) = \mathcal{C}\mathcal{R}\mathcal{C}^{-1} \\ \text{where } \widehat{s} &= \sum_{n=1}^{N} -\ \epsilon_n\ \mu_n(z_{2n-1}\partial_{2n-1} - z_{2n}\partial_{2n})\ . \end{aligned} \tag{4.54}$$

The hyperbolic spiral presents a special hurdle. It is analogous to degenerate perturbation theory in quantum mechanics. Therefore we leave its analysis as an exercise which should not distract the reader. Because the exercise forces us to dip our noses into resonance analysis, it is not a bad idea to come back to it when studying nonlinear resonances.

Exercise 4-4

Without loss of generality, the following discussion can be restricted to the plane of the collision. Taking the point of view of Figure 4.6, we can imagine that prior to the collision, when the eigenvalues are still on the unit circle, the eigenvectors are given by the usual resonance basis: $(r_{1+}, r_{1-}, r_{2+}, r_{2-})$, *i.e.,* $(x + ip_x, x - ip_x, y + ip_y, y - ip_y)$.

1. *At the collision, the map* $\mathcal{R}$ *can be written as*

$$\mathcal{R} = \exp\left(\mu_x : J_y - J_x :\right). \tag{4.55}$$

Find the two additional quadratic functions which are invariant under $\mathcal{R}$. *Answer:*

$$\begin{aligned} c_1 &= \frac{r_{1+}r_{2+} + r_{1-}r_{2-}}{2} = 2\sqrt{J_x J_y}\cos\left(\phi_x + \phi_y\right) \\ s_1 &= \frac{r_{1+}r_{2+} - r_{1-}r_{2-}}{2i} = -2\sqrt{J_x J_y}\sin\left(\phi_x + \phi_y\right). \end{aligned} \tag{4.56}$$

2. *Show that* $[c_1, s_1] \neq 0$.

3. *Hence,* $: J_x - J_y : = \frac{: r_{1+} r_{1+} - r_{2+} r_{2-} :}{2}$ *and* $\alpha : c_1 : + \beta : s_1 :$ *form a complete set of commuting operators. Show that this set can be simplified to* $(: J_x - J_y : , : c_1 :)$ *by a simple canonical transformation. Hint: it must leave* $: J_x - J_y :$ *invariant.*

4. *Using 4-4.3, it is fair to assume that* $\mathcal{R}$ *after the Krein collision can be written as*

$$\begin{aligned} \mathcal{R} &= \exp\{: -\alpha(J_x - J_y) + \beta\, c_1 :\} \\ &= \exp\{-\alpha : J_x - J_y :\} \exp\{\beta : c_1 :\}, \end{aligned} \tag{4.57}$$

and therefore, find eigenfunctions which simultaneously diagonalize the two operators. Answer:

$$\begin{aligned} |\pm i, 1\rangle &= \mp i\, r_{1\pm} + r_{2\mp} \\ |\pm i, -1\rangle &= \pm i\, r_{1\pm} + r_{2\mp}. \end{aligned} \tag{4.58}$$

5. *Show that*

$$\begin{aligned} \mathcal{R}\, |\pm i, 1\rangle &= \exp(\mp i\,\alpha + \beta)\, |\pm i, 1\rangle \\ \mathcal{R}\, |\pm i, -1\rangle &= \exp(\mp i\,\alpha - \beta)\, |\pm i, -1\rangle \\ &\text{and thus, } \alpha = \mu_x. \end{aligned} \tag{4.59}$$

These eigenvalues are precisely as in Figures 4.2 and 4.6.

We conclude that in all the cases described in this section, the map $\mathcal{L}$ can be put into the above normal forms by a symplectic real transformations $\mathcal{A}$, i.e.,

$$\exists \mathcal{A} \text{ such that } \mathcal{R} = \mathcal{A}\mathcal{L}\mathcal{A}^{-1}, \tag{4.60}$$

and that one can put the map into a complete Jordan normal form using a complex resonance basis. The prescription to find the eigenvectors is outlined in Sect. 2.4.2.

4.5.2 The Dissipative Linear Map

We normally think of accelerator theory as a branch of Hamiltonian dynamics. Indeed, neglecting collective effects, it is almost Hamiltonian. However, a charged particle in an electro-magnetic field radiates energy. Quite obviously this effect is very small even for electrons. In electron rings, a particle will loose typically between no more than 10^{-3} and 10^{-4} of its energy per turn. From the point of view of a single-pass system (spectrometer, electron microscope, etc...), a loss of 10^{-4} per passage, which is probably on the order of 10^{-6} per bending magnet, appears totally unmeasurable and therefore unphysical. Indeed, our colleagues in the "single pass" community often wonder why ring physicists neglect large fringe fields (whose dynamical effects are measurable on a laboratory bench) and take into account minuscule contributions to the one-turn map. The answer to this question is what distinguishes periodic structures from single-pass structures. This question is fundamental because it is connected to the most important question of all: what is measurable in a ring? In a single pass system, we can hope to measure accurately the position of a single particle or a pencil beam as it goes through the entire system. From such measurements we deduce the transfer map. In fact, the performance of a single pass system is often reduced to criteria on the size of certain terms of the transfer map. Typically, in spectrometers, one would talk about the $\langle 5 \; |1 \; 5\rangle$ term in the transfer matrix, to use the TRANSPORT notation popularized by Karl Brown, a famous designer and inventor of the code TRANSPORT[17]. One then compares the size of $\langle 5 \; |1 \; 5\rangle$ in relation to other terms, and usually this suffices.

Rings are a different matter because we are interested in stability. Let us set up a gedanken experiment with a "pencil beam". Suppose we inject it in a machine with a small amount of radiation. One can show that it is possible to adjust the control parameters of the system so that Figure 4.1 is modified into Figure 4.7. Of course, we exaggerated the departure from the symplectic condition for illustration purposes. For a truly or mildly linear system, it should be clear that the final destination of this pencil beam is the origin of phase space. If we send a distribution of particles of a given size, we can predict its future size on the basis of the real part of the eigenvalues. Nothing else matters. If in addition, we assume the existence of quantum fluctuations in the radiative process, we can show that, at a given beam size, the damping due to the map $\mathcal{L}$ will be matched by the quantum fluctuation due to the granularity of the photon emission. It

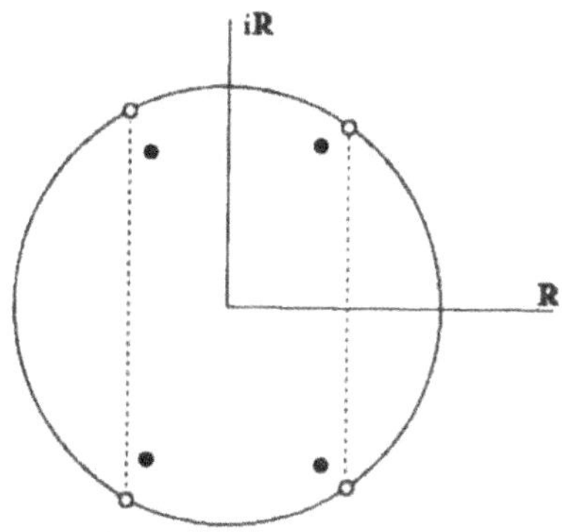

Figure 4.7: Damped Case: Electron Ring.

is remarkable that these effects are microscopic and unmeasurable at the single magnet level. Nevertheless, they determine the final beam size in an electron ring. This example illustrates the different frame of mind one must adopt dealing with rings.

In the case of a dissipative system, the Lie operators for the map $\mathcal{R}$ cannot be Poisson bracket operators. The normal form $\mathcal{R}$ is given by $\mathcal{R} = \exp\left(\widehat{h}\right)$ where $\widehat{h}$ is just

$$\begin{aligned}\widehat{h} &= \sum_{i=1}^{N} -\mu_n : \frac{z_{2n-1}^2 + z_{2n}^2}{2} : + \alpha_n \left\{ z_{2n-1}\partial_{2n-1} + z_{2n}\partial_{2n} \right\} \\ &= - : \vec{\mu} \cdot \vec{J} : + \sum_{n=1}^{N} \alpha_n \widehat{d}_n. \end{aligned} \tag{4.61}$$

As usual the map can be put into a canonical Jordan normal form by going into the resonance basis as in Equation (4.47):

$$\begin{aligned} &\mathcal{S} = \exp\left(\widehat{s}\right) = \mathcal{C}\mathcal{R}\mathcal{C}^{-1} \\ &\text{where} \\ &\widehat{s} = \sum_{n=1}^{N} -i\mu_n (z_{2n-1}\partial_{2n-1} - z_{2n}\partial_{2n}) + \alpha_n \left\{ z_{2n-1}\partial_{2n-1} + z_{2n}\partial_{2n} \right\} \\ &\quad = \sum_{n=1}^{N} (-i\mu_n + \alpha_n) z_{2n-1}\partial_{2n-1} + (i\mu_n + \alpha_n) z_{2n}\partial_{2n}. \end{aligned} \tag{4.62}$$

Exercise 4-5

Combine all the cases we just studied excluding the Krein collision. In other words, the first $N-1$ degrees of freedom are a collection of rotations and integer hyperbolic points. The Nth degree of freedom has a real double root eigenvalue equal to unity.

1. *Write down the operator $\mathcal{R}$.*
2. *Write down the complex eigenbasis as well as the operator $\mathcal{S}$.*

This exercise might appear completely academic. However, in writing a computer library capable of normalizing most situations of interest , it is useful to combine all the cases into one grand case. This is what the author has done for his compositional map library based on automatic differentiation.
Answer: See Interlude VIII.

INTERLUDE VII
EXTENSION OF THE STATE VARIABLES

Moments and Radiation

In this book we restrict our discussion to single particle dynamics which normally implies a six-dimensional phase space. However, the idea of a map can be extended to other phenomena, most notably spin. Not only is spin benefiting from a map approach but also from the introduction of Lie operators. In fact, for many students, their first contact with Lie algebras is probably in the context of angular momentum in quantum mechanics. Other phenomena can make good usage of maps. For example, Chattopadhyay in his CERN lectures[18] on stochastic cooling used a map propagator approach and even Lie operators (i.e., Liouville operators as they are known in non-equilibrium thermodynamics).

In this interlude, we introduce maps to a topic which seems even more remote to single particle tracking: stochastic differential equations. The author is not a specialist on the subject, however there is one critical application in beam dynamics: the computation of "synchrotron integrals." We have already alluded to the importance of quantum fluctuations in determining the final beam sizes of an electron beam. As usual, if one looks for the approach used by ring theorists to compute these quantities, one finds the usual confusion between local and global concepts. And therefore a general state of confusion exists when it is time to extend the usual "synchrotron integrals" to more complex machines. Actually, there is a small core of physicists who have been using the modern approach. These pioneers are Hirata and Ruggiero[19]. It will not surprise the reader to learn that this work is mostly ignored and people still swear by Sands "synchrotron integrals." Again, there is nothing wrong with perturbation theory and Sands formulas to give us some insight. The trouble resides in not distinguishing good formulation from useful formulas. The purpose of this interlude is to present a good formulation.

Let us assume that we have a distribution, ρ, of particles in phase space. Let us compute the quadratic moments of this distribution (around the fixed point):

$$\begin{aligned}\Sigma_{ij} &= \langle z_i z_j \rangle \\ &= \int \rho(\vec{z})\; z_i z_j \; d\vec{z}.\end{aligned} \tag{4.63}$$

One turn later, the moment $\langle z_i z_j \rangle$ has evolved into $\langle \zeta_i \zeta_j \rangle$ where $\vec{\zeta}$ is the one-turn map. This allows us to express these new moments in terms of the initial moments:

$$\Sigma_{ij}^{final} = \langle \zeta_i \zeta_j \rangle = \int \rho(\vec{z})\; \zeta_i(\vec{z})\zeta_j(\vec{z})\; d\vec{z}, \tag{4.64}$$

which in the case of a linear map becomes

$$\begin{aligned}\Sigma_{ij}^{final} &= \int \rho(\vec{z})\; L_{ia} L_{jb} z_a z_b \; d\vec{z} \\ &= L_{ia} L_{jb} \Sigma_{ab} \Rightarrow \Sigma^{final} = L \Sigma L^{\mathsf{T}}.\end{aligned} \tag{4.65}$$

The quadratic moments transform almost like quadratic polynomials under the action of the compositional map $\mathcal{L}$.

Exercise 4-6

1. *Show that the linear compositional map $\mathcal{L}^{\mathsf{T}}$ associated with the linear map whose matrix is L^{T}, transforms the function $\sigma(\vec{z}) = \Sigma_{ab} z_a z_b$ the same way $\mathcal{L}$ acts on the moments Σ_{ab}, i.e.,*

$$\begin{aligned}\forall \vec{z} \quad (\mathcal{L}^{\mathsf{T}} \sigma)(\vec{z}) &= \vec{z} \cdot (L \Sigma L^{\mathsf{T}} \vec{z}) \\ &= z_i L_{ia} \Sigma_{ab} L_{jb} z_j.\end{aligned} \tag{4.66}$$

2. *In the general nonlinear case, consider a vector $\vec{\Sigma}$ containing all the possible moments to order k. Using a truncation scheme*

analogous to that of the k-jets, show that the transformation of this vector under the map $\vec{\zeta}$ is given by

$$\vec{\Sigma}^{final} = \mathbf{M}\vec{\Sigma}. \tag{4.67}$$

where $\mathbf{M}$ *is the transpose of the matrix* $\mathbf{M}^{\intercal}$ *for the compositional map in the monomial basis. See Sect. 4.3 where I introduced* $\mathbf{M}^{\intercal}$.

In a deterministic situation, the dynamics of moments is controlled entirely by the map as one would expect. It is however a very different type of dynamical system. In passing we want to point out that the relationship between the Courant-Snyder invariant in a ring and the "emittance," for a distribution i.e., $\langle x^2\rangle\langle p^2\rangle - \langle xp\rangle^2$ is far more subtle than most people think. In fact, the Courant-Snyder invariant is a dynamical invariant: a function which is left invariant by the map or, equivalently, an eigenfunction of the compositional map. The quantity $\langle x^2\rangle\langle p^2\rangle - \langle xp\rangle^2$ is a kinematical invariant of linear maps in general. Its existence is not necessarily related to invariant tori. Notice that it is an object belonging to the tensor product of the moment space. It is very similar to the case of angular momentum in quantum mechanics where two spin-1/2 particles can combine into a spin zero (i.e., an invariant).

In any event, let us go back to stochastic processes. We imagine that a random event occurs at the one position in the ring which we will choose to be the point of observation. Furthermore, let us assume that this process has the following properties:

$$\begin{gathered}\vec{z} \to \vec{z} + \delta\vec{z} \\ \langle\delta\vec{z}\rangle = 0 \text{ and } \langle\delta z_i \delta z_j\rangle = d\Phi_{ij}.\end{gathered} \tag{4.68}$$

Here we assume that the stochastic force has a zero average, but it has a finite second order correlation which does not depend on $\vec{z}$ (or only very slightly). It is allowed to depend on the location around the ring. This situation is precisely what would happen if we add the granular nature of radiation on top of the classical calculation. Let us work out what happens to the moments Σ in this case:

$$\Sigma \longmapsto \underbrace{L\Sigma L^{\intercal}}_{\substack{\text{Deterministic} \\ \text{one - turn map}}} + \underbrace{d\Phi}_{\substack{\text{Stochastic} \\ \text{Fluctuation}}} . \tag{4.69}$$

In Equation (4.69), the moments $\vec{\Sigma}$ travel through the lattice under the deterministic compositional map $\mathbf{L}$. Then they are "kicked" stochastically by the additive term $d\vec{\Phi}$.

It is remarkable that this map may have a fixed point:

$$\Sigma_\infty = L\Sigma_\infty L^\intercal + d\Phi = \mathbf{L}\vec{\Sigma} + d\vec{\Phi}$$
$$\Downarrow$$
$$\Sigma_\infty = (\mathbf{I} - \mathbf{L})^{-1} d\vec{\Phi}. \tag{4.70}$$

The moments Σ_∞ are the equilibrium moments. When the matrix L has the spectrum of Figure 4.7, a distribution of particles settles into an equilibrium distribution with moments given by Σ_∞.

The important points to remember are:

1. The computation of the stochastic term $d\vec{\Phi}$ is local and thus will be discussed in Part III.

2. The extraction of the equilibrium beam sizes is a fixed point problem. It belongs conceptually to Part II.

In Interlude IX, we will come back later to this topic when we introduce the "conceptual" ring as we did in Sect. 2.5. The final discussion about radiation is in Chapter 15.

Exercise 4-7 (One-turn Envelope Map)
Suppose that stochastic events happen at arbitrary positions s in the ring. Show that the one-turn moment map at $s = 0$, under the conditions of Equation (4.68), *is given by*

$$\Sigma^{final} = L\Sigma L^\intercal + D\Phi$$
$$\textit{where } D\Phi = \oint_0^C L_{s\to C} d\Phi_s L_{s\to C}^\intercal. \tag{4.71}$$

END OF INTERLUDE

This chapter was about the classification of linear maps relevant to accelerator physics. The reader can consult books on dynamics if he is interested in more exhaustive classifications[20, 21]. We will now jump into the nonlinear régime.

CHAPTER 5
FROM LINEAR TO NONLINEAR MAPS

The reader will now learn about the global properties of the ring in the symplectic case. Later we will come back to the radiative ring. Owing to the importance of the Hamiltonian case, we prefer to temporarily neglect nonsymplectic and non-deterministic maps which one encounters in the treatment of synchrotron radiation. The general framework and dichotomous philosophy of this book will apply to these topics nevertheless.

The train of thought in rings starts with the existence of a closed orbit, i.e., a fixed point. We could consider this to be a zeroth order approximation of the motion. Then we ask questions about the linear motion around this fixed point. A great part of accelerator design is to produce the blue print of a ring motivated primarily in terms of its linear properties. Nonlinearities are important and they will introduce quantitative as well as qualitative changes. Therefore, the theory must be developed further. A good theory is one where the intuitive concepts of the linear theory can be carried over to the nonlinear régime at least formally. Otherwise a totally new apparatus must be developed—an apparatus where the linear tools look ugly and complicated for the sake of compatibility with the nonlinear treatment. The map approach to rings succeeds in avoiding this problem. And, when coupled with the local treatment of Part III, it provides a complete theory.

Standard treatments fail at least on two critical issues:

1. Even in the linear case, they fail to see Bethe's point about Bloch functions (see the preface), i.e., Floquet variables. Like every conceptual sickness, it has a symptom: the need to introduce Hill's equation for the ring to justify the Courant-Snyder ansatz of Equation (1.1). Typically, this has a side effect: one must focus on a particular Hamiltonian and tons of approximations are introduced prematurely.

2. Good intuitive first-order treatments of nonlinearities involve summing the so-called "phasors"[5.1] and lumping their effects at one location in the ring.[5.2] Finally, a discretized equation is used (i.e., replace $\frac{d}{ds}$ by $\frac{1}{C}\frac{d}{dn}$) to study the iterates of the map.[5.3] Then, typically, the reader thirsty for more is told to look into Hamiltonian perturbation theory.

One very good example is an introductory book by Edwards and Syphers [4]. This book is remarkable in the sense that the first four chapters of the book follow precisely the pattern I just described. The authors describe intuitive tools which they use in the context of leading order perturbation theory: it is a map theory in disguise. Then they invite the more daring reader to consult the "advanced" treatment based on s-dependent Hamiltonian perturbation theory. That is a tragic mistake. Again, there is nothing wrong with s-dependent perturbation theory; the issue is whether or not it is the most appropriate way to look at an accelerator. Their first four chapters are an eloquent albeit accidental praise to a map-based theory. Their incapability to do anything additional with Hamiltonians should convince the reader of the gulf between the introductory and the "advanced" treatments. One must keep in mind that the purpose of a map-based theory is to avoid this conceptual ocean between the "applied" and the theoretical, the linear and the nonlinear, the coupled and the uncoupled. In fact, our treatment bridges the formal aspects of perturbation theory between classical and quantum physics, with links to the Fokker-Planck equation and with the theory of "synchrotron integrals" through the beam envelope formalism.

5.1 STANDARD SOLVABLE PROBLEMS IN RINGS

We have already indicated that perturbation theory (i.e., k-jet dynamics) is in itself a meaningless enterprise. In general, k-jet dynamics is the beginning of a divergent series, and thus, unless re-interpreted as an actual map, it has little meaning. The re-interpretation can be viewed as a process where one compares the actual system gotten from (numerical) experiment with a simpler (solvable) system. Perturbation theory is then a way to create solvable maps systematically using the information in the k-jet.

[5.1] Read: Lie operators in normalized variables.

[5.2] Read: computing a one-turn map.

[5.3] Read: $\mathcal{M}^n$ is studied with the help of jets.

In addition, the usage of maps will permit us to develop well-defined algorithms which are recursive and conceptually trivial to extend to high orders. In the case of k-jets produced by automatic differentiation libraries, we can truly perform calculations to arbitrary orders. At this stage it is not necessary to distinguish analytical perturbation theory from a computation on a k-jet whose coefficients are known numerically. Again the reader has to believe that the extraction of such k-jets (i.e., power series) is possible from any simulation code using automatic differentiation (AD). In general, most calculations are mixtures of analytical and numerical (AD) manipulations. In analytical calculation, the Dragt-Finn factorization could be in terms of a smallness parameter and the Lie transforms need not be polynomials. One application is the beam-beam interaction where it is possible to use perturbation theory on the actual beam-beam force and derive useful results.

So the question is: what problem do we know how to solve? Of course, by definition, all the linear problems which were catalogued in Sect. 4.5 are solvable. We must look at their possible nonlinear extensions.

5.1.1 Amplitude Dependent Rotations

First and foremost is the case of a rotation. We are given a system where the one-turn map is just

$$\mathcal{R} = \exp\left(: h(\vec{J}) : \right). \tag{5.1}$$

The phase space map induced by $\mathcal{R}$ is given by Equation (4.41). The motion is fully integrable and confined to the intersection of N tori

$$J_i = \frac{z_{2i-1}^2 + z_{2i}^2}{2} = \text{constant}. \tag{5.2}$$

The tune in the ith plane is given by $-\frac{\partial h}{\partial J_i}$. It is possible to show that this function h does not depend on the point of observation, i.e., on the surface of section where the map is computed.

The analogous Hamiltonian statement is that

$$\exists\ K(\vec{J}; s) \text{ such that } h = -\oint_{s_0}^{s_0+C} K(\vec{J}; s)\, ds, \tag{5.3}$$

and that h is unique—it does not depend on s_0. However, the s-dependent function K is not unique. This is related to the phase advance and was illustrated in Sect. 2.6.3.

5.1.2 Amplitude Dependent Rotations for a Coasting Beam

This is a simple extension of the case in Sect. 5.1.1. The one-turn map is given by

$$\mathcal{R} = \exp\left(: h(\vec{J};\delta) : \right)$$
$$\vec{J} = (J_1, \cdots, J_{N-1}) \text{ and } \delta = z_{2N-1}. \tag{5.4}$$

The phase space map induced by $\mathcal{R}$ is again given by Equation (4.41) for the first $N-1$ degrees of freedom. In the Nth degree of freedom, the variable z_{2N} advances by a fixed amount

$$z_{2N}^{final} = z_{2N} + \frac{\partial h}{\partial \delta}. \tag{5.5}$$

If we expand the Lie transform h in a power series, one can identify some famous global quantities:

$$h = -2\pi\vec{\nu}\cdot\vec{J} \; - \; 2\pi\vec{\nu}'\cdot\vec{J}\delta \; + \; C\frac{\alpha_c}{2}\delta^2 \; - \; \pi a_{ij}J_iJ_j + \cdots. \tag{5.6}$$

Using the action-angle representation as in Equation (4.40), we get for the transverse phase space

$$\begin{aligned}\phi_i^{final} - \phi_i &= -\frac{\partial h}{\partial J_i} \\ &= 2\pi \left\{ \underbrace{\nu_i}_{\text{tune}} + \underbrace{\nu_i'}_{\text{chromaticity}} \delta \right\} + 2\pi \underbrace{a_{ij}J_j}_{\substack{\text{tune shift}\\ \text{with amplitude}}} + \cdots,\end{aligned} \tag{5.7}$$

and, for the longitudinal plane

$$\begin{aligned}z_{2N}^{final} - z_{2N} &= \frac{\partial h}{\partial \delta} \\ &= \underbrace{-2\pi\vec{\nu}'\cdot\vec{J}}_{\substack{\text{Amplitude dependent}\\ \text{time of flight}}} + \underbrace{C\alpha_c}_{\substack{\text{Linear Momentum}\\ \text{Compaction}}} \delta + \cdots.\end{aligned} \tag{5.8}$$

One notices immediately that the amplitude-dependent momentum compaction and the chromaticities are related. In fact, the symplectic condition links all the δ-dependence of the transverse motion with the (x, p_x, y, p_y)-dependence of the time-flight. This fact is an immediate consequence of the Hamiltonian nature of the flow. In many standard treatments of accelerator physics, the longitudinal motion is isolated from the transverse at the very start[5.4]. While it is true, even in the presence of a cavity, that the dynamics of the longitudinal plane has some qualitative difference with transverse planes, it is important to realize that these differences are mostly due to the parameter régime in which actual accelerators are run. It is not a fundamental difference.

As before, we can give a Hamiltonian equivalent for the full ring:

$$\exists\ K(\vec{J},\delta;s)\ \text{such that}\ \ h = -\oint_{s_0}^{s_0+C} K(\vec{J},\delta;s)\, ds, \tag{5.9}$$

And once more K is not unique because of the s-dependence.

5.1.3 One-Resonance Map

In the prologue of his book[11], Michelotti introduces the pendulum as an example of a simple system whose dynamics does not reside on a Cartesian space but on the Cartesian product of a sphere and the real axis, i.e., $S^1 \times \mathbf{R}$. Michelotti emphasizes a paradigm shift from the old and the new classical mechanics which started with Poincaré. No one tries to answer long term stability questions by getting the solution of a differential equation (i.e., the history of the system), but by determining the shapes and properties of the surface on which this history might be played out. Poincaré's work gave birth to topology in mathematics and eventually to breakthroughs in chaos and fluid dynamics. As we will now see, a one-resonance map is connected topologically to the standard map and the pendulum. The purpose of this section is to see if it is sometimes possible to understand figures such as Figure 5.1. I want to point out as does Michellotti (see reference [11] pages 198-199) that resonances are not caused by "small denominators." These small denominators are the symptom of a doomed attempt to find global invariant torii which in reality do not exist.

A map might be "resonating" whenever the tune of some trajectories satisfies

$$\vec{m}_r \cdot \vec{\mu}_r = 2\pi p, \quad \text{where } p \in \mathbf{Z}. \tag{5.10}$$

[5.4] In reference [22] the author quotes a totally wrong result: the amplitude-dependent path length is the chromaticity of the quadrupole alone. The reader can look at the original work in reference [23] and try to spot the error. Hint: $\langle x \rangle \neq 0$ to second degree in x and p.

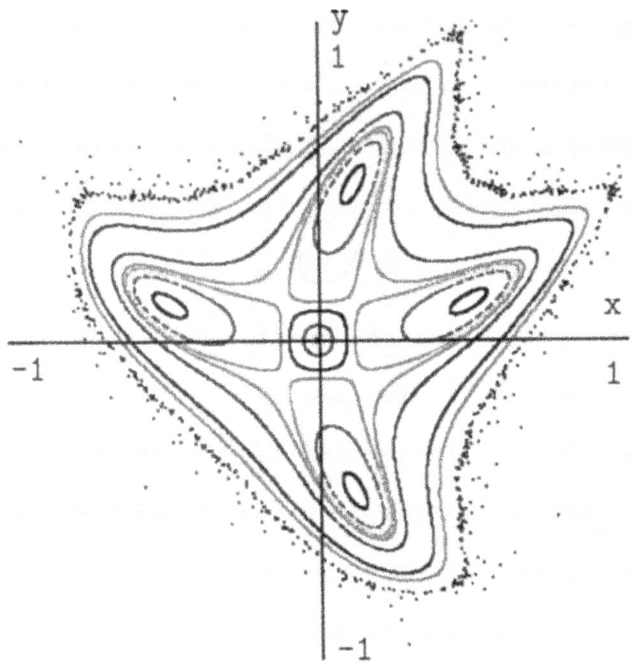

Figure 5.1: Quadratic Map of Exercise 5-1 for $\nu = \frac{1}{4} + 0.001$.

Then, for generic maps, we expect areas of phase space to be locked on the resonance if the map is nonlinear.[5.5] In one degree of freedom this was depicted in Exercise 2-1. The connection to the linear part of the tune was investigated in Exercise 2-2.

The one-resonance map in normalized form is invariant under a discrete group of planar rotations in the plane of the resonance. For example, a planar map which contains only driving terms of the third integer resonance will be invariant under a 120° rotation. This will allow for the immediate computation of $N-1$ invariant tori perpendicular to the plane of the resonance, and in the plane of the resonance, the map is similar to the standard map. Therefore it may not be

[5.5] Of course a map which is topologically a rotation, like the fully normalized map, never "locks" on resonances over finite areas of phase space. The islands have no width.

integrable and chaos could exist on a large scale. In many cases, however, the map will appear integrable in the plane of the resonance. Using k-jet manipulation we will be able to take the one-resonance map and produce an invariant in the plane of the resonance without affecting the existing $N-1$ tori. In other words, we will approximate the standard map by the pendulum!

It is interesting to make an analogy with the full normalizations of the previous sections. In a full normalization, the final map $\mathcal{R}$ is a rotation. The map is invariant under any rotation (or phase advance). Therefore it must commute with the $: J_i :$'s which are the generators of rotation for each plane. It follows that the J_i's are invariants. The map is fully integrable. In the one-resonance normalization, everything we just said is valid in the $N-1$ planes "perpendicular" to the resonance. In the plane of the resonance, the map is invariant only under the discrete subgroup [5.6] C_k of the planar rotation group $SO(2)$. This invariance is consistent with the presence of islands, i.e., bubbles with fixed tunes and separatrices. But, it does not insure integrability. In his approach to the one-resonance normalization, Turchetti[2, 25] imposes these symmetries directly on the Taylor representation of the map. The reader recalls that Turchetti and collaborators develop a map-based theory of ring dynamics which is fundamentally like ours.

A one-resonance map $\mathcal{N}$, in its non-integrable form, has the property that the Lie transforms n_i of its Dragt-Finn factorization are invariant under a discrete rotation group in the plane of the resonance. Going into the resonance basis, we can write down the Dragt-Finn factorization:

$$\mathcal{N} = \exp\left(: -\vec{\mu}\cdot\vec{J} :\right) \exp\left(: n_3 :\right) \cdots \exp\left(: n_k :\right) \cdots$$

where

$$n_k = \sum A_{\vec{m}} \left|\vec{m}^+, \vec{m}^-\right\rangle$$

$$\left|\vec{m}^+, \vec{m}^-\right\rangle = r_{1+}{}^{m_1^+} r_{1-}{}^{m_1^-} \cdots r_{N-}{}^{m_N^-}$$

$$\text{with} \quad \vec{m}^+ - \vec{m}^- = \lambda\vec{m}_r \text{ and } \lambda \in \mathbb{Z}. \tag{5.11}$$

Here, the notation is the same as in Equation (4.36). Of course the one-resonance map $\mathcal{N}$ is related to the original map $\mathcal{M}$ by the relation

$$\mathcal{M} = \mathcal{B}^{-1}\mathcal{N}\mathcal{B} \tag{5.12}$$

[5.6] This is Hamermesh's notation for this subgroup[24].

where $\mathcal{B}$ is computed so as to remove the driving terms of other resonances. Let us prove here a few important properties of the map $\mathcal{N}$.

Theorem 1 *The set $S_{\vec{m}_r}$ of Lie transforms generated by $|\vec{m}^+, \vec{m}^-\rangle$ with the properties enunciated in Equation* (5.11) *forms a closed set under the Poisson bracket.*

It is easiest to prove this theorem by writing $|\vec{m}^+, \vec{m}^-\rangle$ in action-angle variables:

$$|\vec{m}^+, \vec{m}^-\rangle = \prod_{k=1}^{N} (2J_k)^{\frac{m_k^+ + m_k^-}{2}} \times \exp\left(-i\lambda \vec{m}_r \cdot \vec{\phi}\right). \quad (5.13)$$

The Poisson bracket of two such objects will look like

$$\begin{aligned}
&\left[|\vec{m}^+, \vec{m}^-\rangle, |\vec{n}^+, \vec{n}^-\rangle\right] \\
&= \left[\prod_{k=1}^{N} (2J_k)^{\frac{m_k^+ + m_k^-}{2}} e^{-i\lambda \vec{m}_r \cdot \vec{\phi}}, \prod_{k=1}^{N} (2J_k)^{\frac{n_k^+ + n_k^-}{2}} e^{-i\lambda' \vec{m}_r \cdot \vec{\phi}}\right] \\
&= \prod_{k=1}^{N} \prod_{j=1}^{N} \left[(2J_k)^{\frac{m_k^+ + m_k^-}{2}} e^{-i\lambda \vec{m}_r \cdot \vec{\phi}}, (2J_j)^{\frac{n_j^+ + n_j^-}{2}} e^{-i\lambda' \vec{m}_r \cdot \vec{\phi}}\right] \\
&\propto \exp\left(-i(\lambda + \lambda')\vec{m}_r \cdot \vec{\phi}\right).
\end{aligned} \quad (5.14)$$

Of course $\lambda + \lambda' \in \mathbb{Z}$ holds. So, the set is closed.

Theorem 2 *There exists $N-1$ invariants of the compositional map $\mathcal{N}$. They are the $N-1$ functions $\vec{a}^i \cdot \vec{J}$ where $\vec{a}^i$ obeys $\vec{a}^i \cdot \vec{m}_r = 0$ and the $\vec{a}^i$'s form an orthogonal set.*

This is proved by bracketing $\vec{a} \cdot \vec{J}$ with an arbitrary Lie transform belonging to $S_{\vec{m}_r}$.

$$\begin{aligned}
\left[\vec{a} \cdot \vec{J}, |\vec{m}^+, \vec{m}^-\rangle\right] &= \left[\vec{a} \cdot \vec{J}, \prod_{k=1}^{N} (2J_k)^{\frac{m_k^+ + m_k^-}{2}} \exp\left(-i\lambda \vec{m}_r \cdot \vec{\phi}\right)\right] \\
&= \prod_{k=1}^{N} (2J_k)^{\frac{m_k^+ + m_k^-}{2}} \left[\vec{a} \cdot \vec{J}, \exp\left(-i\lambda \vec{m}_r \cdot \vec{\phi}\right)\right] \\
&= \left\{\prod_{k=1}^{N} (2J_k)^{\frac{m_k^+ + m_k^-}{2}}\right\} i\lambda \vec{m}_r \cdot \vec{a} \exp\left(-i\lambda \vec{m}_r \cdot \vec{\phi}\right)
\end{aligned}$$

$$
\begin{aligned}
&= i\lambda \vec{m}_r \cdot \vec{a} \left| \vec{m}^+, \vec{m}^- \right\rangle \\
&= 0. \because \vec{m}_r \cdot \vec{a} = 0.
\end{aligned} \tag{5.15}
$$

It is clear from linear algebra that $N-1$ orthogonal vectors to $\vec{m}_r$ can be constructed. The set of vectors $\{\vec{a}^1, \cdots, \vec{a}^{N-1}, \vec{m}_r\}$ forms an orthogonal basis in $\mathbb{R}^N$.

Theorem 3 *The map $\mathcal{N}$ can be re-factorized as*

$$
\mathcal{N} = \exp\left(: -\frac{p2\pi}{\|m_r\|^2} \vec{m}_r \cdot \vec{J} : \right) \mathcal{N}_c, \tag{5.16}
$$

where the map $\mathcal{N}_c$ is near the identity. Therefore, it can potentially be rewritten with a single Lie transform (k-jet wise). Also, because $\mathcal{N}_c$ and $\mathcal{N}$ commute, the polynomial associated with this single Lie transform is also an invariant of $\mathcal{N}$. This is the missing invariant which possesses the island chain topology. By analogy to Hamiltonian theory, we called $\mathcal{N}_c$ a co-moving map.

The proof proceeds in two steps. First, we notice that all the Lie polynomials of $\mathcal{N}_c$ still belong to $S_{\vec{m}_r}$. In fact, the Dragt-Finn factorization of $\mathcal{N}_c$ is

$$
\mathcal{N}_c = \exp\left(: (\frac{p2\pi}{\|m_r\|^2} \vec{m}_r - \vec{\mu}) \cdot \vec{J} : \right) \exp(: n_3 :) \cdots \exp(: n_k :) \cdots
$$

if $\vec{m}_r \cdot \vec{\mu} = 2\pi p + \epsilon$,

$$
\begin{aligned}
&= \exp\left(: \{ -\sum_{n=1}^{N-1} \frac{\vec{\mu} \cdot \vec{a}^i}{\|a^i\|^2} \vec{a}^i \cdot \vec{J} \} + \frac{-\epsilon}{\|m_r\|^2} \vec{m}_r \cdot \vec{J} : \right) \\
&\quad \times \exp(: n_3 :) \cdots \exp(: n_k :) \cdots
\end{aligned} \tag{5.17}
$$

In Equation (5.17), one notices that the linear part of the map is the exponential of two Lie transforms

$$
\underbrace{-\sum_{n=1}^{N-1} \frac{\vec{\mu} \cdot \vec{a}^i}{\|a^i\|^2} \vec{a}^i \cdot \vec{J}}_{N-1 \text{ commuting planes}} \quad \text{and} \quad \underbrace{\frac{-\epsilon}{\|m_r\|^2} \vec{m}_r \cdot \vec{J}}_{\text{Plane of the resonance}} , \tag{5.18}
$$

where the first term commutes with any part of the map, and the second term is the generator of rotations in the plane of the resonance. One notices that the co-moving map is near the identity in the plane of the resonance since ϵ is assumed to be relatively small.

Therefore, we rewrite $\mathcal{N}_c$ as a single Lie exponent:

$$\mathcal{N}_c = \exp\left(: h_c :\right), \tag{5.19}$$

it is clear that $h_c \in S_{\vec{m}_r}$ using Theorem 1.

In the next step of the proof, we notice that h_c must also be an invariant of $\mathcal{N}$:

$$\begin{aligned}
\mathcal{N} h_c &= \exp\left(: -\frac{p2\pi}{\|m_r\|^2}\vec{m}_r \cdot J :\right)\mathcal{N}_c h_c \\
&\overset{\because\, \mathcal{N}_c h_c = h_c}{=} \exp\left(: -\frac{p2\pi}{\|m_r\|^2}\vec{m}_r \cdot J :\right) h_c \\
&= \exp\left(: -\frac{p2\pi}{\|m_r\|^2}\vec{m}_r \cdot J :\right) \sum_{\vec{m}^+ - \vec{m}^- \in S_{\vec{m}_r}} h_{\vec{m}} \left|\vec{m}^+, \vec{m}^-\right\rangle \\
&\overset{\text{As in Eq. (5.15)}}{=} \sum_{\vec{m}^+ - \vec{m}^- \in S_{\vec{m}_r}} h_{\vec{m}} \underbrace{e^{: -\frac{p2\pi}{\|m_r\|^2}\vec{m}_r \cdot J :} \left|\vec{m}^+, \vec{m}^-\right\rangle}_{\exp(-i2\pi p \lambda_{\vec{m}})\left|\vec{m}^+,\vec{m}^-\right\rangle = \left|\vec{m}^+,\vec{m}^-\right\rangle} = h_c\,.
\end{aligned} \tag{5.20}$$

Conclusion: the one-resonance map, when viewed as a k-jet, has N invariants. The first $N-1$ are tori. They are the functions

$$\vec{a}^i \cdot \vec{J} = \sum_{j=1}^{N} a^i_j \frac{x_j^2 + p_j^2}{2}, \tag{5.21}$$

where $\vec{m}_r \cdot \vec{a}^i = 0$ for all the $\vec{a}^i$'s. The last invariant, which has the topology of an island chain, is given by h_c. Of course, the invariants of the original map $\mathcal{M}$ are gotten by the usual formula

$$\begin{aligned}
&\mathcal{B}^{-1}\vec{a}^i \cdot \vec{J} = \vec{a}^i \cdot \mathcal{B}^{-1}\vec{J} \text{ and } \mathcal{B}^{-1} h_c, \\
&\text{where } \underbrace{\mathcal{M}}_{\text{Original map}} = \mathcal{B}^{-1} \underbrace{\mathcal{N}}_{\text{One-resonance map}} \mathcal{B}\,.
\end{aligned} \tag{5.22}$$

The treatment presented here is absolutely complete from a k-jet point of view. The calculation can be carried to arbitrary order, and indeed, with the help of automatic differentiation , it is carried to high order in numerical (AD) cases. This is to be contrasted with the usual one-resonance treatments found in accelerator physics. Typically, they are based on the recognition that the Hamiltonian

$$H = T(J) + f(J)\cos\left(m\phi - \frac{2\pi}{C}ns\right) \tag{5.23}$$

is solvable exactly (see [26], p.53, section 7). This is seen by going into a co-moving frame using the mixed-variable generating function $F_2 = (\phi - \frac{2\pi}{mC}ns)\overline{J}$. The new Hamiltonian $\overline{H}$ is

$$\overline{H} = T(\overline{J}) - \frac{2\pi}{mC}n\overline{J} + f(J)\cos\left(m\overline{\phi}\right). \tag{5.24}$$

The Hamiltonian of Equation (5.23) is very special. It has only one transverse driving term and it has only one Fourier mode in the time-like variable s.

While it is possible to produce a Hamiltonian theory equivalent to our one-resonance map approach, it is usually not done. It is deemed too difficult or simply not understood.

⌜ *Dear Reader, if you are of the old school of classical mechanics, you might have tried to follow the algebra of this section too closely losing track of the purpose. Let me try to anticipate the questions you might have:*

◎*1. Is the procedure applicable to a map which is far away from the resonance?*

◎*2. And, if yes, what is the meaning of all of this?*

◎*3. When should we do one-resonance normalizations versus the full normalizations of Sects.* 5.1.1 *and Sect.* 5.1.2*?*

If you are asking yourselves these questions (or similar ones), you are probably drifting away from the topological point of view. I invite the reader to look at Figure 5.2 *or Figure* 5.4 *for example. One sees clearly triangular shapes, hyperbolic and elliptic fixed points of the third iterate of the map, etc.*

Of course, I am repeating myself. As I said, there is little meaning in a jet normalization. If we do not see any resonances in the phase plot [5.7] *then we may as well assume that the real map can be normalized into a rotation. Again, we need to remember that our goal is to produce exactly solvable solutions which compare favorably to exact tracking in a relevant region of phase space. If we completely normalize the map there is no hope of producing an island topology. At best, we will describe the motion up to the separatrix and no further.* ⌟

[5.7] Even though one does not see a resonance, the expansion of the normalization transformation $\mathcal{A}$ in a Taylor series can have a small radius of convergence. The islands of the "complexified" map control the convergence.

Let us pause and study a simple map which generates Figure 5.2.

Exercise 5-1

Consider the compositional map

$$\mathcal{M} = \exp\left(: -2\pi(\frac{1}{3} + \epsilon)J :\right) \exp\left(: \frac{1}{3}x^3 :\right). \tag{5.25}$$

1. *Write down the map in phase space* ($\mathbf{R}^2 \longrightarrow \mathbf{R}^2$).
2. *Track a few turns and produce Figure* 5.2 *for* $\epsilon = 0.01$.
3. *Write the compositional map in resonance basis. Answer:*

$$\mathcal{M} = \mathcal{R}\exp\left(\frac{1}{24} : r_+{}^3 + r_-{}^3 + 3r_+{}^2r_- + 3r_-{}^2r_+ :\right),$$
$$\text{where } \mathcal{R} = \exp\left(: -2\pi(\frac{1}{3} + \epsilon)J :\right). \tag{5.26}$$

4. *In Equation* (5.26), *the Lie transform proportional to* $r_+{}^2r_- + r_-{}^2r_+$ *belongs to the integer resonance. Therefore the map* $\mathcal{M}$ *is not a one-resonance map. Using a canonical transformation of the form* $\mathcal{B}_2 = \exp\left(: ar_+{}^2r_- + a^*r_-{}^2r_+ :\right)$ *one can remove this resonance to leading order. Compute the complex number "a" and rewrite the Lie exponents in terms of real functions.*
5. *Suppose that the new one-resonance map* $\mathcal{N}_2$ *is rewritten with a single Lie exponent:*

$$\begin{aligned}\mathcal{N}_2 &= \mathcal{B}_2\mathcal{M}\mathcal{B}_2^{-1}\\ &= \mathcal{R}\exp\left(\frac{1}{24} : r_+{}^3 + r_-{}^3 :\right)\\ &= \exp\left(: -2\pi(\frac{1}{3} + \epsilon)J + br_+{}^3 + b^*r_-{}^3 :\right).\end{aligned} \tag{5.27}$$

Give topological reasons why, in the best possible situation, the single Lie operator of Equation (5.27) *is the beginning of a series which, if convergent, is valid only inside the separatrix and cannot be prolongated any further. Answer: At the separatrix and beyond the motion induced by the map jumps from one open island to the next one. This motion cannot be a stroboscopic view of a time-independent Hamiltonian which is precisely what a single exponent representation provides.*

6. *Compute the function b. Show that it has a divergence of the form* ϵ^{-1}. *This is the analytic symptom for the divergence at the separatrix. The reader can see that the topological reason is far more convincing.*

7. *Compute* $\mathcal{N}_2^3$ *in the Dragt-Finn representation and argue that for sufficiently small* ϵ, *the 3-turn map can be written as:*

$$\mathcal{N}_2^3 \approx \exp\left(: -6\pi\epsilon J + \frac{1}{8} r_+{}^3 + \frac{1}{8} r_-{}^3 : \right). \tag{5.28}$$

   This map can potentially describe the motion past the separatrix. Why? Using this map locate the fixed points of the map by rewriting the Lie transform in action-angle variables. This is "super naïve" theory found in elementary texts[4].

It is important to remember the topological frame of mind of this book. One first suspects the presence of a resonance through an analysis of the tracking results and then one tries to put the approximate map into that form.

In the next section, we examine in more detail the one-resonance map.

5.2 ABOUT NORMALIZATION ALGORITHMS

In Sect. 5.1, we have described two standard solvable situations. The reader should master the concept of a one-resonance map and the algorithm which leads to it in the k-jet algebra. Considering the practical importance of resonant extraction, just to name one application, the mastering of the one-resonance normal form should be part of the "ABC" of accelerator theory. For this reason, we will reserve one full section for this topic.

It is remarkable that the one-resonance map is poorly understood in accelerator physics. Still today, one can hear a collection of statements which are false and widespread in their belief. I will list a few:

1. Sextupoles can only produce third order and first order resonances. More generally, unless an expansion of the Hamiltonian in the resonance basis explicitly displays a given driving term for a resonance, then the given resonance is forbidden. *Nothing could be more false.*

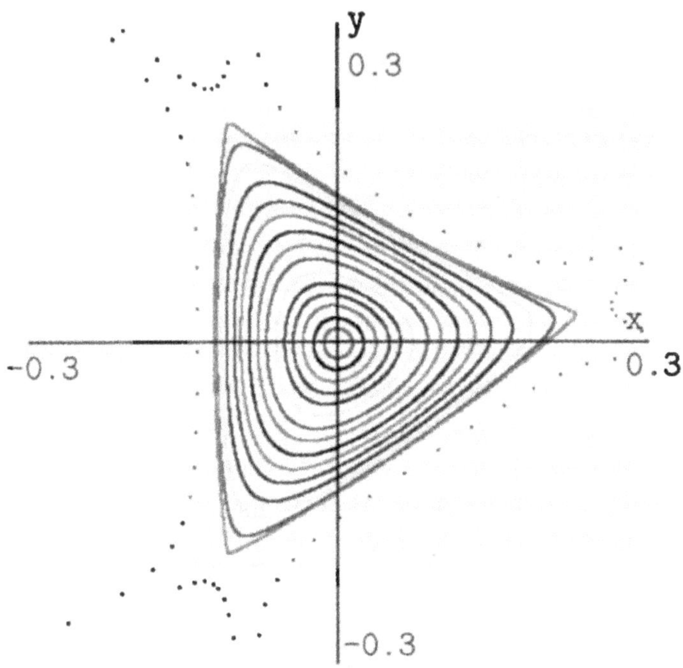

Figure 5.2: Quadratic Map for $\nu = \frac{1}{3} + 0.01$.

2. Corollary of 1, is the belief that the map (or worse the Hamiltonian) must contain the driving term of the resonance in the leading order if it is supposed to reproduce this resonance accurately.

3. Close enough to a resonance, the topological behavior in phase space will be described by the leading order term in perturbation theory.

4. In a map or Hamiltonian, the leading order term for, let us say, the $3\mu = k2\pi$ resonance, must be of the form $f(J)\cos(3\phi + \phi_0)$.

In this section, we illustrate the theory by the study of a "pet" problem. Before we start I will "theorize" on the source of the falsehoods.

First of all, if one considers the map of Exercise 5-1 and tracks it near a tune of 1/4, the reader will see that four islands appear. This was displayed also in Exercise 2-1 where five, four, and three islands where produced by a quadratic map. This is also seen in Figure 5.1 at the beginning of this chapter. Therefore, the appearance of higher-order resonances in a sextupole map (quadratic Hamiltonian) is an experimental fact. One could try to rescue 1 by stating that perturbation theory is always incapable of reproducing these results, but that would also be false. What is true however, is that it takes *more than first-order* perturbation theory to get four island out of a quadratic map. The source of these misconceptions is grounded in Hamiltonian perturbation theory which is rarely carried out beyond first order. Typically, introductory texts are often misleading. For example, in [27] the author of a lecture on "Non-linearities and Resonances" says:

> "The order, n, is related to the order of the multipole which drives the resonance. For example, fourth-order resonances are driven by multipoles with 2n poles, i.e., octupoles. Multipoles can drive resonance of lower-order; octupoles drive fourth, second-order; sextupoles third and first-order, etc., but here we only consider the highest order driven."

The highest order driven is potentially ∞! Thus permanent brain damage has been inflected! Personally I do not put the blame on any particular author. Most of them, educated in the 1960's or even earlier, try their best to teach a subject vaster than what is examined here. Younger physicists, raised in the computer era, do not speak loud enough. That is the problem!

Falsehood 2 can be viewed as a corollary of 1. In fact, because a pure sextupolar field produces five islands, then any "epsilontic" quintic term added to it will not perturb the motion in the limit of epsilon going to zero. Hence, the five islands are correctly described by a map without a fifth order driving term, as in Figure 2.2.

Remarkably Edwards and Syphers[4] start their chapter on resonances with a figure (4.1, p.109) displaying our sextupole map (Exercise 5-1) with a tune of 0.42. As expected, five islands emerge. While they warn the reader that they will not attempt to unravel all the features of this figure, they manage to spoil this introduction with a statement, eight pages later, that sextupoles in two degrees of freedom have four sum resonances belonging to the first and third integer! In a sense, even when confronted with the factual and undeniable statement of a figure, we relapse into the usual nonsense of standard accelerator theory. The more I try to understand the origin of these misconceptions, the more I realize that they are ingrained in the culture, and thus, one cannot blame individuals for making what are ultimately contradictory statements. A colleague once told me that he had to "unlearn" the perversions which accompany many introductory treatments.

Falsehood 3 is more subtle. For example, again in [4], the authors assume that near 1/3-integer resonance, the first order effects of sextupoles dominate. This is not necessarily true as we will see.

Finally falsehood 4 is simple to understand. Consider a map such as

$$\exp\left(: -2\pi(\frac{1}{3} + \epsilon)J : \right) \exp\left(: x^6 : \right). \tag{5.29}$$

It will resonate near $\epsilon = 0$, producing 2 sets of 3 islands. Therefore, the leading order driving term of the third order resonance is of the form $J^3 \cos(6\phi + \phi_0)$.

Exercise 5-2

Plot the phase space of the map

$$\exp\left(: -2\pi(\frac{1}{3} + \epsilon)J : \right) \exp\left(: \delta x^3 + x^6 : \right), \tag{5.30}$$

for $\epsilon = 0.01/3$ *and values of* δ *ranging from* 0 *to* 0.1. *You will see islands grouped in pairs. They undergo pair collision as* δ *approaches* 1/50. *The local topology of this collision is that of the half-integer resonance.*

All these facts are obtainable from simple numerical (or even real) experiments. I always thought that it was the prerogative of theorists, some would say terrorists, like myself, to ignore simulations and experiments! But, as I said in the beginning of this chapter, even decent introductory books go in a sort of panic mode when approaching the time to discuss nonlinear phenomena.

Here we will keep our cool and look at the problem of third order sextupole resonance to second-order in sextupole strength. The map we will use is rich enough to display the features ignored in the first order treatments such as that of Edwards and Syphers.

5.2.1 Bringing the Map into a One-Resonance Map

Our pet example will be the following map:

$$\mathcal{M} = \mathcal{R}_1 \exp\left(: -8x^3 :\right) \mathcal{R}_2 \exp\left(: k_s 8x^3 + 16k_o x^4 :\right),$$
$$\text{where } \mathcal{R}_i = \exp\left(- : \mu_i J :\right). \tag{5.31}$$

This map represents two sextupoles which are μ_2 apart; the second sextupole being equipped with an "octupole winding" of strength k_o.

The steps I will describe here are characteristic of an analytical calculation. A k-jet calculation done with an automatic differentiation (AD) library would not reveal as much. We will come back to the distinctions between analytical and AD later when we define the Floquet ring. Let us start the analysis of this map assuming that the total tune $\mu = \mu_1 + \mu_2$ is near $\frac{2\pi}{3}$.

5.2.1.1 Lumping the perturbations

As we will see later in Sect. 9.2.4, the pet problem we are now trying to solve has far more generality then one can imagine. The reader will discover that what we will now describe is in fact a second-order analytical calculation for an arbitrary distribution of sextupoles (in one degree of freedom). For pedagogical reasons, we will provide a correspondence with Edwards and Syphers in Exercise 5-5.

In general, we can think of the maps $\mathcal{R}_1$ and $\mathcal{R}_2$ as the unperturbed propagators (fully normalized) between the two perturbations (sextupoles in this case). The first step consists in lumping all these perturbations at one location in the ring.

$$\begin{aligned}\mathcal{M} &= \mathcal{R}_1 \exp\left(: -8x^3 :\right) \mathcal{R}_2 \exp\left(: k_s 8x^3 + 16k_o x^4 :\right) \\ &= \mathcal{R}_1 \mathcal{R}_2 \mathcal{R}_2^{-1} e^{:-8x^3:} \mathcal{R}_2 e^{:k_s 8x^3 + 16k_o x^4:} \\ &= \mathcal{R} \exp\left(: -\mathcal{R}_2^{-1} 8x^3 :\right) \exp\left(: k_s 8x^3 + 16k_o x^4 :\right).\end{aligned} \tag{5.32}$$

In Equation (5.32) all the equalities are strictly true. The last line can be interpreted as follows: the perturbed one-turn map is equal to the original one-turn normalized propagator $\mathcal{R}$ followed by the first sextupole whose Lie transform is phase-shifted by the appropriate map (i.e., $\mathcal{R}_2^{-1}$); the action of the second sextupole-octupole system comes immediately afterwards. These phase-shifted perturbations are called "phasors" in the jargon of intuitive accelerator theory.

The next step consists in lumping all the perturbations at exactly the same location. For this, one may use the BCH formula which we already mentioned in Sect. 3.3.2 (see also Sect. 6.1.2).

$$\begin{aligned}\mathcal{M} &= \mathcal{R}\exp\left(: -\mathcal{R}_2^{-1}8x^3 :\right)\exp\left(: k_s 8x^3 + 16k_o x^4 :\right)\\ &= \mathcal{R}\exp\left(: 8\{-\mathcal{R}_2^{-1}x^3 + k_s x^3\} + 16k_o x^4\right.\\ &\qquad \left. -32k_s\left[\mathcal{R}_2^{-1}x^3, x^3\right] + \cdots :\right). \end{aligned} \tag{5.33}$$

This calculation is now approximate, but unlike the intuitive treatment of Edwards and Syphers, the Poisson bracket between the sextupoles contains all the potentially important second-order terms of the phasors.

5.2.1.2 Going into resonance basis

The map $\mathcal{R}$ is fully diagonal in the complex resonance basis of Equation (4.53). So, we express Equation (5.33) in this basis—we first have

$$\mathcal{R} = \exp\left(: -\frac{\mu}{2}r_+r_- :\right),$$

followed by

$$\begin{aligned}8\mathcal{R}_2^{-1}x^3 &= \mathcal{R}_2^{-1}(r_+ + r_-)^3,\\ &= \underbrace{e^{i3\mu_2}r_+^3 + e^{-i3\mu_2}r_-^3}_{\text{1/3 resonance}} + \underbrace{3e^{i\mu_2}r_+^2r_- + 3e^{-i\mu_2}r_+r_-^2}_{\text{integer resonance}},\end{aligned}$$

and finally, the octupole term is simply

$$\begin{aligned}16k_o x^4 &= 16k_o(r_+ + r_-)^4,\\ &= k_o\left\{\underbrace{r_+^4 + r_-^4}_{\text{1/4 resonance}} + \underbrace{4r_+^3r_- + 4r_+r_-^3}_{\text{1/2 resonance}} + 6\underbrace{r_+^2r_-^2}_{\text{tune shift}}\right\}.\end{aligned}$$

Using this basis, we can work out the bracket which represents the interaction of the two sextupoles to leading order:

$$\begin{aligned}-32k_s\left[\mathcal{R}_2^{-1}x^3, x^3\right] &= k_s\left\{9ie^{i\mu_2}(e^{i2\mu_2} - 1)r_+^4 - 9ie^{-i\mu_2}(e^{-i2\mu_2} - 1)r_-^4\right\}\\ &\quad + k_s\left\{-36e^{i\mu_2}\sin(2\mu_2)\,r_+^3r_- - 36e^{-i\mu_2}\sin(2\mu_2)\,r_+r_-^3\right\}\\ &\quad + k_s\left\{-18\sin(3\mu_2) - 54\sin(\mu_2)\right\}r_+^2r_-^2.\end{aligned} \tag{5.34}$$

Therefore the map $\mathcal{M}$ can be written as:

$$\mathcal{M} = \mathcal{R} \exp\left(: f_3 + f_4 :\right),$$

$$f_3 = f_3^3 + f_3^1 \text{ and } f_4 = f_4^4 + f_4^2 + f_4^0,$$

$$\begin{aligned}
f_3^3 &= (k_s - e^{i3\mu_2})r_+^3 + (k_s - e^{-i3\mu_2})r_-^3 \\
f_3^1 &= 3(k_s - e^{i\mu_2})r_+^2 r_- + 3(k_s - e^{-i\mu_2})r_+ r_-^2 \\
f_4^4 &= \left\{k_o + k_s 9ie^{i\mu_2}(e^{i2\mu_2} - 1)\right\} r_+^4 \\
&\quad + \left\{k_o - k_s 9ie^{-i\mu_2}(e^{-i2\mu_2} - 1)\right\} r_-^4 \\
f_4^2 &= \left\{4k_o - k_s 36e^{i\mu_2} \sin(2\mu_2)\right\} r_+^3 r_- \\
&\quad + \left\{4k_o - k_s 36e^{-i\mu_2} \sin(2\mu_2)\right\} r_+ r_-^3 \\
f_4^0 &= \left\{6k_o - k_s 18 \sin(3\mu_2) - k_s 54 \sin(\mu_2)\right\} r_+^2 r_-^2 \,. \qquad (5.35)
\end{aligned}$$

The reader will recognize in Equation (5.35) the "phasors" of the first-order theory. The polynomial f_3^1 contains the phasors of the integer resonance while f_3^3 is made up of third integer phasors. The second order expression is more complex. It contains the phasors corresponding to all possible pairs of sextupoles interacting with one another. The octupole terms are again trivial: they are the first order phasors of an octupole insertion. These statements are rigorous and extend, as we will later discover, to arbitrary distributions of sextupoles and octupoles. As I have been saying in this book (and elsewhere), the Lie polynomials in resonance basis are *rigorous and extendable generalizations* of the intuitive concepts of phasors.

The map in Equation (5.35) is not a "one-resonance map". In the naïve theory, the entire contribution from the phasors belonging to other resonances is ignored, thus failing, in some cases, to reproduce the correct global topology even as ϵ approaches zero—no matter how close we get to the third-order resonance. The stable and unstable topologies of Figures 2.4 and 5.2 are both possible as the linear tune approaches one-third; generically these cases are resolved by performing a second-order calculation.

5.2.1.3 Canonical transformation

To get the map into one-resonance, we act on it with a canonical transformation of the form:

$$\mathcal{B} = \exp\left(: F_4 :\right) \exp\left(: F_3 :\right),$$

$$
\begin{aligned}
F_3 &= \cancel{F_3^3} + F_3^1, \\
F_4 &= F_4^4 + F_4^2 + \cancel{F_4^0}.
\end{aligned}
\tag{5.36}
$$

Notice that I crossed out two terms. First, the canonical transformation should not contain any terms related to the 1/3 resonance. Secondly, the F_4^0 term of F_4 is a tune shift term which is not removable from any map. Let us proceed:

$$
\begin{aligned}
\mathcal{N} &= \mathcal{BMB}^{-1} \\
&= e^{:F_4:}e^{:F_3:}\mathcal{R}\exp\left(: f_3 + f_4 :\right)e^{:-F_3:}e^{:-F_4:}, \\
&= \mathcal{R}\exp\left(: \mathcal{R}^{-1}F_4 :\right)\exp\left(: \mathcal{R}^{-1}F_3 :\right)\exp\left(: f_3 + f_4 :\right) \\
&\qquad \times \exp\left(: -F_3 :\right)\exp\left(: -F_4 :\right), \\
&= \mathcal{R}\exp\left(: n_3 + n_4 :\right), \\
n_3 &= \mathcal{R}^{-1}F_3 - F_3 + f_3, \\
n_4 &= \mathcal{R}^{-1}F_4 - F_4 + f_4 + \frac{1}{2}\left[\mathcal{R}^{-1}F_3, f_3\right] \\
&\qquad + \frac{1}{2}\left[\mathcal{R}^{-1}F_3 + f_3, -F_3\right].
\end{aligned}
\tag{5.37}
$$

The expression in Equation (5.37) is the correct second-order expression for any type of transformation. If we focus on our problem, we get to sextupole order:

$$
\begin{aligned}
F_3 = F_3^1 &= \left(\mathcal{I} - \mathcal{R}^{-1}\right)^{-1} f_3^1 \\
&= \frac{3(k_s - e^{i\mu_2})}{1 - e^{i\mu}} r_+^2 r_- + \frac{3(k_s - e^{-i\mu_2})}{1 - e^{-i\mu}} r_+ r_-^2 .
\end{aligned}
\tag{5.38}
$$

Exercise 5-3
Here we analyze the Poisson bracket terms.

1. *Show that the second order cross term can be written as*

$$
\begin{array}{ccc}
\frac{1}{2}\left[\mathcal{R}^{-1}F_3, f_3\right] & & + \frac{1}{2}\left[\mathcal{R}^{-1}F_3 + f_3, -F_3\right] \\
\downarrow & & \downarrow \\
\frac{1}{2}\left[\mathcal{R}^{-1}F_3^1, f_3^1\right] + \frac{1}{2}\left[\mathcal{R}^{-1}F_3^1, f_3^3\right] & + & \frac{1}{2}\left[f_3^3, -F_3^1\right]
\end{array}
\tag{5.39}
$$

2. *Show that the tune shift coming from these bracket originates in*

$$
\frac{1}{2}\left[\mathcal{R}^{-1}F_3^1, f_3^1\right].
$$

3. Hence the normalized map of Equation (5.37) *is given by*

$$
\begin{aligned}
n_3 &= n_3^3 = (k_s - e^{i3\mu_2})r_+^3 + (k_s - e^{-i3\mu_2})r_-^3, \\
n_4 &= n_4^0 = f_4^0 + \frac{1}{2}\left[\mathcal{R}^{-1}F_3^1, -F_3^1\right] \qquad (5.40)\\
&= \big(\ \{6k_o - k_s 18\sin(3\mu_2) - k_s 54\sin(\mu_2)\} \\
&+\ 27\cot(\mu/2)\left\{k_s^2 + 1 - 2k_s\cos(\mu_2)\right\}\ \big)\, r_+^2 r_-^2 .
\end{aligned}
$$

For our purposes, the actual form of $F_4 = F_4^4 + F_4^2$ is not important. We simply notice that it is possible to remove the driving terms for the 1/4 and 1/2 resonances.

5.2.1.4 Let us pause and think

Normally, the purpose of a one-resonance normalization is to understand and quantify the features of a simulation or experiment. In Figure 5.2, the phase space is a triangle with unstable hyperbolic fixed points at the vertices. Let us assume that the tune shift terms vanish in Equation (5.40), then following the ideas of Exercise 5.2, we cube the map. If μ is almost $2\pi/3$, the three-turn map is given by

$$
\begin{aligned}
\mathcal{M}^3 &= \mathcal{B}^{-1}\mathcal{N}^3\mathcal{B} \\
&= \mathcal{B}^{-1}\left(\mathcal{R}\exp\left(: n_3^3 :\right)\right)^3\mathcal{B} \\
&= \mathcal{B}^{-1}\mathcal{R}^3\exp\left(: \mathcal{R}^{-2}n_3^3 + \mathcal{R}n_3^3 + n_3^3 :\right)\mathcal{B} \\
\text{if } \mu = \frac{2\pi(1+\epsilon)}{3} &\Rightarrow \mathcal{R}^k n_3 = 3n_3 + O(\epsilon) \text{ and } \mathcal{R}^3 = \exp\left(: -2\pi\epsilon J :\right) \\
&\Downarrow \\
\mathcal{M}^3 &= \mathcal{B}^{-1}\exp\left(: -2\pi\epsilon J :\right)\exp\left(: 3n_3^3 :\right)\mathcal{B} \\
&= \mathcal{B}^{-1}\exp\left(: -2\pi\epsilon J + 3n_3^3 + O(\epsilon)\ :\right)\mathcal{B}. \qquad (5.41)
\end{aligned}
$$

For small ϵ it represents correctly the motion in the neighborhood of the origin. If fact, even if the tune shifts do not vanish, since they are one degree higher than the resonance term n_3^3, one is tempted to say that close enough to the resonance, the topology of the map of Equation (5.41) is always representative of the general case. This is actually false.

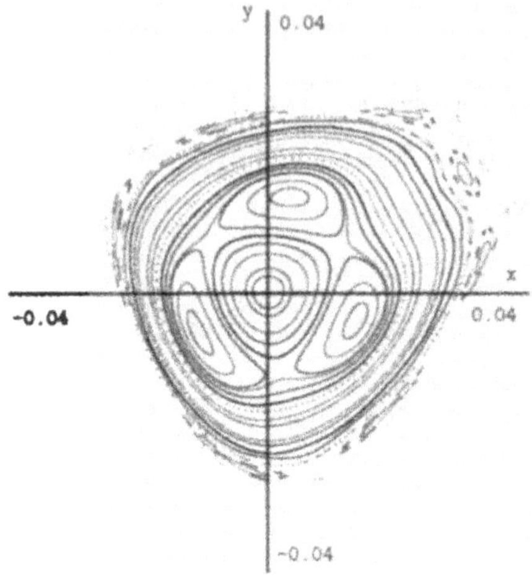

Figure 5.3: $\epsilon = 0.03$.

Exercise 5-4
Show that the map of Equation (5.41) *has three unstable fixed points on either side of the resonance, but it does not have stable elliptic points besides the origin.*

In Figures 5.3 and 5.4, the three islands do not disappear as we approach $\epsilon = 0$. This is realized by setting $k_s = 1$, $k_o = 0$ and $\mu_2 = -0.64\ \pi$.

If we imagine trying to extract a beam using a septum at $x = 0.01$, then this procedure will fail. This is not just an "academic" disaster. In the late 1980's, I had the chance to examine a design of the Kaon factory project at the TRIUMF laboratory in Vancouver, British Columbia. In their initial design, it was not possible to extract the beam below or above the third integer. In a ring, the sextupoles are usually set to fixed values so as to reduce the chromaticities. Then, extraction is done by moving the tunes using quadrupoles. The entire operation amounts to a change of ϵ. So, what is happening in Figures 5.3 and 5.4? The map is actually undergoing a nonlinear stabilization: a good thing unless you are extracting the beam. This is due to the "tune shift" term in the one-resonance map, i.e., n_4^0 in Equation (5.40). Therefore, it is actually important to carry the calculation beyond first order. There are actually pathological cases where one needs to carry

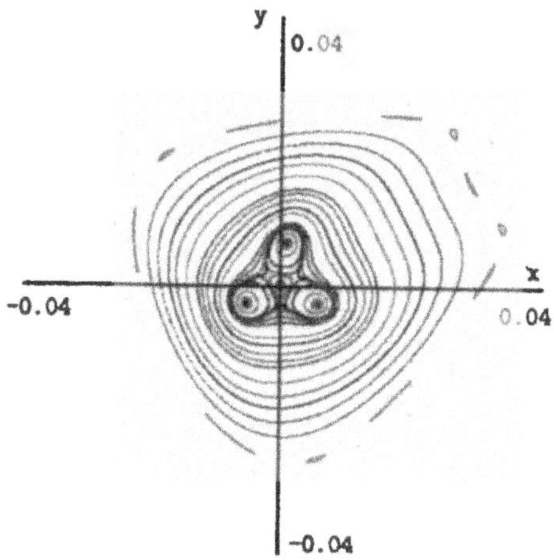

Figure 5.4: $\epsilon = 0.003$.

the calculation all the way to fourth order in sextupole strength in order to get the correct answer! This is the case of Figure 5.1: it was produced by using the map $\exp\left(-2\pi\left(\frac{1}{4}+0.001\right)J\right)\exp\left(:8x^3:\right)$.

Now, let us go back to the calculation of the invariant in the plane of the resonance.

5.2.1.5 The co-moving map

The idea of "cubing" a map near the third integer in order to get a map near the identity was proposed by Dragt in his early lectures on Lie methods[28]. Clearly the map $\mathcal{M}^3$ must have the same invariant as $\mathcal{M}$. However, while it is clear that this procedure will always work for resonances involving a single plane (one-dimensional resonance), we must use the theory of Sect. 5.1.3 in the general case.

We apply Equation (5.17) to the map in Equation (5.40):

$$\begin{aligned}\mathcal{N}_c &= \exp(:\frac{2\pi}{3}J:)\mathcal{N} \\ &= \exp(:-\frac{2\pi}{3}\epsilon J:)\exp\left(:n_3^3+n_4^0:\right). \qquad (5.42)\end{aligned}$$

The challenge consists in writing the map $\mathcal{N}_c$ in terms of a single exponent to all orders in the linear part. In analytical calculations, this is best achieved by temporarily normalizing $\mathcal{N}_c$ into a rotation

$$\begin{aligned}\mathcal{R}_c &= \mathcal{C}_{1/\epsilon}\mathcal{N}_c\mathcal{C}_{1/\epsilon}^{-1}\\ &= \exp(:-\frac{2\pi}{3}\epsilon J + n_4^0 + \Delta n_4^0 :\ ),\end{aligned} \tag{5.43}$$

and then using the transformation $\mathcal{C}_{1/\epsilon}$ to reconstruct the map $\mathcal{R}_c$ as a single exponent map. The answer is simply

$$\begin{aligned}\mathcal{N}_c &= \mathcal{C}_{1/\epsilon}^{-1}\mathcal{R}_c\mathcal{C}_{1/\epsilon}\\ &= \exp(:-\frac{2\pi}{3}\epsilon\mathcal{C}_{1/\epsilon}^{-1}J + \mathcal{C}_{1/\epsilon}^{-1}n_4^0 + \mathcal{C}_{1/\epsilon}^{-1}\Delta n_4^0 :\ )\\ \Rightarrow h_c &= -\frac{2\pi}{3}\epsilon\mathcal{C}_{1/\epsilon}^{-1}J + \mathcal{C}_{1/\epsilon}^{-1}n_4^0 + \mathcal{C}_{1/\epsilon}^{-1}\Delta n_4^0.\end{aligned} \tag{5.44}$$

In Equation (5.44), the single exponent h_c is an invariant of the map $\mathcal{N}_c$ as well as $\mathcal{N}$. I remind the reader of the $1/\epsilon$ divergence with the notation $\mathcal{C}_{1/\epsilon}$. We will see explicitly how this divergence cancels out in the expression for the invariant h_c, thus permitting good behavior past the separatrix and the three islands.

5.2.1.6 Computation of $\mathcal{C}_{1/\epsilon}$ and h_c

As usual we write $\mathcal{C}_{1/\epsilon}$ in terms of its "inverse" Dragt-Finn representation:

$$\mathcal{C}_{1/\epsilon} = \exp\left(: c_4^4 + c_4^2 :\right)\exp\left(: c_3^3 :\right) = \exp\left(: c_3^3 :\right). \tag{5.45}$$

One notices immediately that there are no quartic Lie transforms for the map $\mathcal{C}_{1/\epsilon}$. This is because the Poisson bracket of r_+^3 with r_-^3 is just $-18i\ r_+^2 r_-^2 = -72i\ J^2$, which contributes only to the tune shift term Δn_4^0. The result is

$$\begin{aligned}&\text{Given } n_3^3 = \underbrace{\left(k_s - e^{i3\mu_2}\right)}_{a} r_+^3 + \underbrace{\left(k_s - e^{i3\mu_2}\right)}_{a^*} r_-^3\\ &\Rightarrow c_3^3 = \frac{a}{1-\exp\left(i2\pi\epsilon\right)}\, r_+^3 + \frac{a^*}{1-\exp\left(-i2\pi\epsilon\right)}\, r_-^3\\ &\Delta n_4^0 = 36\left|a\right|^2\cot\left(\pi\epsilon\right)J^2.\end{aligned} \tag{5.46}$$

Using Equation (5.46), we can evaluate h_c to leading order:

$$
\begin{aligned}
h_c &= -\frac{2\pi}{3}\epsilon C_{1/\epsilon}^{-1} J + C_{1/\epsilon}^{-1} n_4^0 + C_{1/\epsilon}^{-1} \Delta n_4^0 = h_{c2}^0 + h_{c3}^3 + h_{c4}^0 \\
&= -\frac{2\pi}{3}\epsilon \left(J - [c_3^3, J] + \frac{1}{2}\,[c_3^3, [c_3^3, J]] \right) + n_4^0 + \Delta n_4^0 \\
h_{c2}^0 &= -\frac{2\pi}{3}\epsilon J \\
h_{c3}^3 &= \frac{-i2\pi\epsilon a}{1-\exp(i2\pi\epsilon)} r_+^3 + \frac{i2\pi\epsilon a^*}{1-\exp(-i2\pi\epsilon)} r_-^3 \\
h_{c4}^0 &= \left\{ 36\cot(\pi\epsilon) - \frac{72\pi\epsilon}{1-\cos(2\pi\epsilon)} \right\} |a|^2 J^2 + n_4^0.
\end{aligned} \tag{5.47}
$$

The tune shift term written the long way is just

$$
\begin{aligned}
h_{c4}^0 = J^2 \Bigg(& 36 \left\{ \cot(\pi\epsilon) - \frac{2\pi\epsilon}{1-\cos(2\pi\epsilon)} \right\} \left\{ k_s^2 + 1 - 2k_s \cos(3\mu_2) \right\} \\
& + 108 \cot\left(\frac{\mu}{2}\right) \left\{ k_s^2 + 1 - 2k_s \cos(\mu_2) \right\} \\
& + 4 \left\{ 6k_o - k_s 18 \sin(3\mu_2) - k_s 54 \sin(\mu_2) \right\} \Bigg).
\end{aligned} \tag{5.48}
$$

Thus we can rewrite the original map $\mathcal{M}$ as

$$
\begin{aligned}
\mathcal{M} &= \mathcal{B}^{-1} \exp\left(-\frac{2\pi}{3} : J : \right) \exp\left(: h_c : \right) \mathcal{B} \\
&= \exp\left(-\frac{2\pi}{3} : \mathcal{B}^{-1} J : \right) \exp\left(: \mathcal{B}^{-1} h_c : \right).
\end{aligned} \tag{5.49}
$$

The reader can appreciate that the co-moving phase

$$
\exp\left(-\frac{2\pi}{3} : \mathcal{B}^{-1} J : \right)
$$

in the original variables cannot be easily guessed. It is in general a rotation distorted by a nonlinear diffeomorphism (i.e., $\mathcal{B}^{-1}$).

5.2.1.7 Discussion: limiting cases

For the readers who have a copy of Edwards and Syphers' book, here is a little exercise.

Exercise 5-5

The purpose of this exercise is to derive Equation (4.37), page 118, in reference [4]. Consider a one-dimensional sextupole map where a thin lens sextupole is located at an arbitrary location "s."

$$\mathcal{M} = \exp\left(- : \Delta\mu_s J :\right) \exp\left(: -\frac{\widetilde{k}_s}{3} x^3 : \right) \exp\left(- : (\mu - \Delta\mu_s) J :\right),$$

$$\text{where } \widetilde{k}_s = \frac{B_s''}{2Brho} \beta_s^{3/2}. \tag{5.50}$$

We can define an amplitude "a" as follows

$$x = \sqrt{2J} \cos(\phi) = a \cos(\phi) \ \ \text{and}\ p = -\sqrt{2J} \sin(\phi)$$
$$\text{with } [\phi, J] = 1. \tag{5.51}$$

Show that the change in "a" to leading order is given by

$$\Delta a = \left[-\frac{\widetilde{k}_s}{3} \exp\left(- : \Delta\mu_s J :\right) x^3, a \right]$$
$$= \frac{\widetilde{k}_s a^2}{4} \left\{ \sin(\phi + \Delta\mu_s) + \sin(3\phi + 3\Delta\mu_s) \right\}. \tag{5.52}$$

Clearly to leading order, the change in "a" due to a distribution of sextupoles involves the integral of Δa over s. One realizes that the "s" dependence in the Lie algebraic treatment is carried by the functions $\Delta\mu_s$ and $\widetilde{k}_s$. The action-angle variables are just dummy variables or functions of the initial phase space coordinates (depending on your point of view). This makes the entire treatment a matter of a couple of lines versus four pages in reference [4].

Let us return to the exact second-order treatment of our "pet" problem and see how it reduces to the usual naïve treatments found in the accelerator literature. We will concentrate on the one-resonance invariant h_c which is topologically similar to $\mathcal{B}^{-1}h_c$, the invariant of the real map.

1. **Super Naïve Theory: $k_o \gg \beta k_s^2$ and $\epsilon \approx 0$**

   In this limit, the invariant h_c (as well as $\mathcal{B}^{-1}h_c$) is just

$$h_c = -\frac{2\pi}{3}\epsilon J + \left(k_s - e^{i3\mu_2}\right) r_+^3 + \left(k_s - e^{-i3\mu_2}\right) r_-^3 + 6k_o r_+^2 r_-^2$$
$$= -\frac{2\pi}{3}\epsilon J + 2\left(k_s \cos(3\phi) - \cos(3\{\mu_2 - \phi\})\right)(2J)^{3/2}$$
$$+ 24 k_o J^2. \tag{5.53}$$

Here, tune shifts are ignored unless they appear explicitly in the Hamiltonian due to the presence of octupoles. This is the treatment one finds in Edwards and Syphers[4].

2. **Not so Naïve Theory: $k_o \approx \beta k_s^2$ and $\epsilon \approx 0$**

   For small octupole strength (or no octupole at all), one must keep the effect of the sextupoles to second order. However, as we approach the resonance, the asymptotic value of h_c can be read by setting $\epsilon = 0$ in the nonlinear part of Equation (5.40):

$$\begin{aligned} h_c = & -\frac{2\pi}{3}\epsilon J \; + \; 2\left(k_s \cos\left(3\phi\right) - \cos\left(3\{\mu_2 - \phi\}\right)\right)(2J)^{3/2} \\ & + \frac{108}{\sqrt{3}}\left\{k_s^2 + 1 - 2k_s \cos\left(\mu_2\right)\right\} J^2 \\ & + 4\left\{6k_o - k_s 18 \sin(3\mu_2) - k_s 54 \sin(\mu_2)\right\} J^2 . \end{aligned} \tag{5.54}$$

   Of course, one gets the same result by taking the asymptotic limit of the exact solution.

The reader can verify that the presence of "detuning," namely the terms in J^2 in h_c allow a closure of the separatrix, and thus, the presence of stable islands for values of ϵ above and below zero.

5.2.2 Other normalizations

Other types of symplectic normalization are usually simpler conceptually than the one-resonance normalization. It is clear that we could have elected to remove all the resonances from the map. In fact, this was done in Sect. 5.2.1.6 and the result was given in Equation (5.46). Of course, in a total normalization we take out the resonances all at once, combining $C_{1/\epsilon}$ and $\exp\left(: F_3^1 :\right)$ into one single operator.

The theory will look very similar in the case of a coasting beam normalization since one can treat the variable δ as a system parameter when doing the nonlinear normalization. (Except for the fact that Lie transforms, which are functions of δ alone, must be kept in the map since they produce pure momentum compaction.)

A full unified treatment must involve the (quasi-symplectic) radiative map. The Lie operators of this map cannot be made of Poisson brackets. Therefore it is time to pause for a chapter on general vector fields.

CHAPTER 6
VECTOR FIELDS AND CANONICAL TRANSFORMATIONS

This chapter can be viewed by the reader as a big theoretical interlude. Many readers may find it, to be polite, too abstract or plainly unreadable. Nevertheless it should be read diagonally at the very least.

Here, I intend to achieve four goals:

1. to display the homomorphism which exists between various brackets,

2. to use these homomorphisms to derive the effect of s-dependent canonical transformations on the map equation (Interlude I),

3. thus make contact with Hamiltonian (s-dependent) theory,

4. and finally, to extend the concept of the resonance basis to an arbitrary vector field (i.e., map).

The last item on the list is truly useful in ring calculations. It allows us to treat the case of an electron ring using tools which are as close as possible to the original symplectic formalism. As for the first few items, they allow the reader to make the connection with the s-dependent theory. In fact, we will spend some time working out simple problems different ways in the exercises so as to give ourselves some appreciation of the adequacy of a map based theory in ring dynamics. My hidden goal is to send the studious reader straight to the vomitorium by force feeding him Hamiltonian theory!

6.1 VECTOR FIELDS

Let us recall the equation for the map derived in Interlude I, adding here a smallness parameter ε:

$$\frac{d}{ds}\vec{\zeta}_{s_0\to s} = \varepsilon\vec{F}_s(\vec{\zeta}_{s_0\to s})$$
$$\Downarrow$$
$$\mathcal{M}_{s_0\to s} = \frac{d}{ds}\mathcal{M}_{s_0\to s}\,\varepsilon\vec{F}_s\cdot\vec{\nabla}. \tag{6.1}$$

We can ask some formal questions, i.e., neglecting issues of convergence, about the solutions of Equation (6.1). For example, what is the nature of the following operators

$$\mathcal{M}_{s_0\to s} = \exp\left(\hat{h}\right) \text{ where } \hat{h} = \sum_{n=1}^{\infty}\varepsilon^n\hat{h}_n,$$
$$\overset{\text{or}}{=}\cdots\exp\left(\varepsilon^k\hat{f}_k\right)\cdots\exp\left(\varepsilon\hat{f}_1\right), \tag{6.2}$$

or more precisely, of the Lie operators of these two ε-expansions?

The answer critical to this book is that these operators are linear combinations of the vector field $\vec{F}_s\cdot\vec{\nabla}$ evaluated at different s's, and of multi-fold commutators of the same vector field. This is a generalized Baker-Campbell-Hausdorff theorem (BCH). Secondly, and of equal importance, is the fact that the commutator of two vector fields is a vector field. Thus all the operators of Equation (6.2) are vector fields. This implies that a homomorphism exists between the commutator of two vector fields and a bracket of the vector functions defining the fields themselves. Finally, in the case of Hamiltonian dynamics, this bracket will reduce to the Poisson bracket. We now intend to prove these fundamental results.

6.1.1 The Homomorphisms

Since the homomorphisms are central to our discussion, we will start with their proofs. Let us compute the commutator of two vector fields $\vec{F}\cdot\vec{\nabla}$ and $\vec{G}\cdot\vec{\nabla}$, by letting the commutator act on an arbitrary function f:

$$\forall f\in C_\infty \to \left\{\vec{F}\cdot\vec{\nabla},\vec{G}\cdot\vec{\nabla}\right\}f = \vec{F}\cdot\vec{\nabla}\left(\vec{G}\cdot\vec{\nabla}f\right) - \vec{G}\cdot\vec{\nabla}\left(\vec{F}\cdot\vec{\nabla}\right)f$$
$$\text{but } \vec{F}\cdot\vec{\nabla}\left(\vec{G}\cdot\vec{\nabla}f\right) = F_a\left(\partial_a G_b\right)\partial_b f + F_a\left(G_b\partial_a\partial_b f\right),$$

exchanging F and G,

$$\left\{\vec{F}\cdot\vec{\nabla},\vec{G}\cdot\vec{\nabla}\right\} = \left(F_a\left(\partial_a G_b\right) - G_a\left(\partial_a F_b\right)\right)\partial_b = \vec{H}\cdot\vec{\nabla}$$
$$\text{where } \vec{H} = \left\langle\vec{F},\vec{G}\right\rangle \equiv F_a\left(\partial_a\vec{G}\right) - G_a\left(\partial_a\vec{F}\right). \tag{6.3}$$

Therefore, we have defined a bracket between the vector functions $\vec{F}$ and $\vec{G}$ denoted by $\left\langle\vec{F},\vec{G}\right\rangle$.

The next important homomorphism concerns the special case of a Hamiltonian flow. In that case there exists two functions f and g such that

$$\vec{F} = \mathrm{J}\vec{\nabla}f \text{ and } \vec{G} = \mathrm{J}\vec{\nabla}g. \tag{6.4}$$

Exercise 6-1

1. *Show that*

$$\left\langle\mathrm{J}\vec{\nabla}f,\mathrm{J}\vec{\nabla}g\right\rangle = \mathrm{J}\vec{\nabla}\left[g,f\right] \tag{6.5}$$

is true.

2. *The Poisson bracket operator is defined by Dragt as* $: f : g = [f,g]$*, show that this implies that*

$$: -f := \left(J\vec{\nabla}f\right)\cdot\vec{\nabla}, \tag{6.6}$$

and therefore, it follows from Equation (6.5) *that*

$$\{: -f :,: -g :\} = \ :[-f,-g]:$$
or simply
$$\{: f :,: g :\} = \ :[f,g]:. \tag{6.7}$$

In the Hamiltonian theory, we have established a homomorphism between the commutator $\{\ ,\ \}$, the vector function bracket $\langle\ ,\ \rangle$ and the Poisson bracket $[\ ,\ ]$. Why are the homomorphisms important? Firstly, in the case of Hamiltonian dynamics, the homomorphism between the Poisson bracket and vector field bracket tells us that we can deal with functions instead of vector functions. That, of course, is the essence of the Hamiltonian formalism. Secondly, we can ask what happens if the operators involved in the map equation have a small non-Hamiltonian part, due to radiation for example. If we rewrite the vector field of Equation (6.1) as $\vec{F}_s \cdot \vec{\nabla} =: -H_s : + \delta\vec{F}_s \cdot \vec{\nabla}$ and use the still unproven BCH theorem, then we know that the solution for the compositional map must involve commutators

$$\left\{\vec{F}_s \cdot \vec{\nabla}, \vec{F}_{s'} \cdot \vec{\nabla}\right\} = \left\{: -H_s : + \delta\vec{F}_s \cdot \vec{\nabla}, : -H_{s'} : + \delta\vec{F}_{s'} \cdot \vec{\nabla}\right\}$$
$$=: [-H_s, -H_{s'}] : + O(\delta) \cdots \qquad (6.8)$$

Hence the map is Hamiltonian to order δ. But, more important is the issue of a canonical transformation on the map: if the Lie operators of such a transformation are computed from the Lie operators of the map (as we did for the symplectic case Chapters 4 and 5), then the resulting normalized map (whatever it is) will also be Hamiltonian to order δ. In general, if arbitrary non-canonical transformations are used this is not necessarily the case.

6.1.2 The BCH Theorem

The standard version of the BCH theorem states

$$\exp\left(\widehat{C}(\varepsilon)\right) = \exp\left(\varepsilon\widehat{A}\right)\exp\left(\varepsilon\widehat{B}\right) \qquad (6.9)$$
$$\text{where } \widehat{C} = \varepsilon\widehat{A} + \varepsilon\widehat{B} + \frac{\varepsilon^2}{2}\left\{\widehat{A}, \widehat{B}\right\}$$
$$+ \frac{\varepsilon^3}{12}\left\{\widehat{B}, \{\widehat{B}, \widehat{A}\}\right\} + \frac{\varepsilon^3}{12}\left\{\{\widehat{B}, \widehat{A}\}, \widehat{A}\right\} + \cdots,$$

and $\widehat{C}$ is made out of commutators involving $\widehat{A}$ and $\widehat{B}$ only. It is possible to compute this series to high order. For example, Koseleff carries this calculation to order ε^9 in his thesis[29]. Here, we will prove a more general result. We intend to show that the map of Equation (6.1) is made out of commutators of $\vec{F}_s \cdot \vec{\nabla}$ taken at different values of s. The proof will rely heavily on the homomorphisms displayed in Sect. 6.1.1.

The proof starts with rewriting the equation for the map

$$\frac{d}{ds}\mathcal{M} = \mathcal{M}\varepsilon\vec{F}_s \cdot \vec{\nabla}$$
$$\mathcal{M}^{-1}\left\{\frac{d}{ds}\mathcal{M}\right\} = \varepsilon\vec{F}_s \cdot \vec{\nabla}, \qquad (6.10)$$

and then writing the map $\mathcal{M}$ as $\exp\left(\alpha\widehat{C}\right)$ where α is a smallness parameter which we will later put equal to one. By differentiating, we get a differential equation in α for the map $\mathcal{K} = \mathcal{M}^{-1}\frac{d}{ds}\mathcal{M}$:

$$\frac{d}{d\alpha}\mathcal{K} = \frac{d}{d\alpha}\left\{e^{(-\alpha\widehat{C})}\left\{\frac{d}{ds}e^{(\alpha\widehat{C})}\right\}\right\} = \left\{-\widehat{C},\mathcal{K}\right\} + \frac{d}{ds}\widehat{C}. \quad (6.11)$$

We can now defined an adjoint operator in terms of the commutator:

$$\mathrm{ad}\widehat{C}\ \widehat{A} = \left\{\widehat{C},\widehat{A}\right\}. \quad (6.12)$$

The differential equation for $\mathcal{K}$ becomes

$$\begin{aligned} &\frac{d}{d\alpha}\mathcal{K} = -\mathrm{ad}\widehat{C}\ \mathcal{K} + \frac{d}{ds}\widehat{C}, \\ &\mathcal{K}_{\alpha=0} = 0. \end{aligned} \quad (6.13)$$

This operator equation can be solved by "variation of parameters" ("Variation of maps" is more accurate here!). First we solve the homogeneous equation:

$$\begin{aligned} &\frac{d}{d\alpha}\mathcal{K}_h = -\mathrm{ad}\widehat{C}\ \mathcal{K}_h \\ &\Rightarrow \mathcal{K}_h(\alpha) = \exp\left(-\alpha\ \mathrm{ad}\widehat{C}\right)\mathcal{K}_0. \end{aligned} \quad (6.14)$$

In the next step, as in the usual variation of parameters, one lets the map $\mathcal{K}_0$ depend on α in an attempt to get a particular solution to Equation (6.13). This leads to a simple quadrature for $\mathcal{K}_0(\alpha)$

$$\frac{d}{d\alpha}\exp\left(-\alpha\ \mathrm{ad}\widehat{C}\right)\mathcal{K}_0(\alpha) = -\mathrm{ad}\widehat{C}\exp\left(-\alpha\ \mathrm{ad}\widehat{C}\right)\mathcal{K}_0 + \frac{d}{ds}\widehat{C}$$

$$\Downarrow$$

$$\frac{d}{d\alpha}\mathcal{K}_0(\alpha) = \exp\left(\alpha\ \mathrm{ad}\widehat{C}\right)\frac{d}{ds}\widehat{C}$$

$$\Downarrow$$

$$\mathcal{K}_0(\alpha) = \frac{\exp\left(\alpha\ \mathrm{ad}\widehat{C}\right)}{\mathrm{ad}\widehat{C}}\frac{d}{ds}\widehat{C}. \quad (6.15)$$

The final solution is constructed as the sum of the particular solution (Equation (6.15)) and the homogeneous solution (Equation (6.14))

$$\mathcal{K}(\alpha) = \exp\left(-\alpha \,\mathrm{ad}\widehat{C}\right)\mathcal{K}_0 + \frac{1}{\mathrm{ad}\widehat{C}}\frac{d}{ds}\widehat{C} \tag{6.16}$$

where $\mathcal{K}_0$ is selected so as to ensure $\mathcal{K}(\alpha=0) = 0$. Therefore, the solution (setting $\alpha = 1$) is

$$\mathcal{K} = \frac{1-\exp\left(-\mathrm{ad}\widehat{C}\right)}{\mathrm{ad}\widehat{C}}\frac{d}{ds}\widehat{C} \tag{6.17}$$

which we now substitute in Equation (6.10), with the result

$$\frac{1-\exp\left(-\mathrm{ad}\widehat{C}\right)}{\mathrm{ad}\widehat{C}}\frac{d}{ds}\widehat{C} = \varepsilon \vec{F}_s \cdot \vec{\nabla}. \tag{6.18}$$

We can defined a function $\Upsilon(x) = \frac{x}{1-e^{-x}}$ and integrate Equation (6.18) to get a formula for $\widehat{C}$:

$$\widehat{C}(s) = \int_0^s d\sigma\ \Upsilon(\mathrm{ad}\widehat{C}(\sigma))\varepsilon \vec{F}_\sigma \cdot \vec{\nabla}. \tag{6.19}$$

If one rewrites $\widehat{C}(s)$ as an expansion in ε i.e., $\widehat{C} = \sum_{n=1}^{\infty} \varepsilon^i \widehat{C}_i$, then it is easy to check by induction on the powers of ε that Equation (6.19) will only generate integrals over commutators of $\vec{F}_s \cdot \vec{\nabla}$ with itself at different values of s. Hence, the operator $\widehat{C}$ can be written as $\vec{C} \cdot \vec{\nabla}$. This proves the BCH theorem.

Using the homomorphisms of Sect. 6.1.1, we can write the following sequence of equalities:

$$\begin{gathered}
\vec{C} \cdot \vec{\nabla} = \int_0^s d\sigma\ \Upsilon(\mathrm{ad}\vec{C} \cdot \vec{\nabla})\varepsilon \vec{F}_\sigma \cdot \vec{\nabla} \\
\Downarrow \\
\vec{C} = \int_0^s d\sigma\ \Upsilon(:\vec{C}:)\varepsilon \vec{F}_\sigma \\
\Downarrow \text{ if } \exists \vec{F}_\sigma = -J\vec{\nabla} H_\sigma \Rightarrow \vec{C} = -J\vec{\nabla} c \\
c = \int_0^s d\sigma\ \Upsilon(-:c:)\varepsilon H_\sigma.
\end{gathered} \tag{6.20}$$

The operator $:\vec{C}:$ is defined by analogy to the Poisson bracket operator as

$$:\vec{C}:\vec{F} = \left\langle \vec{C}, \vec{F} \right\rangle . \tag{6.21}$$

Exercise 6-2

1. *Using the operator $\mathcal{K}$, show that the operator $\widehat{C}$ of Equation* (6.9) *can be written as*

$$\widehat{C}(\epsilon) = \int_0^\epsilon d\epsilon \; \epsilon \; \Upsilon(\mathrm{ad}\widehat{C}(\epsilon)) \left(e^{-\epsilon \; \mathrm{ad}\widehat{B}} \widehat{A} + \widehat{B} \right) . \tag{6.22}$$

2. *Derive the terms displayed in Equation* (6.9) *using Equation* (6.22).

The topic of formal Lie algebras can get rather involved and is relevant to analytic perturbation theory, the derivation of symplectic integrators and fast Taylor series algorithms using Lie operators[30]. In this book, we will not go into abstract mathematics, but the reader who would like to know more about this topic is invited to look at reference [29]. Nevertheless we need these homomorphisms to make contact with canonical transformations. In the next section, we examine the effect of a canonical transformation on the Hamiltonian.

6.2 LIE TRANSFORMATIONS ON VECTOR FIELDS

We first briefly review the case of the mixed generating function and its first order connection with the Lie transform.

6.2.1 The Mixed Generating Function: $\vec{q} \cdot \vec{p}^{\,f} + w$

It is well-known to students of classical mechanics that it is possible to transform a Hamiltonian by a canonical transformation expressed in terms of a mixed generating function. For example, consider the generating function

$$S(\vec{q}, \vec{p}^{\,f}) = \vec{q} \cdot \vec{p}^{\,f} + w(\vec{q}, \vec{p}^{\,f}; s), \tag{6.23}$$

then the final variables $\vec{z}^f = (q_1^f, p_1^f, \cdots, q_N^f, p_N^f)$ are related to the initial variables $\vec{z}$ by the formula

$$\begin{aligned} q_i^f &= q_i + \frac{\partial w}{\partial p_i^f} \\ p_i^f &= p_i - \frac{\partial w}{\partial q_i}. \end{aligned} \tag{6.24}$$

By solving these equations one gets a canonical transformation $\vec{s}$:

$$\mathbf{R}^{2N} \xrightarrow{\vec{s}} \mathbf{R}^{2N}$$

$$\vec{z} \longmapsto (q_1^f, p_1^f, \cdots, q_N^f, p_N^f). \tag{6.25}$$

If we rewrite this formula using a more rigorous notation, we will see immediately the connection with Poisson brackets to first order:

$$\vec{z}^f = \vec{z} - [w, \vec{z}] \circ \vec{s}_p. \tag{6.26}$$

In this equation $\vec{z}$ is again our sloppy notation for the identity transformation in $\mathbf{R}^{2N}$, and the other entities are

$$\mathbf{R}^{2N} \xrightarrow{\vec{s}_p} \mathbf{R}^{2N}$$

$$\vec{z} \longmapsto \left(z_1, p_1^f(\vec{z}), \cdots, z_N, p_N^f(\vec{z})\right) \tag{6.27}$$

where $\vec{p}^f = (z_2^f, \cdots, z_{2N}^f)$ is the map relating the final momentum to the initial position and momentum. The quantity w in Equation (6.26) is the generator viewed as a function—thus taking any possible argument. It follows from Equation (6.26) that the compositional map associated with $\vec{s}$ is to leading order

$$\mathcal{S} \approx \exp\left(: -w :\right). \tag{6.28}$$

Finally, the effect of $\mathcal{S}$ on a Hamiltonian H_s can be written as follows in our rigorous [6.1] notation:

$$\begin{aligned} H_s \longmapsto K_s &= H_s \circ \vec{s}^{-1} + \partial_s S \circ \vec{s}_z^{-1} \\ &= H_s \circ \vec{s}^{-1} + \partial_s w \circ \vec{s}_z^{-1} \\ &= \mathcal{S}^{-1} H_s + \partial_s w \circ \vec{s}_z^{-1}. \end{aligned} \tag{6.29}$$

[6.1] I am trying to write more precisely the usual formula $K = H + \frac{\partial S}{\partial s}$.

The map $\vec{s}_x$ is the obvious counterpart of $\vec{s}_p$: the momenta are left unchanged and the final positions are expressed in terms of their initial values. Again, to leading order in perturbation, one can write

$$K_s \approx \mathcal{S}^{-1} H_s + \partial_s w = H_s + [w, H_s] + \partial_s w. \tag{6.30}$$

From our result, it follows that the compositional transformation $\mathcal{S}^{-1}$ is the same to leading order as

$$\mathcal{S}^{-1} = \mathcal{A} = \exp(: w :) . \tag{6.31}$$

It is interesting to notice that the passive point of view is the natural choice for Lie transforms while one takes an active standpoint with generating functions. This is not surprising since generating functions are coordinate changes (i.e., generators of maps in $\mathbf{R}^{2N}$) while Lie transformations act on our function space. Of course it is a matter of perspective, but it is unsettling for someone used to generating functions. One particular symptom of the confusion is the definition of the function J. The "Lie transform person" tends to think of J as a simple function in phase space i.e., half the square radius, he then sees a canonical transformation as a distortion of the charts in which the Hamiltonian (or map) is locally expressed. In the distorted chart of a complete normalization, the Hamiltonian (or the single Lie generator of the map) is a function of J only. The generating function person sees the action-angle variables J and ϕ changing into a new set I and ψ whose dependence on the original chart is complex. However the motion of the new action-angle variables is simple. In a way, the emphasis in the Lie case is on the manifold itself while it is on the coordinate change in the usual formalism.

Now, we would like to derive the rigorous effect of an s-dependent Lie transformation on an arbitrary Hamiltonian.

6.2.2 Canonical Transformation on H: a Map Perspective

For an arbitrary form of the Lie transformation, the proof involves the homomorphisms of Sect. 6.1.1. However, if we specialize to a transformation of the form $\exp(: w :)$ this can be bypassed by extending phase space to include the time-like variable s and using the usual time-independent formalism.

Let us start with the definition of an s-dependent transformation. Consider a phase space map $\vec{\zeta}_{0\to s}$ which propagates a phase space vector at 0 to its image at s. A canonical transformation $\vec{\alpha}_s$ is applied at 0 and s. The new map $\vec{\xi}_{0\to s}$ is given by the formula

$$\vec{\xi}_{0\to s} = \vec{\alpha}_s^{-1} \circ \vec{\zeta}_{0\to s} \circ \vec{\alpha}_0. \tag{6.32}$$

This is a definition. It simply says that, to propagate in the new phase space, one must insert the ray in the original space at $s = 0$, then it propagates in the original frame, and finally it is converted to the new variables at s. In general, this transformation must be s-dependent to be useful and periodic if it is used to normalize the Hamiltonian of the ring. For example, in rings the Courant-Snyder transformation is the simplest example of such transformations—the normalized position x is scaled by the square-root of the beta function at location 0 and then scaled back by $1/\sqrt{\beta_s}$ at location s. What does this imply for the compositional maps associated with all these transformations? It follows simply from the definition of a compositional map:

$$\text{Definitions} \Rightarrow \ \forall f \in C_\infty \ \begin{cases} \mathcal{N}_s f = f \circ \vec{\xi}_{0\to s} \\ \mathcal{M}_s f = f \circ \vec{\zeta}_{0\to s} \\ \mathcal{A}_s f = f \circ \vec{\alpha}_{0\to s} \end{cases}$$

$$\begin{aligned}
\mathcal{N}_s f &= f \circ \vec{\xi}_{0\to s} = f \circ \vec{\alpha}_s^{-1} \circ \vec{\zeta}_{0\to s} \circ \vec{\alpha}_0 \\
\mathcal{N}_s f &= \mathcal{A}_0 \left(f \circ \vec{\alpha}_s^{-1} \circ \vec{\zeta}_{0\to s} \right) \\
&= \mathcal{A}_0 \mathcal{M}_s \left(f \circ \vec{\alpha}_s^{-1} \right) \\
&= \mathcal{A}_0 \mathcal{M}_s \mathcal{A}_s^{-1} f \\
\Rightarrow \mathcal{N}_s &= \mathcal{A}_0 \mathcal{M}_s \mathcal{A}_s^{-1}.
\end{aligned} \tag{6.33}$$

Let us take the derivative of the map $\mathcal{N}_s$ with respect to s (denoted by a dot) assuming that $\mathcal{M}_s$ obeys Equation (6.1):

$$\begin{aligned}
\frac{d}{ds}\mathcal{N}_s &= \mathcal{A}_0 \dot{\mathcal{M}}_s \mathcal{A}_s^{-1} + \mathcal{A}_0 \mathcal{M}_s \dot{\mathcal{A}}_s^{-1} \\
&= \mathcal{N}_s \left\{ \mathcal{A}_s \varepsilon \vec{F}_s \cdot \vec{\nabla} \mathcal{A}_s^{-1} + \mathcal{A}_s \dot{\mathcal{A}}_s^{-1} \right\} \\
&= \mathcal{N}_s \left\{ \varepsilon \, \overrightarrow{\mathcal{A}_s \vec{F}_s} \cdot \vec{\nabla} + \mathcal{A}_s \dot{\mathcal{A}}_s^{-1} \right\} \\
&= \mathcal{N}_s \left\{ \varepsilon \, \overrightarrow{\mathcal{A}_s \vec{F}_s} \cdot \vec{\nabla} + \Delta \vec{F}_s \cdot \vec{\nabla} \right\}.
\end{aligned} \tag{6.34}$$

In Equation (6.34), we are using the BCH theorem to assert that the operator $\mathcal{A}_s \dot{\mathcal{A}}_s^{-1}$ must be a vector field operator whose vector is denoted by $\Delta\vec{F}_s$. Of course, if we specialize to Hamiltonian systems Equation (6.34) becomes

$$\frac{d}{ds}\mathcal{N}_s = : -\varepsilon\mathcal{A}_s H_s - \delta H_s : . \tag{6.35}$$

In these equations, the second term vanishes if the canonical transformation is s-independent. This is analogous to the second term in Equation (6.29), i.e., the partial derivative of the mixed generating function with respect to s. The question remains: how to express $\Delta\vec{F}_s$ or δH_s in terms of the Lie transform generating $\mathcal{A}_s$?

The general answer to this question can be expressed compactly for a general Lie transformation using all the homomorphisms of Sect. 6.1.1. We quote it here, but we will not derive it. First, let us assume that the map $\mathcal{A}_s$ is generated by a σ-dependent vector field:

$$\frac{d}{d\sigma}\mathcal{A}_{s;\sigma} = \mathcal{A}_{s;\sigma}\vec{A}_{s;\sigma}\cdot\vec{\nabla} \quad \text{with} \quad \begin{cases} \mathcal{A}_{s;0} = \mathcal{I} \\ \mathcal{A}_{s;1} = \mathcal{A}_s \end{cases} . \tag{6.36}$$

The canonical transformation $\mathcal{A}_s$ is generated by integrating a σ-dependent vector field $\vec{A}_{s;\sigma}$ from $\sigma = 0$ to $\sigma = 1$. Therefore it is very general. The final answer for the vector map $\mathcal{A}_s\dot{\mathcal{A}}_s^{-1}$ is given by

$$\begin{aligned} \mathcal{A}_s\dot{\mathcal{A}}_s^{-1} &= -\int_0^1 d\sigma\ \mathcal{A}_{s;\sigma}(\mathrm{ad}\vec{A}_{s;\sigma}\cdot\vec{\nabla})\ \frac{\partial}{\partial s}\vec{A}_{s;\sigma}\cdot\vec{\nabla} \\ \Rightarrow \Delta\vec{F}_s &= -\int_0^1 d\sigma\ \mathcal{A}_{s;\sigma}(\because\vec{A}_{s;\sigma}\because)\ \frac{\partial}{\partial s}\vec{A}_{s;\sigma} \Leftarrow\ \because\vec{A}\because\vec{B} = \left\langle\vec{A},\vec{B}\right\rangle \\ \overset{\text{If Hamiltonian}}{\Longrightarrow}\ \delta H_s &= \int_0^1 d\sigma\ \mathcal{A}_{s;\sigma}(: w_{s;\sigma} :)\ \frac{\partial}{\partial s}w_{s;\sigma}. \end{aligned} \tag{6.37}$$

As promised we concentrate on the special case of a σ-independent $\mathcal{A}_s$, then the integral in Equation (6.37) can be carried out:

$$\delta H_s = \frac{\exp(: w_s :) - 1}{: w_s :}\frac{\partial}{\partial s}w_s. \tag{6.38}$$

The derivation of equation Equation (6.38) is actually very simple and transparent using the usual phase space extension trick. Let us consider functions of $2N+1$ variables:

$$\vec{z} \text{ is replaced by } \vec{\tilde{z}} = (\vec{z}, s). \tag{6.39}$$

Equation (6.1) can be rewritten as

$$\frac{d}{d\sigma}\mathcal{M}_{\sigma_0\to\sigma} = \mathcal{M}_{\sigma_0\to\sigma}\left(\vec{F}\cdot\vec{\nabla}+\partial_s\right). \tag{6.40}$$

The reader can check that the original dynamics is contained in Equation (6.40). The phase space is now extended to include time which is indicated by dropping the subscript s from the vector field and the Hamiltonian. Why do we need this trick? Let us restrict this to the canonical case: the operator ∂_s is related to the canonical momentum associated with the new variable s.

$$\frac{d}{d\sigma}\mathcal{M}_{\sigma_0\to\sigma} = \mathcal{M}_{\sigma_0\to\sigma}\left(: -H - p_s :\right). \tag{6.41}$$

The effect of the canonical transformation $\exp(: w :)$ is now trivial because it does not formally depend on the new time σ. We will denote by $: w :_\sigma$ the Lie transform of the extended space. Let us proceed:

$$\begin{aligned} K &= \exp(: w :_\sigma)\,(H + p_s) \\ &= \exp(: w :)\,H_s + \sum_{n=0}^{\infty}\frac{: w :_\sigma{}^n}{n!}p_s \\ \Rightarrow \delta H_s &= \sum_{n=1}^{\infty}\frac{: w_s :^{n-1}}{n!}\partial_s w = \frac{\exp(: w_s :)-1}{: w_s :}\partial_s : w_s : . \end{aligned} \tag{6.42}$$

One notices that the result of Equation (6.42) agrees to leading order with Equation (6.30).

6.3 REVISITING HAMILTONIAN NORMALIZATION

In this book I have been proclaiming (*ad vitam eternam*) the virtues of a map-based theory. It is worth emphasizing again that this statement is relevant to rings: it is not a blanket affirmation. We will illustrate this once more by revisiting the example we used in the one-resonance normalization. If we set k_s and k_o to zero for simplicity, the map of Equation (5.31) can be generated by the following s-Hamiltonian:

$$\begin{aligned} H &= \frac{\mu}{2\pi}J + \delta_p\left(s - \frac{\mu_1}{\mu}2\pi\right)\varepsilon V \\ V &= 8x^3 \end{aligned} \tag{6.43}$$

where ε is a smallness parameter. The Hamiltonian has periodicity 2π and the one-turn map from $s = 0$ to $s = 2\pi$ is simply

$$\mathcal{M}_{0\to 2\pi} = \exp\left(: -\mu_1 J :\right) e^{:-\varepsilon V:} \exp\left(: -\mu_2 J :\right). \qquad (6.44)$$

Of course, we have already normalized this map to order f_4 in Sect. 5.2.1; and please look at footnote 2.8 for a proper interpretation of the delta function in computing the map. The insanity of standard accelerator physics is in failing to exploit the extremely localized nature of the perturbing potentials resulting in an orgy of algebra. This is more apparent in some standard references such as Guignard's work[31]. Others, such as Ando and later Bengtsson[32, 33], have successfully taken into account the localized nature of the sextupoles in a second order normalization into a rotation. Ando did this in the usual normal form theory with characteristic functions and Bengtsson in a rigorous approach to time-dependent perturbation theory of the map itself. Nevertheless, by failing to realize that they were dealing with maps between delta functions, both formalisms are unnecessarily complicated.

In Sect. 6.3.1 we derive the canonical transformation which removes all the leading order effects in ε. In Sect. 6.3.2, we take another look at the so-called "isolated resonance" theory.

6.3.1 Computation of w à la Guignard

Here we will take the extreme point of view found in Guignard's work. Typically in Guignard's work one finds the Hamiltonian expanded in Fourier modes in s. Just to give a feeling to the reader we now proceed in that manner. First, let us expand the delta function:

$$\delta_p\left(s - \frac{\mu_1}{\mu} 2\pi\right) = \frac{1}{2\pi} \sum_{k=-\infty}^{\infty} \exp\left(ik\left\{s - \frac{\mu_1}{\mu} 2\pi\right\}\right). \qquad (6.45)$$

Secondly, we write the potential V in resonance basis using the action-angle variables:

$$H = \frac{\mu}{2\pi} J + \frac{\varepsilon}{2\pi} \sum_m \sum_{k=-\infty}^{\infty} \exp\left(ik\left\{s - \frac{\mu_1}{\mu} 2\pi\right\} + im\phi\right) V_m$$

$$V_m(J) = \begin{cases} (2J)^{3/2} \text{ for } m = 3, -3 \\ 3\,(2J)^{3/2} \text{ for } m = 1, -1 \\ 0 \text{ for } m \neq 3, -3, 1, -1 \end{cases} \qquad (6.46)$$

Let us expand the Lie generator w of the canonical transformation $\exp(\varepsilon : w :)$ in a Fourier series:

$$w = \sum_m \sum_{k=-\infty}^{\infty} \exp\left(i\{ks + m\phi\}\right) w_{mk}. \tag{6.47}$$

We can now solve for the generator w to leading order in ε using Equation (6.30):

$$\begin{aligned} K_s^1 &= \frac{\mu}{2\pi} J + \varepsilon \frac{\mu}{2\pi}\,[w, J] + \varepsilon \partial_s w + \varepsilon V + O(\varepsilon^2) \\ &\Downarrow \\ 0 &= i(\frac{m\mu}{2\pi} + k) w_{mk} + \frac{1}{2\pi} \exp(-ik\frac{\mu_1}{\mu} 2\pi) V_m. \end{aligned} \tag{6.48}$$

From Equation (6.48), we deduce the final form for the Lie transform w:

$$w = \sum_m \sum_{k=-\infty}^{\infty} -\frac{\exp\left(-k\frac{2\pi\mu_1}{\mu}\right)}{i\,(m\mu + 2\pi k)} \exp\left(i\{ks + m\phi\}\right) V_m. \tag{6.49}$$

When looking at an expression for w in Equation (6.49), it is far from evident that one can actually perform the summation in k and produce a closed form expression. Of course, if we know the map theoretical approach, we know *a priori* that this object has a simple closed form expression. This is the topic of the next exercise.

Exercise 6-3

In this exercise, we compute the one-turn map $\mathcal{M}_{s\to s+2\pi}$ for various values of s. Let us parameterize the one-turn map as

$$\mathcal{M}_{s\to s+2\pi} = \exp\left(-\varepsilon : P_s :\right) \exp\left(: -\mu J :\right). \tag{6.50}$$

1. *Show, using map theory, that the function P_s is given by*

$$\begin{aligned} P_s &= \begin{cases} \exp\left(: -\left\{\mu_1 - \frac{\mu}{2\pi}s\right\} J :\right) V \text{ if } 0 < s < s_1 = \frac{2\pi}{\mu}\mu_1 \\ \exp\left(: -\left\{\mu + \mu_1 - \frac{\mu}{2\pi}s\right\} J :\right) V \text{ if } s_1 < s < 2\pi \end{cases} \\ &= \left\{(e^{:-\mu J:} - 1)\Theta(s - s_1) + 1\right\} e^{:-\{\mu_1 - \frac{\mu}{2\pi}s\} J:} V. \end{aligned} \tag{6.51}$$

The function $\Theta(s)$ is the usual step function taking the value zero for negative arguments and the value one for positive arguments.

2. *Normalize the map $\mathcal{M}_{s\to s+2\pi}$ with a canonical transformation* $\exp(\varepsilon : w_s :)$ *into a new map* $e^{\varepsilon : w_s :}\mathcal{M}_{s\to s+2\pi}e^{-\varepsilon : w_s :}$. *Select w so as to remove the function P_s completely. Answer:*

$$\begin{aligned} w_s &= \frac{1}{1-\exp(:-\mu J:)}P_s \\ &= \sum_m \frac{\left\{(e^{im\mu}-1)\Theta(s-s_1)+1\right\}e^{im\{\mu_1-\frac{\mu}{2\pi}s\}}}{1-\exp(im\mu)}e^{im\phi}V_m. \end{aligned} \tag{6.52}$$

3. *Equation* (6.52) *was derived using the map method. Show directly that this P_s is identical to the expression in Equation* (6.49). *Hint: Fourier transform Equation* (6.52), *i.e., multiply by $\frac{e^{-iks}}{2\pi}$ and integrate around the ring.*

We will not pursue this very far since it can get ugly pretty fast. Obviously a second order calculation can be very hard if one does not use Equation (6.52).

6.3.2 "Isolated Resonance Theory"

I like very much the locution "Isolated Resonance" in reference to the topological features one might see in a tracking study. However in the literature, under the title "Isolated Resonance Theory," it refers to the isolation of one term in the Hamiltonian. This theory is equivalent to the "super naïve approximation" derived from the correct one-resonance calculation in Sect. 5.2.1.7.

Let us assume that the tune $\mu = \frac{2\pi(1+\epsilon)}{3}$ is near a third. That is $\epsilon \approx 0$. As we did in Sect. 5.2.1, we will leave the driving terms of this resonance in the Hamiltonian. Hence, the canonical transformation w is given by

$$w = \underbrace{F_3^1}_{\text{Eq. 5.38}} = \sum_{\substack{k=-\infty \\ m=\pm 1}}^{\infty} -\frac{e^{-k\frac{2\pi\mu_1}{\mu}}}{i(m\mu+2\pi k)}e^{i\{ks+m\phi\}}V_m. \tag{6.53}$$

The resulting Hamiltonian to first order is

$$\begin{aligned} H_1^3 &= \frac{\mu}{2\pi}J + \frac{\varepsilon}{2\pi}\sum_{m=3,-3}\sum_{k=-\infty}^{\infty}\exp\left(ik\{s-\frac{\mu_1}{\mu}2\pi\}+im\phi\right)V_m + O(\varepsilon^2) \\ &= \frac{\mu}{2\pi}J + \varepsilon\delta_p(s-\frac{\mu_1}{\mu}2\pi)\sum_{m=3,-3}\exp(im\phi)V_m + O(\varepsilon^2). \end{aligned} \tag{6.54}$$

Because of the Dirac delta function, the map theorist immediately integrates Equation (6.54) and obtains the map $\mathcal{N}$ of Equation (5.37). He then follows the procedure of Sect. 5.1.3 if he desires an exact k-jet algorithm. **Period**.

The Hamiltonian person usually enters an s-dependent dilemma at this point. Clearly, the Hamiltonian of Equation (6.54) is a function of s and therefore it is not an invariant. However, given a bit of wishful thinking (see Exercise 6-4), the main contribution to the resonance, in this huge summation, is generically s-independent. Therefore one aims at that term. To get it we imagine what happens if ϵ tends towards zero. If one attempted a total normalization, the denominator $m\frac{\mu}{2\pi} + k = \frac{m}{3}(1+\epsilon) + k$ vanishes if $\epsilon = 0$ and $(m,k) = (\pm 3, \mp 1)$. Therefore, let us isolate these terms:

$$H^3_{1\epsilon} = \frac{(1+\epsilon)}{3} J + \frac{\epsilon \sqrt[3]{2J}}{2\pi} \{e^{-i\{s-3\mu_1-3\phi\}} + e^{i\{s-3\mu_1-3\phi\}}\}.$$

Then, we go into a co-moving frame by defining a new phase $\overline{\phi} = -\frac{s}{3} + \phi$. This transformation, produced by the mixed generating function $S(\phi, \overline{J}) = \{-\frac{s}{3} + \phi\}\overline{J}$, leads to the new Hamiltonian

$$\begin{aligned} H^3_{1\epsilon} &= \frac{\epsilon}{3}\overline{J} + \frac{1}{2\pi}\left(2\overline{J}\right)^{3/2}\left\{e^{i3\mu_1 + i3\overline{\phi}} + e^{-i3\mu_1 - i3\overline{\phi}}\right\} \\ &= \frac{\epsilon}{3}\overline{J} + \frac{1}{\pi}\left(2\overline{J}\right)^{3/2}\cos\left(3\overline{\phi} - 3\mu_2\right). \end{aligned} \tag{6.55}$$

We used the fact that $\epsilon \approx 0$ implies $3\mu_1 \approx 2\pi - 3\mu_2$ where μ_2 is the phase advance from the sextupole to the end of the ring as in Equation (6.44).

The reader can compare this result with the results of Sect. 5.2.1.7. In fact, Equation (6.55) is none other than the "super naïve theory." The map-based theory shows clearly that the "non-resonant" terms in Equation (6.54) are in reality potential "island makers." This implies that they should not be removed by canonical transformations. The standard "isolated resonance theory" sweeps them under the carpet. The map based approached gives a clear prescription for this "sweeping under the carpet" procedure. Again, we emphasize that part of these conceptual difficulties arise in conventional treatments because one forgets the topological reality of a resonance and replaces it by a pair of algebraic terms in the Hamiltonian, which becomes the "isolated resonance."

Exercise 6-4

In this exercise, we illustrate the incomplete nature of the "isolated resonance" calculation. The map we will now study has perfect 120° *symmetry.*

1. *Plot the map*

$$\mathcal{M} = \exp\left(: -\frac{2\pi}{3} J : \right) \exp\left(: -\frac{2\pi}{60} J - 2\alpha(x^3 - 3xp^2) : \right) \\ \times \exp\left(: 2(x^3 - 3xp^2) : \right) \tag{6.56}$$

for α ranging from zero to one. For this exercise, you may need some high order integrator, a lot of integrations steps, or an automatic differentiation package.

2. *For $\alpha = 1$, show that the "isolated resonance" theory gives J as the one-resonance invariant. This should be in flagrant disagreement with your simulations in 6-4.1.*

3. *Using the correct one-resonance theory, show that the first order invariant is given by:*

$$H = -\Delta\mu\ J + \left\{ \left(\frac{i3\Delta\mu}{e^{i3\Delta\mu} - 1} - \alpha \right) r_+^3 + \left(\frac{-i3\Delta\mu}{e^{-i3\Delta\mu} - 1} - \alpha \right) r_-^3 \right\} \\ \text{where } \Delta\mu = \frac{2\pi}{60}. \tag{6.57}$$

It is clear that when α is unity, the leading term in $\Delta\mu$ vanishes. In that case, the full expression of Equation (6.57) *must be used.*

4. *Show that, when $\alpha = 1$, the triangle is tilted by* 30° *for small $\Delta\mu$. This should match your simulations.*

6.3.3 Hamiltonian and Map: Equivalence

Of course all our computations start with some Hamiltonian. Strictly speaking there is no true map theory since the one-turn map is the result of integrating an infinitesimal map around the ring described by H_s. The difference between the Hamiltonian and the map approach resides in the (lack of) exploitation of the localized nature of the forces acting on the particle. Many authors do not follow the extreme point of view presented in the last section. For example, Courant, Ruth and Weng[34] in their section 5 on canonical perturbation theory, only mention in passing that it is sometimes useful to Fourier expand in the time-like variable. Unfortunately, in their treatment of the "isolated resonance" they do not give even a hint of the procedure needed to go beyond the Guignard treatment. So, if one reads between the lines and is very attentive to the global aspects of Hamiltonian theory, we can see treatments in the literature which are more "map-like" than others. Here, we present the ultimate "map-like" computation of the Green's function of perturbation theory starting from the Hamiltonian. But, first let us review the pure "map-style" derivation. As usual if a map has a single subscript s it denotes the one-turn map from s back to s.

6.3.3.1 The Green's function: map-wise

Let us consider a perturbation $\delta_p(s-\sigma)V$ of the ring at $s=\sigma$. The perturbed one-turn map at some location s can be written as:

$$\mathcal{M}_s = \mathcal{U}_{s\to\sigma} \exp\left(:-V:\right)\mathcal{U}_{\sigma\to s}. \tag{6.58}$$

Because the location label "s" is a modulo C number where C is the "circumference," we must carefully define the map $\mathcal{U}_{s\to s'}$ to ensure that it always propagates for a turn or less. First always choose s and s' between 0 and C (modulo C number) and interpret the resulting map as follows:

$$\mathcal{U}_{s\to s'} = \begin{cases} \mathcal{U}_{s\to s'} & s < s' \\ \mathcal{U}_{s\to C}\mathcal{U}_{0\to s'} & s' < s \end{cases}. \tag{6.59}$$

As usual, we perform a canonical transformation on $\mathcal{M}_s$:

$$\mathcal{N}_s^1 = \exp\left(:w_s:\right)\mathcal{U}_{s\to\sigma} \exp\left(:-V:\right)\mathcal{U}_{\sigma\to s} \exp\left(:-w_s:\right). \tag{6.60}$$

To leading order in the perturbation, we can derive an equation for $:w_s:$ the usual way:

$$\begin{aligned} \mathcal{N}_s^1 &= \mathcal{U}_s \exp\left(:\mathcal{U}_s^{-1}w_s:\right) \exp\left(:-\mathcal{U}_{\sigma\to s}^{-1}V:\right) \exp\left(:-w_s:\right) \\ &\approx \mathcal{U}_s \exp\left(:\{\mathcal{U}_s^{-1}-\mathcal{I}\}w_s - \mathcal{U}_{\sigma\to s}^{-1}V:\right)\cdots. \end{aligned} \tag{6.61}$$

Let us define a projection operator $\mathcal{P}_s$ whose purpose is to project out the terms in V which are untouched by the normalization:

$$w_s = \left\{\mathcal{U}_s^{-1}-\mathcal{I}\right\}^{-1}\mathcal{P}_s\mathcal{U}_{\sigma\to s}^{-1}V. \tag{6.62}$$

For example, in a complete normalization, $\mathcal{P}_s$ projects out the part of $\mathcal{U}_{\sigma\to s}^{-1}V$ which belongs to the range of $\mathcal{U}_s^{-1} - \mathcal{I}$, that is to say $\{\mathcal{U}_s^{-1} - \mathcal{I}\}(\mathcal{I} - \mathcal{P}_s)\mathcal{U}_{\sigma\to s}^{-1}V = 0$. Assuming a distributed perturbation V_s, we can write the total Lie transform W_s as

$$W_s = \oint_0^C \{\mathcal{U}_s^{-1} - \mathcal{I}\}^{-1}\mathcal{P}_s\mathcal{U}_{\sigma\to s}^{-1}V_\sigma d\sigma. \tag{6.63}$$

It should be noted that in general, we have a gauge freedom because an element of the kernel of $\mathcal{U}_s^{-1} - \mathcal{I}$ can be added to the Green's function w_s:

$$\begin{aligned} W_s &= \kappa_s + \oint_0^C \left\{ \{\mathcal{U}_s^{-1} - \mathcal{I}\}^{-1}\mathcal{P}_s\mathcal{U}_{\sigma\to s}^{-1}V_\sigma \right\} d\sigma \\ \kappa_s &= \mathcal{U}_s^{-1}\kappa_s. \end{aligned} \tag{6.64}$$

As we have seen in Sect. 2.6.3, this gauge freedom leads to different phase advances.

6.3.3.2 The Green's function: Hamiltonian-wise

Our goal is to take advantage of the delta function and rederive the result of Sect. 6.3.3.1 starting with the equation for the new Hamiltonian. So, we start with Equation (6.30) which gives us the new first order Hamiltonian K_s^1:

$$K_s^1 = H_s + \delta_p(s - \sigma)V + [w_s, H_s] + \partial_s w_s. \tag{6.65}$$

Because we expect the actual potential distribution to be closer to a sum of delta functions than a smooth distribution, we elect *not* to expand the delta function. Therefore, let us examine the propagation of the Green's function in the potential-free region. One reasonable choice is to insist that the Hamiltonian be unchanged outside the delta function:

$$\begin{aligned} &[w_s, H_s] + \partial_s w_s = 0 \\ &\qquad \Rightarrow \partial_s w_s = :H_s: w_s. \end{aligned} \tag{6.66}$$

Equation (6.66) can be solved by using the ansatz

$$w_s = \mathcal{N}_{s_0 \to s} w_{s_0} \qquad . \tag{6.67}$$

which, when substituted in Equation (6.66), gives us an equation for $\mathcal{N}_{s_0}$:

$$\partial_s \mathcal{N}_{s_0 \to s} = : H_s : \mathcal{N}_{s_0 \to s}. \tag{6.68}$$

This equation is actually the equation for the inverse of the unperturbed propagator $\mathcal{U}$ (see Exercise 2-24):

$$\mathcal{N}_{s_0 \to s} = \mathcal{U}^{-1}_{s_0 \to s}. \tag{6.69}$$

Therefore, provided we do not cross over the delta function, we have

$$w_s = \mathcal{U}^{-1}_{s_0 \to s} w_{s_0}. \tag{6.70}$$

Of course, in the absence of any perturbation we would set w_{s_0} to zero and the Hamiltonian stays unchanged. It remains to find out what exactly happens as we go across the delta function. To answer this, we integrate Equation (6.65) from $\sigma_- = \sigma - \delta$ to $\sigma_+ = \sigma + \delta$ where δ is a quantity tending towards zero. This is a classical trick.

$$K^1_+ - K^1_- = V + w_+ - w_-. \tag{6.71}$$

We then notice that the propagation between σ_+ and σ_- is given by Equation (6.70). Thus we can rewrite Equation (6.71)

$$K^1_+ - K^1_- = V + w_+ - \underbrace{\mathcal{U}_\sigma^{-1} w_+}_{w_-}$$

$$\Downarrow$$

$$K^1_+ - K^1_- = (\mathcal{I} - \mathcal{U}_\sigma^{-1}) w_+ + V. \tag{6.72}$$

Again, we choose w_+ in the range of $\mathcal{U}_\sigma^{-1} - \mathcal{I}$:

$$w_+ = (\mathcal{U}_\sigma^{-1} - \mathcal{I})^{-1} \mathcal{P}_\sigma V. \tag{6.73}$$

If we evaluate Equation (6.62) at $s = \sigma$, we reproduce the above result. To get w_s from Equation (6.73), we simply use the free propagation once more:

$$\begin{aligned} w_s &= \mathcal{U}_{\sigma\to s}^{-1} w_+ \\ &= \mathcal{U}_{\sigma\to s}^{-1} (\mathcal{U}_\sigma^{-1} - \mathcal{I})^{-1} \mathcal{P}_\sigma V \\ &= \mathcal{U}_{\sigma\to s}^{-1} (\mathcal{U}_\sigma^{-1} - \mathcal{I})^{-1} \mathcal{U}_{\sigma\to s} \underbrace{\mathcal{U}_{\sigma\to s}^{-1} \mathcal{P}_\sigma \mathcal{U}_{\sigma\to s}}_{\mathcal{P}_s} \mathcal{U}_{\sigma\to s}^{-1} V \\ &= (\mathcal{U}_s^{-1} - \mathcal{I})^{-1} \mathcal{P}_s \mathcal{U}_{\sigma\to s}^{-1} V. \end{aligned} \tag{6.74}$$

We obtain the same result as before.

Exercise 6-5

The gauge freedom in choosing w_s was reflected in the map theory by the addition of a function κ_s belonging to the kernel of $\mathcal{U}_s^{-1} - \mathcal{I}$. The purpose of this exercise is to show that this term is not zero in the Courant-Snyder transformation. We use the Courant-Snyder invariant as a pseudo-Hamiltonian which is to be normalized. In Exercise 8-7, the reader will study the same effect on a time-dependent problem.

1. *Show that the transformation represented by the matrix A of Equation (2.75) has the following single Lie transform representation:*

$$\begin{aligned} \mathcal{A} &= \exp\,(: w :) \\ \textit{where } w(x,p) &= \frac{\alpha \ln\left(\sqrt{\beta}\right)}{1-\beta} x^2 - \ln\left(\sqrt{\beta}\right)\, xp. \end{aligned} \tag{6.75}$$

2. *If we take the map point of view, the map to diagonalize is given by $\exp\left(- : \frac{\mu}{2}\varepsilon :\right)$ where ε is the Courant-Snyder invariant. This map can be generated by the time-independent Hamiltonian $\frac{1}{2}\varepsilon$ integrated for a "time" of $C = \mu$. Splitting this Hamiltonian as*

$$\begin{aligned} H &= \underbrace{\frac{x^2+p^2}{2}}_{J} + \underbrace{\frac{(\gamma-1)}{2}x^2 + \frac{(\beta-1)}{2}p^2 + \alpha xp}_{V} \\ &= J + V, \end{aligned} \tag{6.76}$$

solve for W using Equation (6.64). Hint: remember that $U_{s\to\sigma}$ is just $\exp(:-(\sigma-s)J:)$ and thus integrate Equation (6.64) formally.

3. *If you could not do Exercise 6-5.2, then go back to basics and use Equation (6.65) with a time independent perturbation V. Show that the answer for Exercise 6-5.2 must be:*

$$W = :J:^{-1} \underbrace{PV}_{\text{Projects out the kernel}} . \tag{6.77}$$

The solution of Exercise 6-5.2 is:

$$w = \frac{\gamma-\beta}{4}xp - \frac{\alpha}{4}(x^2-p^2). \tag{6.78}$$

4. *We can compare the approximate solution (Equation (6.78)) with the exact solution (Equation (6.75)) by assuming the that $\gamma \approx 1$, $\beta \approx 1$, and $\alpha \approx 0$. Do it! Show that the two expressions differ by a rotation: compute the exact κ appearing in Equation (6.64).*

In this exercise we have seen that perturbation theory does not necessarily pick up the Courant-Snyder choice for the gauge term κ. In general, two algorithms carried beyond first order in perturbation, will differ by a rotation. In the map based theory, we always insist that the choice of the canonical transformation only depends on the one-turn map. This is certainly the case for the Courant-Snyder transformation. It is not the case of Hamiltonian perturbation theory à la Guignard since the choice of the transformation depends on the local properties, i.e., derivatives of H with respect to the "time s."

6.4 RESONANCE BASIS FOR VECTOR FIELDS

In practical cases, the map may not be Hamiltonian (symplectic). In the presence of radiation we know that the linear part is characterized by a normal form of the type illustrated in Figure 4.7. At this point, the reader should (re-)read Sect. 4.5 and in particular Sect. 4.5.2 before going any further.

6.4.1 Transforming into the Resonance Basis

Let us assume that the linear part of the map is given by Equation (4.61). We know that a transformation to the resonance variables (given by Equation (4.47) or (6.84)) will diagonalize the linear part. What does it do to the nonlinear part of the map? To answer this we look back at Sect. 4.4.2 Equations (4.28) and (4.29): given a nonlinear vector field $\vec{F}$, the new transformed vector field $\overrightarrow{C\vec{F}}$ is given by

$$\begin{aligned} F^{\dagger}_{2k-1} &= \left.\overrightarrow{C\vec{F}}\right|_{2k-1} = \left(\vec{F}\cdot\vec{\nabla}c^{-1}_{2k-1}\right)\circ\vec{c} \\ &= (F_{2k-1} + iF_{2k})\circ\vec{c} \\ &= CF_{2k-1} + iCF_{2k}, \end{aligned} \tag{6.79}$$

and thus

$$F^{\dagger}_{2k} = CF_{2k-1} - iCF_{2k}. \tag{6.80}$$

Let us illustrate this by a simple symplectic example. Consider the planar vector field generated by the operator $:x^3:$, i.e., the usual sextupole kick. We can immediately write this in the resonance basis

$$\begin{aligned} :Cx^3: &= -\frac{i}{4} : x^3 + p^3 + 3x^2p + 3xp^2 : \\ &= -\frac{i}{4}\left(3\{x+p\}^2\{\partial_p - \partial_x\}\right) \\ &= \begin{pmatrix} \frac{3i}{4}\{x+p\}^2 \\ -\frac{3i}{4}\{x+p\}^2 \end{pmatrix}\cdot\vec{\nabla}\ . \end{aligned} \tag{6.81}$$

In Equation (6.81) we use "x" and "p" to represent the projection functions, that is to say the components of the identity map $\vec{I}$. The same result can be obtained by using Equations (6.79) and (6.80). The original operator $:x^3:$ has the following vector field representation:

$$:x^3: = 3x^2\partial_p = \begin{pmatrix} 0 \\ 3x^2 \end{pmatrix}\cdot\vec{\nabla}\ . \tag{6.82}$$

It follows that $F_1 = 0$ and $F_2 = 3x^2$. We substitute this in Equations (6.79) and (6.80):

$$\begin{aligned} F^{\dagger}_1 &= CF_1 + iCF_2 \\ &= i3x^2\circ\vec{c} = i3\left(\frac{x+p}{2}\right)^2 = \frac{3i}{4}(x+p)^2\ . \end{aligned} \tag{6.83}$$

INTERLUDE VIII
TRANSFORMING INTO THE RESONANCE BASIS: REAL AXIS TUNES

In Figure 4.4, we discussed the possibility of having some unstable planes with tunes on the real axis. For example, one may want to compute the shape of a separatrix for the purpose of extraction using a third integer resonance. Numerically, one extracts a three-turn map around the unstable fixed point closest to the septum using automatic differentiation. Using an "unstable" normal form, one then computes the shape of the separatrix as well as the rate of escape along this separatrix. Such a calculation, judged very difficult in the standard theory, can be done routinely with a map-based approach.

In general, we can combine Equations (4.50), (4.54), and (4.62) into one case. Physically, it does not mean much to have a coasting plane in the presence of damping: however this generalization comes in very handy in the writing of a library based on automatic differentiation[35].

To get the transformational rules, we first define the canonical transformation $\vec{c}^{-1}$ as we did in Equation (4.47):

$$(z_{2j-1}, z_{2j}) \overset{\vec{c}^{-1}}{\longmapsto} \left(\underbrace{z_{2j-1} + \epsilon_k z_{2j}}_{r_{j+}(\vec{z})}, \underbrace{z_{2j-1} - \epsilon_k z_{2j}}_{r_{j-}(\vec{z})} \right). \qquad (6.84)$$

We remind the reader that ϵ_k is one if the tunes are on the real axis. Otherwise this quantity is equal to the imaginary number i. In the coasting plane (the last one), we simply set $\epsilon_N = 0$. The map $\mathcal{R}$ is then transformed into $\mathcal{S}$:

$$\begin{aligned} \mathcal{S} &= \exp\left(\hat{s}\right) = \mathcal{CRC}^{-1} \\ \text{where } \hat{s} &= \sum_{n=1}^{N-1} \left\{-\epsilon_n \mu_n + \alpha_n\right\} z_{2n-1}\partial_{2n-1} + \left\{\epsilon_n \mu_n + \alpha_n\right\} z_{2n}\partial_{2n} \\ &+ \underbrace{C\alpha_c}_{\substack{\text{Momentum} \\ \text{compaction}}} z_{2N-1}\partial_{2N} \end{aligned} \qquad (6.85)$$

It follows that an arbitrary vector field $\vec{F}$ is transformed under $\vec{c}^{-1}$ into

$$F^{\dagger}_{2k-1} = \left.\overrightarrow{CF}\right|_{2k-1} = \left(\vec{F} \cdot \vec{\nabla} c^{-1}_{2k-1}\right) \circ \vec{c}$$

$$
\begin{aligned}
&= (F_{2k-1} + \epsilon_k F_{2k}) \circ \vec{c} \\
&= CF_{2k-1} + \epsilon_k CF_{2k}.
\end{aligned} \tag{6.86}
$$

END OF INTERLUDE

6.4.2 Eigenvector Fields

Because of the homomorphism between symplectic vector fields and Poisson bracket operators, we know that the eigen-operators of a rotation are simply monomials in the resonance basis. Here, we intend to check that the eigen-operators of $\mathcal{S}$ (as given by Equation (4.62)) are just

$$P_{\vec{m}} \partial_{2k-\delta} \text{ where } \delta = 0, 1. \tag{6.87}$$

Here $P_{\vec{m}}$ is a monomial (as defined in Equation (4.10)) and $\partial_{2k-\delta}$ is the derivative operator for the $(2k-\delta)^{th}$ plane. If we use the map associated with $\mathcal{S}$

$$
\vec{s}(\vec{z}) = S\vec{z} = \begin{pmatrix} \lambda_1 & 0 & & & \\ 0 & \lambda_1^* & & & \\ & & \vdots & & \\ & & & \lambda_N & 0 \\ & & & 0 & \lambda_N^* \end{pmatrix} \begin{pmatrix} z_1 \\ z_2 \\ \vdots \\ z_{2N-1} \\ z_{2N} \end{pmatrix},
$$

$$\text{where } \lambda_k = \exp\left(-i\mu_k + \alpha_k\right), \tag{6.88}$$

and the transformational properties of a vector field, we get

$$
\begin{aligned}
\mathcal{S}\left\{P_{\vec{m}}\partial_{2k-\delta}\right\}\mathcal{S}^{-1} &= e^{(-1)^{\delta+1} i\mu_k - \alpha_k} \\
&\quad \times e^{-i\left\{\vec{m}^+ - \vec{m}^-\right\}\cdot\vec{\mu} + \left\{\vec{m}^+ + \vec{m}^-\right\}\cdot\vec{\alpha}} \left\{P_{\vec{m}}\partial_{2k-\delta}\right\}.
\end{aligned} \tag{6.89}
$$

As an example, we can go back to the operator $: x^3 :$. Using Equation (6.89), we can immediately extract the eigen-operators out of Equation (6.81):

$$F^\dagger = \begin{pmatrix} \frac{3i}{4}\{x+p\}^2 \\ -\frac{3i}{4}\{x+p\}^2 \end{pmatrix}$$

$$= \frac{3i}{4} \begin{pmatrix} \overbrace{x^2}^{-\nu=k} + \overbrace{p^2}^{3\nu=k} + 2\overbrace{xp}^{\nu=k} \\ -\underbrace{x^2}_{-3\nu=k} - \underbrace{p^2}_{\nu=k} - 2\underbrace{xp}_{-\nu=k} \end{pmatrix}. \tag{6.90}$$

6.4.3 Section 4.4.2 Revisited

In Sect. 4.4.2 we looked at the recursive normalization algorithm. Starting at Equation (4.34), we illustrated the procedure using the case of a symplectic map being brought into a rotation. Here we simply retrace the steps of that section assuming the general case of Interlude VIII. For simplicity, we turn off the coasting beam case because physically it is always associated with a symplectic map. Thus all the N planes are "oscillating." Let us write the equivalent of Equation (4.34) in the resonance basis:

$$\vec{N}^{[j]} = \left\{ \exp\left(\mathbf{:} -\vec{N}^{[1]} \mathbf{:} \right) - \mathcal{I} \right\} \vec{A}^{[j]} + \vec{F}^{[j]}_{(j-1)}$$
$$\text{given } \mathcal{S} = \exp\left(\mathbf{:} \vec{N}^{[1]} \mathbf{:} \right). \tag{6.91}$$

Again we use the vector field bracket operator $\mathbf{:}\vec{F}\mathbf{:}$ as defined in Equation (6.37). Next, we write the equivalent of Equation (4.36):

$$\text{if} \quad \left\{ \begin{array}{l} \vec{N}^{[j]} = \displaystyle\sum_{|\vec{m}^+|+|\vec{m}^-|=j} \vec{N}_{\vec{m}}\, z_1^{m_1^+} z_2^{m_1^-} \cdots z_N^{m_N^-} \\ \text{where } \vec{m} = (\vec{m}^+, \vec{m}^-) \text{ similarly for } \vec{A}^{[j]} \text{ and } \vec{F}^{[j]}_{(j-1)} \end{array} \right.$$

$$\Rightarrow N_{\vec{m};2k-\delta} = \left\{ e^{(-1)^\delta i\mu_k + \alpha_k} e^{i\vec{\mu}\cdot(\vec{m}^+ - \vec{m}^-) - (\vec{m}^+ + \vec{m}^-)\cdot\vec{\alpha}} - 1 \right\} A_{\vec{m};2k-\delta}$$
$$+ F_{\vec{m};2k-\delta}. \tag{6.92}$$

The value of the parameter δ is defined in Sect. 6.4.2. As we explained before the factor $e^{(-1)^\delta i\mu_k + \alpha_k}$ represents the transformation of the vector field itself while the other factor is related to the transformation of the functional form of each component. This equation, as Equation (4.37), can be solved:

$$A_{\vec{m};2k-\delta} = \frac{F_{\vec{m};2k-\delta}}{1 - e^{(-1)^\delta i\mu_k + \alpha_k} e^{i\vec{\mu}\cdot(\vec{m}^+ - \vec{m}^-) - (\vec{m}^+ + \vec{m}^-)\cdot\vec{\alpha}}}$$

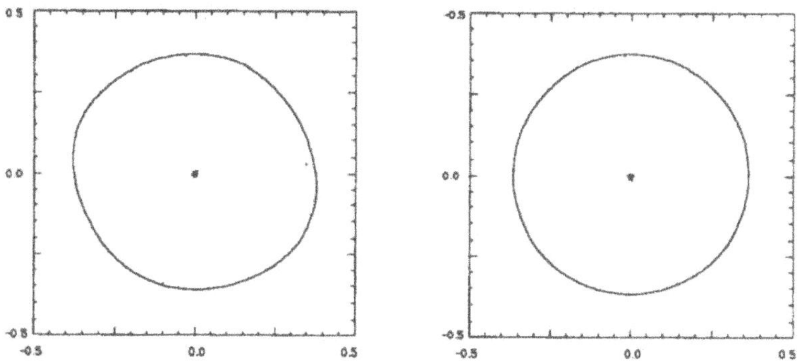

Figures 6.1 and 6.2: Limit cycle in regular and and normalized variables.

$$\text{where } \vec{m} \in \{\text{Set to be removed}\}. \tag{6.93}$$

One notices that, in theory, due to damping, it is possible to remove all the resonances and the tune shifts from the map. This simply reflects the fact that close enough to the origin all trajectories are attracted or repelled. However, in typical rings, the nonsymplectic effects are very small. Thus, we must usually use methods which keep us near the symplectic case avoiding the removal of tune shifts.

I would like to close this section with an exercise involving strong damping. Our goal will be to compute the limit cycle of a map.

Exercise 6-6

Consider a rotation $\vec{\rho}$ of angle μ in phase space. Let us construct the following oddity:

$$\begin{aligned}\zeta_1(x,p) &= (1+a-bx^2)\rho_1(x,p)\\ &= (1+a-bx^2)\left(\cos(\mu)\,x+\sin(\mu)\,p\right)\\ \zeta_2(x,p) &= \rho_2(x,p) = \cos(\mu)\,p-\sin(\mu)\,x.\end{aligned} \tag{6.94}$$

1. *For small a, argue that a limit cycle (circle) is found at a radius of*

$$r = \sqrt{\frac{4a}{b\,\{2+\cos(2\mu)\}}}. \tag{6.95}$$

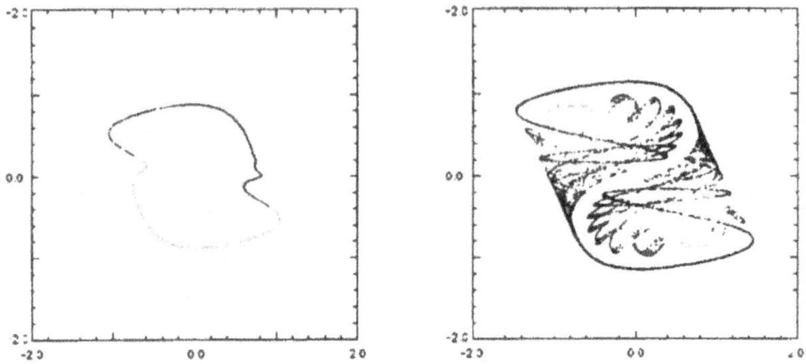

Figures 6.3 and 6.4: From the Limit Cycle to the Strange Attractor.

2. *Plot the limit cycle for* $\mu = 266\pi/300$, $a = 1/10$ *and* $b = 1$. *See Figure* 6.1.

3. *Keep increasing "a" until the limit cycle turns into a strange attractor.*

In Figure 6.2 the map of Figure 6.1 is normalized to 7th order in the variables (x, p). In this normalization we left the tune shift with amplitude in the map as the spirals in Figure 6.2 indicate. In Figure 6.3, the parameter a is cranked up to 6/10. Finally, in Figure 6.4, a is now 9/10. A strange attractor has now appeared covering a large section of phase space.

INTERLUDE IX
NORMAL FORM FOR THE STOCHASTIC MOMENT MAP

We have seen in Interlude VII that the beam sizes are components of the fixed point of the map $\vec{\Sigma}^{final} = \mathbf{L}\vec{\Sigma} + D\vec{\Phi}$. Actually, it is remarkable that we can use the theory presented in this section to find this fixed point in an elegant way. The actual computation of $D\vec{\Phi}$, a local quantity, belongs to Part III. Here we assume that it is known and that the linear map $\mathcal{L}$ has a normal form of the type of Figure 4.7—three pairs of complex conjugate eigenvalues just inside the unit circle.

Using the results of Exercise 4-6, we can write the stochastic map in terms of a compositional map:

$$\begin{aligned}\sigma^{final} &= \mathcal{L}^{\mathsf{T}}\sigma + d\\ \text{where } \sigma(\vec{z}) &= \Sigma_{ab} z_a z_b\\ \text{and } d(\vec{z}) &= D\Phi_{ab} z_a z_b.\end{aligned} \tag{6.96}$$

We assume that the compositional map can be diagonalized

$$\mathcal{R}^{\mathsf{T}} = \mathcal{A}^{\mathsf{T}-1}\mathcal{L}^{\mathsf{T}}\mathcal{A}^{\mathsf{T}}, \tag{6.97}$$

where $\mathcal{R}$ corresponds to Equation (4.61).

$$\mathcal{S}^{*} = \mathcal{C}\mathcal{A}^{\mathsf{T}-1}\mathcal{L}^{\mathsf{T}}\mathcal{A}^{\mathsf{T}}\mathcal{C}^{-1}. \tag{6.98}$$

The eigenvalues of the matrix S^{*} are the complex conjugates of those of S. Equivalently, as in Equation (4.62) for $\mathcal{S}$, the map $\mathcal{S}^{*}$ in terms of Lie transforms is just

$$\begin{aligned}\mathcal{S}^{*} &= \exp\left(\widehat{s}^{*}\right)\\ \text{where}&\\ \widehat{s}^{*} &= \sum_{n=1}^{N}(i\,\mu_n + \alpha_n)z_{2n-1}\partial_{2n-1}\\ &\quad +(-i\,\mu_n + \alpha_n)z_{2n}\partial_{2n}.\end{aligned} \tag{6.99}$$

We proceed as usual by letting the map $\mathcal{C}\mathcal{A}^{\mathsf{T}-1}$ act on Equation (6.96):

$$\begin{aligned}\mathcal{C}\mathcal{A}^{\mathsf{T}-1}\sigma^{final} &= \mathcal{C}\mathcal{A}^{\mathsf{T}-1}\mathcal{L}^{\mathsf{T}}\sigma + \mathcal{C}\mathcal{A}^{\mathsf{T}-1}d\\ \widetilde{\sigma}^{final} &= \mathcal{S}^{*}\widetilde{\sigma} + \widetilde{d}.\end{aligned} \tag{6.100}$$

The tilde represents the function in resonance basis: $\widetilde{f} = \mathcal{C}\mathcal{A}^{\mathsf{T}-1}f$. The operator $\mathcal{S}^{*}$ is diagonal is the monomial basis. Therefore we can solve for the fixed point $\widetilde{\sigma}^{\infty}$ by setting $\widetilde{\sigma}^{final} = \widetilde{\sigma}$. The answer is

$$\begin{aligned}\widetilde{\sigma}^{\infty} &= \{\mathcal{I} - \mathcal{S}^{*}\}^{-1}\widetilde{d}\\ &= \sum_{|\vec{m}|=2}\frac{\widetilde{d}_{\vec{m}}P_{\vec{m}}}{1-\exp\left(i(\vec{m}^{+}-\vec{m}^{-})\cdot\vec{\mu} + (\vec{m}^{+}+\vec{m}^{-})\cdot\vec{\alpha}\right)}\\ \vec{m} &= (m_1^{+}, m_1^{-}, \cdots, m_N^{+}, m_N^{-}).\end{aligned} \tag{6.101}$$

The dimensionality N of phase space is normally equal to three unless spin is present. The original beam sizes are gotten using the canonical transformation

$$\sigma^{\infty} = \mathcal{A}^{\mathsf{T}} \mathcal{C}^{-1} \tilde{\sigma}^{\infty}. \tag{6.102}$$

It is interesting to ask what the consequences are of the usual small damping found in rings. This can be answered immediately using Equation (6.101). Typically, we are far from linear resonances, that is

$$\left\| 1 - \exp\left(i(\vec{m}^{+} - \vec{m}^{-}) \cdot \vec{\mu} \right) \right\| \gg 0 \tag{6.103}$$

from which it follows that the dominant term in Equation (6.101) corresponds to the "so-called" emittances $|\vec{m}| = 0$. Therefore, the equilibrium sizes in resonance basis are:

$$\begin{aligned} \tilde{\sigma}^{\infty} &= \{\mathcal{I} - \mathcal{S}^{*}\}^{-1} \tilde{d} \\ &= \sum_{j=1}^{N=3} \frac{\tilde{d}_j x_j p_j}{-2\alpha_j} + O(\|\vec{\alpha}\|^0). \end{aligned} \tag{6.104}$$

Of course, to leading order, the canonical transformation $\mathcal{A}$ is the symplectic transformation prior to the introduction of radiative effects. This is why we talk about "equilibrium emittances" even though they are not defined in a dissipative system. In the real resonance basis (before $\mathcal{C}^{\mathsf{T}}$ is used), the equilibrium emittances are just:

$$\begin{aligned} \varepsilon^{\infty} &= \mathcal{C} \tilde{\sigma}^{\infty} \\ &= \{\mathcal{I} - \mathcal{R}^{\mathsf{T}}\}^{-1} \tilde{d} \\ &= \sum_{j=1}^{N=3} \frac{\tilde{d}_j \left(x_j^2 + p_j^2 \right)}{-2\alpha_j} + O(\|\vec{\alpha}\|^0). \end{aligned} \tag{6.105}$$

Now, we can transform ε^{∞} into the unnormalized variables using the map $\mathcal{A}^{\mathsf{T}}$:

$$\begin{aligned} \Sigma_{ij}^{\infty} z_i z_j &= \mathcal{A}^{\mathsf{T}} \varepsilon^{\infty}(\vec{z}) \\ &= \sum_{j=1}^{N=3} \varepsilon_j \left(\left[\sum_{k=1}^{2N=6} A_{k\ 2j-1} z_k \right]^2 + \left[\sum_{k=1}^{2N=6} A_{k\ 2j} z_k \right]^2 \right) \\ &\quad + O(\|\vec{\alpha}\|^0) \end{aligned} \tag{6.106}$$

where $\varepsilon_j = \dfrac{\tilde{d}_j}{-2\alpha_j}$.

From the definition of the function $\sigma(\vec{z})$, we can read off the equilibrium beams sizes

$$\langle z_k z_m \rangle = \sum_{j=1}^{3} (A_{k\ 2j-1} A_{m\ 2j-1} + A_{k\ 2j} A_{m\ 2j})\, \varepsilon_j. \quad (6.107)$$

One can conclude that for damping small compared to the distance from any linear resonance the equilibrium beam sizes are controlled by three numbers: the equilibrium emittances ε_j. To our knowledge, this was first exploited by Alex Chao in reference [36]. In analytical calculations for ideal machines, in addition to the small damping, one can exploit the fact that often the motion in the "longitudinal" plane (energy and time of flight) is decoupled from the transverse. This leads to the so-called "synchrotron integral" formalism of Sands[37].

Our goal here was to present the diagonalization of the moment matrix as a purely global normalization procedure. Its computation is relegated to Part III. The formalisms of Chao and Sands, like most symbolic calculations, will mix the global and local aspect of the physics involved. They are most useful at a design stage when a premium is put on the analytical form of the formula rather than on its actual evaluation for a given ring.

END OF INTERLUDE

CHAPTER 7
THE RING

All along we have trumpeted the primacy of the one-turn map. Indeed, the kind of normalization one should perform is dictated by the topological features of the one-turn map. This normalization is obtained from an approximate map which can be a linear map, a nonlinear map or a map containing information about stochastic processes as we have seen in Interludes VII and IX.

The ring, as we will define it, is a collection of maps necessary for the description/calculation of physical phenomena that interest the scientist. The beauty of a map based theory is that it keeps the number of maps to an absolute minimum, as dictated by the (gedanken) experimental conditions. If the reader needs some practice with these concepts, he can certainly go back to Sect. 2.5 for a less abstract setting. Here, we will introduce a more complete, but slightly more abstract formalism.

The ring being a collection of maps, it is necessary to remind the reader of the definition of a map. First of all, because of the recent popularity of approximate one-turn maps (Taylor series expansions or fitted maps), it is easy to forget that the most fundamental map is the one obtained by measurements (on the computer or/and real rings). Therefore, the tracking code (integrator) provides the accelerator physicist with the most reliable map for simulating the behavior of a machine. In general, this map propagates a vector $\vec{z} \in \mathbf{R}^6$ from one surface of section to another one. The k-jet itself is an extension of this concept. With the help of automatic differentiation (AD), we can propagate not only a vector $\vec{z}$, but the derivatives of this vector with respect to some initial conditions and some chosen control parameters. Finally, last but not least, we can propagate moments in the presence of stochastic processes. To these state variables, the reader can add spin and other quantities of his choice. For example, one can add interval arithmetic in order to estimate error propagation due to a Taylor series approximation.[7.1]

When taking the point of view of AD, it is important to realize that the k-jet is not a map! It is an extension of the phase space variable. Rays in $\mathbf{R}^6$, k-jets, moments, etc. are all objects which are transformed by the map. It just so happens that from the k-jet one can construct a Taylor series approximation of the usual map in $\mathbf{R}^6$ provided by the simulation code. To sum it up, let us not confuse the word map with the approximate Taylor series which is derived from k-jets.

7.1 RINGS AND SUB-RINGS

The ring is a finite and ordered collection of charts (i.e., surface of sections with a reference frame) with the maps needed to propagate from one chart to the next. In Part III, we will concentrate on the definition of the charts. Here, it suffices to say that we know implicitly what they are. Thus, in this chapter, the emphasis is on an ordered collection of maps and on transformations of the original charts for purposes of analysis. If we label the charts from $i = 0$ to $i = \aleph - 1$, then the ring $Ring_\aleph$ is given by the set:

$$Ring_\aleph = \{\mathcal{M}_{i \to i+1} \mid i = 0, \aleph - 1 \}\,. \tag{7.1}$$

The ring defined in Equation (7.1) is based on a theoretical/experimental understanding of the situation which incites the physicist to select $\aleph$ surfaces of sections. Obviously, if one does not understand what is going on, one's first reaction is to take the maximal ring $\aleph = \infty$. This is Hamiltonian theory. The Hamiltonian is the generator of infinitesimal maps: if you can't pick them judiciously, pick them all! In practice, this means that one is then dealing conceptually with the most complex object. It is the fundamental mistake which this book tries avoid.

Of course, from a ring one can define sub-rings. Consider an integer function $\wp$ from the mod[Ω] integers [7.2] to the mod[$\aleph$] integers where we assume that $\Omega \leq \aleph$:

$$\begin{aligned} \mathbf{Z}/\Omega\mathbf{Z} &\xrightarrow{\wp} \mathbf{Z}/\aleph\mathbf{Z} \\ i &\longmapsto \wp(i)\,. \end{aligned} \tag{7.2}$$

[7.1] Point often made by M. Berz over the years.

[7.2] In a ring, the positions are modulo $\aleph$ numbers because of the obvious periodicity. In a single-pass system, we can also talk of sub-systems. Obviously, in that case, the positions are described by regular integers.

Then a sub-ring of $Ring_{\aleph}$ can be constructed easily with $\wp$:

$$Ring_{\wp(\aleph)} = \left\{ \mathcal{M}_{\wp(i)\to\wp(i+1)} \mid i = 0, \Omega - 1 \right\}$$
$$\mathcal{M}_{\wp(i)\to\wp(i+1)} = \mathcal{M}_{\wp(i)\to\wp(i)+1} \cdots \mathcal{M}_{\wp(i+1)-1\to\wp(i+1)}. \quad (7.3)$$

A sub-ring is a smaller collection of maps built with the "master ring" $Ring_{\aleph}$. The master ring is itself coming from the full Hamiltonian.[7.3] In circular ring theory, there is a ring which we must study: the one-turn map. In fact, it is useful to think of $\aleph$ such rings as in the set

$$\Upsilon = \left\{ \{\mathcal{M}_{0\to\aleph}\}, \cdots, \{\mathcal{M}_{\aleph-1\to2\aleph-1}\} \right\} \quad (7.4.a)$$
$$\mathcal{M}_i = \mathcal{M}_{i\to i+\aleph} = \text{ One turn map at } i. \quad (7.4.b)$$

where[7.4] every element is a ring containing a one-turn map. One might believe naïvely that the set Υ of one-turn maps is equivalent to the ring which generated it ($Ring_{\aleph}$). In fact, it is not equivalent. This will be quite obvious when we make the connection to normalized variables: all the one-turn maps are identical rotations!

When dealing with pure tracking/experimental data, it is quite clear that we "push" from chart to chart the state variables which describe our system. This is experimentally self-evident: we cannot have an infinite number of beam position monitors and diagnostics around a ring. In simulations, any realistic model on the computer will require integration algorithms which have a finite number of integrations steps, chart transformations, etc. Any attempt to resolve the physical processes within an integration step is obviously[7.5] ridiculous. Therefore my challenge is not to convince the reader that one must define a finite number of observation points, but to develop out of this conviction a complete analytical and theoretical framework that uses only these experimental locations. Our goal is to redefine the Courant-Snyder theory (coupled/nonlinear/radiation/Sands/etc.) strictly in terms of the finite maps. The theory should not wander around, fishing for things which the experiment (gedanken or real) does not need. *This is the issue.*

[7.3]We use Hamiltonian in a loose sense: the infinitesimal generator of maps. This could include radiative and stochastic effects.

[7.4]In general, if for example $0 \le i < \aleph$, $i \le j - k\aleph < \aleph$, and $j \ge \aleph$ then the map $\mathcal{M}_{i\to j}$ is clearly interpreted as $\mathcal{M}_i^k \mathcal{M}_{i\to j-k\aleph}$. The reader can dream up other cases.

[7.5]Well, not so fast buster! See footnote 11.2: tracking with "s" or time self-consistently forces one to worry about the inside of an integration step.

7.2 TRANSFORMING THE RING: $\mathcal{A}_i\,[Ring_\aleph]$

We have already studied the effect of canonical transformations on the one-turn map. Indeed, we have discussed all sorts of normalizations whose purpose is to create simple models of the dynamics that help us understand the actual tracking/experimental data. Given a ring containing $\aleph$ maps at $\aleph$ charts, it is reasonable to ask if it is possible to transform the ring into a "normalized or Floquet" ring. For example, consider a ring belonging to the collection Υ, such as the one-turn map at $i = 0$, i.e.,

$$\{\mathcal{M}_{0\to\aleph}\}\,.$$

Let us assume that this map can be transformed into a rotation:

$$\{\mathcal{A}_0\mathcal{M}_{0\to\aleph}\mathcal{A}_0^{-1}\} = \{\mathcal{R}\}\,. \tag{7.5}$$

All we are saying here is that in a different basis, the ring looks like a rotation. If we think in terms of a chart at the point of observation, we can view the transformation $\mathcal{A}_0^{-1}$ as a providing a "pair of glasses" which distorts the chart at $i = 0$ in such a way as to turn the trajectories in phase space into a bunch of circles. If we look at a real map the effect of $\mathcal{A}_0^{-1}$ will be to bring the map close to a rotation near the fixed point; generically, what is said here can be strictly true only for linear maps and k-jets. This has been known since the days of Poincaré who, in 1890, proved the divergence of the series of Lindstedt. Nevertheless normal forms are useful near the origin as Poicaré pointed out himself:

> Ces séries, d'après le théorème de M. Lindstedt, satisfont "formellement" aux équations du mouvement. Si donc elles étaient uniformément convergentes, elles nous donneraient l'intégrale générale de ces équations. "Je dis que cela n'est pas possible."
>
> ...
>
> Les considérations qui précèdent peuvent donc permettre d'établir que les séries habituelles de la Mécanique céleste sont divergentes ; ce n'est pas qu'elles ne puissent néanmoins être utilement employées ; en effet il peut arriver que les termes d'une série décroissent d'abord très rapidement pour croîre ensuite indéfiniment, et par conséquent que cette série, quoique divergente, puisse servir à représenter

une fonction avec une approximation très grande, mais non indéfinie.

In any event we are free to change the chart any way we want and to create a modified "one map" ring at location $i = 0$. This is simply the similarity transformation which we have exploited all along in our attempt to understand one-turn maps. The real question is whether or not this concept can be extended to rings containing an arbitrary number of maps. Actually, the answer to this question is already in Sect. 6.2.2. There we derived the effect of a canonical transformation on the Hamiltonian using a map-based approach. The map $\mathcal{N}_s$ of Equation (6.33) is an element of a transformed ring. Here our approach is to stick to a finite ring and to get the same result.

The answer is very simple and very intuitive. So, let us go back to the usual transfer map in $\mathbf{R}^{2N}$. Simulation codes or real rings propagate rays:

$$\begin{array}{ccc} \mathbf{R}^{2N} & \xrightarrow{\vec{\zeta}_{i\to i+1}} & \mathbf{R}^{2N} \\ \vec{z} & \longmapsto & \vec{\zeta}_{i\to i+1}(\vec{z})\,, \end{array} \tag{7.6}$$

and, in the usual way,

$$\forall f\ \mathcal{M}_{i\to i+1} f = f \circ \vec{\zeta}_{i\to i+1}. \tag{7.7}$$

Now, let us distort the charts at i and $i+1$ by a canonical transformation $\vec{\alpha}$:

$$\begin{aligned} \vec{y} &= \vec{\zeta}_{i\to i+1}(\vec{x}) \\ \Rightarrow\ \underbrace{\vec{\alpha}_{i+1}^{-1}(\vec{y})}_{\vec{z}} &= \vec{\alpha}_{i+1}^{-1}\left(\vec{\zeta}_{i\to i+1}(\vec{x})\right) \\ \Rightarrow \vec{z}\ &= \vec{\alpha}_{i+1}^{-1}\left(\ \vec{\zeta}_{i\to i+1}\left(\ \underbrace{\vec{\alpha}_i(\vec{w})}_{\vec{x}}\ \right)\ \right). \end{aligned} \tag{7.8}$$

Equation (7.8) defines the distortion of a chart: the application of a diffeomorphism $\vec{\alpha}_k$ on the tracking data at chart k. The new map, which connects the initial distorted chart at i (dummy $\vec{w}$) to the final distorted chart at $i+1$ (dummy $\vec{z}$), is $\vec{\alpha}_{i+1}^{-1} \circ \vec{\zeta}_{i\to i+1} \circ \vec{\alpha}_i$. In the language of compositional maps, this is just $\mathcal{A}_i \mathcal{M}_{i\to i+1} \mathcal{A}_{i+1}^{-1}$, which leads to the Floquet ring induced by $\mathcal{A}$:

$$\mathcal{A}\,[Ring_{\aleph}] = \left\{ \mathcal{N}_{i\to i+1} \,\middle|\, \begin{array}{c} i = 0, \aleph - 1 \\ \mathcal{N}_{i\to i+1} = \mathcal{A}_i \mathcal{M}_{i\to i+1} \mathcal{A}_{i+1}^{-1} \end{array} \right\}. \tag{7.9}$$

INTERLUDE X
COORDINATE DEPENDENT OR INDEPENDENT DESCRIPTION

In the definition of the transformed ring $\mathcal{A}\,[Ring_{\aleph}]$ we must carry two things: the maps themselves and the definition of the charts. We remind the reader that it is also true in the case of $Ring_{\aleph}$; there is an implicit definition of the charts. Mathematically, as we will see in Part III, we must have a big Cartesian frame to locate the ring. Then, we need the maps as well as the definition of the local charts. Therefore, even if we do not mention the maps $\mathcal{A}$, one must always remember that they exist. Furthermore, there exists a map which connects an original local chart to the Lorentz frame of the Laboratory. All these transformations must exist and are implicit when not mentioned.

Thus if we want to relate to experiments, we must go back to the usual Cartesian world. This is where the experimentalist lives and conducts his measurements. Therefore, the theorist must always relate any calculation he does to this original frame by using whatever $\mathcal{A}$ he concocted.

For a mathematician, any frame is as good as any other! Just to crystallize the discussion, let us imagine that an actual ring exists and we play God. Suppose that we intercept a particle somewhere in the ring. The future of this particle is obviously "chart" independent: God does not care about the theorist's choice of charts. He will propagate the particle independently of us. This is so obvious that we often forget it. Units do not matter, surface of section can be weird and the measuring frame on them can be weirder. The mathematician and theoretical physicist exploit this fact using the coordinate-free language of theoretical dynamics: the language of differential forms. This language takes advantage of the fact that from a topological point of view no one cares about the frame of reference used! Indeed, the rings $Ring_{\aleph}$ and $\mathcal{A}\,[Ring_{\aleph}]$ are related by a diffeomorphism namely $\mathcal{A}$. Therefore, if the motion is on a torus, it remains topologically a torus for all diffeomorphisms applied to the ring.

In the rarified air of mathematics and theoretical dynamics, one is interested primarily in the topological nature of phase space. This has been true since Poincaré. In this book, we also emphasize geometrical features because we find this to be more powerful for practical and theoretical applications. The experimentalists will see figures being drawn on their screens, not equations or special functions. However, these figures are drawn in the world of Cartesian geometry–in a Lorentz frame, the frame of the control room. We do not want to adopt a totally coordinate-free notation since there is a preferential frame after all and this is why optometrists have a job!

Therefore, we always consider the ring to be a collections of maps and charts whose connection to the control room is of great importance. Despite its look, this is a book on accelerator physics, grounded in the belief that "if it ain't measurable it ain't worth a damn!"
END OF INTERLUDE

7.3 FLOQUET RINGS: A THEORETICAL CONSTRUCT

We have seen that the purpose of normalization algorithms on k-jets is two fold:

1. First the approximate canonical transformation can be used to simplify tracking data for all sorts of reasons such as making ellipses into circles.
2. The normalized approximate map can be solved exactly and compared to the normalized tracking data in an attempt to relate observed features to a simple model.

These two purposes are two sides of the same coin. They require normalization algorithms applied to Taylor series maps. With the advent of automatic differentiation we can routinely perform these calculations on the Taylor series map as well as on the stochastic moment map of Interlude IX. There is another reason to normalize k-jets and transform rings of k-jets: analytical calculations. Consider the famous first-order tune shift formula for quadrupoles which, by the way, assumes mid-plane symmetry:

$$\Delta\nu_m = \frac{(-1)^{m+1}}{4\pi} \int_0^C k(s)\beta_m \, ds. \tag{7.10}$$

This integral, in our theory, is replaced by a summation over the relevant integration steps:

$$\Delta\nu_m = \frac{(-1)^{m+1}}{4\pi} \sum_{n=1}^{N} \Delta L_n \, k_n \, \beta_{m;n} \,. \tag{7.11}$$

In Equation (7.11), the integral is a sum over "thin lens kicks." Assuming that the underlying code is a symplectic integrator, usually the discretization will not introduce any error relative to the code itself. Therefore, the evaluation of Equation (7.10) has been reduced again to the definition of a ring: the set of maps between the relevant integration steps. The number of quadrupoles is finite, so the cardinality $\aleph$ of the needed ring is finite. We will study this topic in Chapter 8.

If one looks more carefully at Equation (7.11), one notices that only the so-called beta function is needed. Therefore, to evaluate this analytical formula, it suffices to analyze $\aleph$ maps with automatic differentiation for the purpose of extracting beta. This is a trivial task with modern tools. Hence, we conclude that transformed rings are necessary for the analysis of the tracking data as well as the evaluation of analytical formulas. (See Chapter 8.)

7.3.1 The Simplest Floquet Ring: Phase Advances

The most important transformation is the deformation of an N-torus into a perfectly circular N-torus. This was the solvable problem of Sect. 5.1.1. The resulting phase space is a nested set of concentric circles in all N-planes. From a k-jet point of view, we perform the following transformation on the ring:

$$\mathcal{A}\left[Ring_{\aleph}\right] = \left\{ \mathcal{R}_{i\to i+1} \middle| \begin{array}{c} i = 0, \aleph - 1 \\ \mathcal{R}_{i\to i+1} = \mathcal{A}_i \mathcal{M}_{i\to i+1} \mathcal{A}_{i+1}^{-1} \end{array} \right\}. \quad (7.12)$$

The maps $\mathcal{R}_{i\to i+1}$ are all amplitude dependent rotations in phase space.

$$\begin{aligned} \mathcal{R}_{i\to i+1} &= \exp\left(- : \omega_{i\to i+1} :\right) \\ &\quad \omega_{i\to i+1} \text{ is a function } \vec{J} \text{ only.} \end{aligned} \quad (7.13)$$

The effect of such a map on the phase space was discussed in Sect. 4.4.2. When acting on the identity map it produces a rotation of angles

$$\mu_{k;i\to i+1} = \frac{\partial}{\partial J_k} \omega_{i\to i+1}. \quad (7.14)$$

Given a ring, we can ask how much of the Floquet ring

$$\mathcal{A}\left[Ring_{\aleph}\right]$$

is dependent on the particular choice of $\mathcal{A}$. We have seen that there is an infinite number of choices for $\mathcal{A}$. In the concrete case of a planar linear map, we gave the so-called Courant-Snyder choice in Equation (2.75) and the "anti-Courant-Snyder" transformation in Equation (2.101). More generally, we pointed out in Equation (2.57) that

$\mathcal{A}$ is known up to a nonlinear rotation. This freedom should make the reader *highly* suspicious of a pseudo-question on the "right" definition of the phase advance. See for example the list of Sect. 2.4.4.

Due to this gauge dependence, we would like to list the properties of the Floquet ring of Equation (7.12) that are gauge independent. If such properties exist, one can refer to them with the definite article "the."

Let us start with the most important gauge independent quantity: the tunes of the one-turn map. Let us first assume that the zeroth order tunes (tunes of the 1-jet) are nonresonant (see the discussion following Equation (4.37)).

7.3.1.1 The tune

The first thing we can construct is the set of one-turn maps associated with this Floquet ring

$$\mathcal{A}[\Upsilon] = \left\{ \{\mathcal{R}\}, \cdots \{\mathcal{R}\}, \cdots, \{\mathcal{R}\} \right\}. \tag{7.15}$$

It contains only one map. Because rotations commute, their product does not depend in the order and thus all the one-turn maps are the same.

In a ring it is natural to look at things that are averages over turns or over a distribution. Hence we define the observables to be turn-averages. Consequently consider a function f on the original phase space which we evaluate after n turns at position i. The answer is given by

$$f(n) = \mathcal{M}^n_{i \to i+N} f. \tag{7.16}$$

Dropping temporarily the label (everything is done at location i), we can express the function f in action-angle variables

$$\mathcal{A}f = \sum F_{\vec{m}}(\vec{J}) \exp\left(i\vec{m} \cdot \vec{\phi}\right). \tag{7.17}$$

Again we expand f in the monomial basis of k-jets as done in Sect. 4.3. Then we compute the function $f(n)$ using Equation (7.17):

$$\begin{aligned}
f(n) &= \mathcal{M}^n f \\
&= \mathcal{A}^{-1}\mathcal{R}^n \mathcal{A} f \\
&= \mathcal{A}^{-1} \sum F_{\vec{m}}(\vec{J}) \mathcal{R}^n \exp\left(i\vec{m} \cdot \vec{\phi}\right)
\end{aligned}$$

$$= \mathcal{A}^{-1} \sum F_{\vec{m}}(\vec{J}) \exp\left(in\, \vec{m} \cdot \vec{\mu}\right) \exp\left(i\vec{m} \cdot \vec{\phi}\right). \tag{7.18}$$

Finally, as would be done in many experiments, we perform a Fourier transform over the turn number of the function f:

$$f(\omega) \doteq \lim_{N \to \infty} \frac{1}{N} \sum_{j=1}^{N} e^{-i\omega j} f(j), \tag{7.19}$$

which leads to the results:

$$f(\omega) = \mathcal{A}^{-1} \sum F_{\vec{m}}(\vec{J}) \exp\left(i\vec{m} \cdot \vec{\phi}\right) \delta_{\left(\vec{m} \cdot \vec{\mu}(\vec{J}), \omega\right)}$$

$$= \sum F_{\vec{m}}(\vec{J}^{\text{inv}}) \exp\left(i\vec{m} \cdot \vec{\phi}^{\text{inv}}\right) \delta_{\left(\vec{m} \cdot \vec{\mu}(\vec{J}^{\text{inv}}), \omega\right)}. \tag{7.20}$$

In Equation (7.20) the invariant action-angles are expressed in the original frame of the measurements

$$\vec{\phi}^{\text{inv}} = \mathcal{A}^{-1} \vec{\phi} \text{ and } \vec{J}^{\text{inv}} = \mathcal{A}^{-1} \vec{J}. \tag{7.21}$$

The function $\delta_{(x,y)}$ is a Kronecker delta on the real numbers x and y. Because of the delta function, sharp peaks will appear whenever

$$\omega = \vec{m} \cdot \vec{\mu}(\vec{J}^{\text{inv}}),$$

where $\vec{z}$ locates the particle in the laboratory frame. The height of this peak is given by

$$F_{\vec{m}}\left(\vec{J}^{\text{inv}}(\vec{z})\right) \exp\left(i\vec{m} \cdot \vec{\phi}^{\text{inv}}(\vec{z})\right)$$

where $\vec{J}^{\text{inv}}(\vec{z})$ is the invariant action expressed as a function of the laboratory frame and $\vec{\phi}^{\text{inv}}(\vec{z})$ is the invariant angle in the same coordinate frame.

The position of these peaks is not a gauge dependent quantity. And, therefore the set of normalized one-turn maps $\mathcal{A}[\Upsilon]$ is unique within a mere relabeling of the planes.

Exercise 7-1

This exercise is about the uniqueness of the map $\mathcal{R}$ in Equation (7.15). *In this exercise, let us concentrate on the nonlinear part of a k-jet.*

1. *If the one-turn Floquet maps differ at position i and j, show that it implies that*

$$\exists \mathcal{A} \text{ such that } \mathcal{A}\mathcal{R}_i\mathcal{A}^{-1} = \mathcal{R}_j. \tag{7.22}$$

2. *For symplectic maps $\mathcal{A}$, show recursively on the order of the jet that $\mathcal{A}$ is a rotation and that $\mathcal{R}_i = \mathcal{R}_j$. Your proof should depend critically on $\mathcal{A}$ being symplectic and the zeroth order tunes being nonresonant.*

Here is another exercise on the topic of "minimal normal form." If one is willing to use noncanonical transformations, it is possible to simplify the form $\mathcal{R}$ further[38, 39, 40].

Exercise 7-2

For planar symplectic maps show that it is possible to find a non-canonical transformation $\mathcal{A}$ such that

$$\exists s \text{ such that } \mathcal{A}\mathcal{M}\mathcal{A}^{-1} = \exp\left(: -\mu J + aJ^s :\right). \tag{7.23}$$

where Equation (7.23) is true for k-jets assuming $k \geq 2s$.

7.3.1.2 The phase advance and matching

I n the definition of the Floquet ring (Equation (7.12)), we can insist that the canonical transformation $\mathcal{A}_i$ at location i depends only on the one-turn map at i and nothing else. This is a natural thing to do and indeed the Courant-Snyder transformation given the matrix

$$\begin{pmatrix} \sqrt{\beta} & 0 \\ -\dfrac{\alpha}{\sqrt{\beta}} & \dfrac{1}{\sqrt{\beta}} \end{pmatrix}$$

depends only on the one-turn matrix through the lattice functions β and α.

In fact, let us define matching. We will say that two locations(charts) of the ring are matched if and only if they have identical one-turn maps. Using rings with canonical transformations depending on the one-turn map alone, we thus obtain:

$$\begin{aligned} &\text{charts } i \text{ and } j \text{ are matched} \\ \iff\ & \mathcal{M}_i = \mathcal{M}_j \\ \iff\ & \mathcal{A}\left[\mathcal{M}_i\right] = \mathcal{A}\left[\mathcal{M}_j\right]. \end{aligned} \tag{7.24}$$

It follows from Equation (7.24) that the phase advance between matched charts does not depend on the choice of canonical transformations. Mathematically, suppose we have a ring $Ring_{\aleph}$ such that

$$\begin{gathered}\mathcal{A}\left[Ring_{\aleph}\right] = \{\mathcal{R}_{i\to i+1} \mid i = 0, \aleph - 1\,\} \\ \text{and} \\ \mathcal{B}\left[Ring_{\aleph}\right] = \left\{\mathcal{R}'_{i\to i+1} \mid i = 0, \aleph - 1\,\right\},\end{gathered} \tag{7.25}$$

if Equation (7.24) is true we must have

$$\mathcal{R}_{i\to j} = \mathcal{R}'_{i\to j}. \tag{7.26}$$

This is true if we are away from any linear resonances. The proof of this is simple. We express the map from i to j in terms of the phase advances:

$$\begin{aligned}\mathcal{M}_{i\to j} &= \mathcal{A}_i^{-1}\mathcal{R}_{i\to j}\mathcal{A}_j \\ &\overset{\text{Eq. 7.24}}{=} \mathcal{A}_i^{-1}\mathcal{R}_{i\to j}\mathcal{A}_i \\ &= \mathcal{B}_i^{-1}\mathcal{R}'_{i\to j}\mathcal{B}_i.\end{aligned} \tag{7.27}$$

We equate those two expressions for the map

$$\begin{gathered}\mathcal{A}_i^{-1}\mathcal{R}_{i\to j}\mathcal{A}_i = \mathcal{B}_i^{-1}\mathcal{R}'_{i\to j}\mathcal{B}_i \\ \Downarrow \\ \mathcal{B}_i\mathcal{A}_i^{-1}\mathcal{R}_{i\to j}\mathcal{A}_i\mathcal{B}_i^{-1} = \mathcal{R}'_{i\to j}.\end{gathered} \tag{7.28}$$

Using, the results of Exercise 7-1, it follows that

$$\mathcal{R}_{i\to j} = \mathcal{R}'_{i\to j} \tag{7.29}$$

and that

$$\mathcal{B}_i\mathcal{A}_i^{-1} \tag{7.30}$$

is actually a rotation.

Exercise 7-3

This is a trivial exercise. Compute the product of matrices $A^{-1}B$ where B is the anti-Courant-Snyder transformation and A is the Courant-Snyder transformation. The answer is of course a rotation:

$$A^{-1}B = \begin{pmatrix} \frac{1}{\sqrt{1+\alpha^2}} & \frac{\alpha}{\sqrt{1+\alpha^2}} \\ -\frac{\alpha}{\sqrt{1+\alpha^2}} & \frac{1}{\sqrt{1+\alpha^2}} \end{pmatrix}. \tag{7.31}$$

Relate this to the results of Exercise 2-26.3.

In fact between matched sections one can talk about "the" phase advance. The concept of a phase advance becomes a gauge independent quantity independent of the perturbation theory scheme which produced the Floquet ring. Therefore it should be possible to imagine a measurement, by a turn average of some sort, of a quantity in the laboratory frame which produces "the" phase advance. Let us go back to the observable f in Sect. 7.3.1.1. Suppose that f is an observable at position 1. We are interested to know how the Fourier transform at 1 compares with the transform at 2 if position 2 is matched. It is easy to show that the peaks are shifted by "the" phase advance between 1 and 2. Hence Equation (7.20) at location 2 is given by

$$\begin{aligned} f_2(\omega) &= A^{-1} \sum F_{\vec{m}}(\vec{J}) \exp\left(i\vec{m}\cdot\vec{\phi} + \vec{\Delta}(\vec{J})\right) \delta_{(\vec{m}\cdot\vec{\mu}(\vec{J}),\omega)} \qquad (7.32) \\ &= \sum F_{\vec{m}}(\vec{J}^{\text{inv}}) \exp\left(i\vec{m}\cdot\left\{\vec{\phi}^{\text{inv}} + \vec{\Delta}(\vec{J}^{\text{inv}})\right\}\right) \delta_{(\vec{m}\cdot\vec{\mu}(\vec{J}^{\text{inv}}),\omega)}. \end{aligned}$$

Here the function $\vec{\Delta}$ is the phase advance between the two observation points. This expression holds for nonlinear matched maps.

7.3.2 The Floquet Ring for Coasting Beams

In the previous sections we discussed the case of a ring in which all planes are oscillating. This case is extremely important because it describes the transverse dynamics in the absence of longitudinal (temporal) focusing. Furthermore, because of the low frequency of the longitudinal oscillations, it is possible to look at the full 6-dimensional problem with a cavity in the adiabatic limit. Many global quantities which are defined in a coasting beam formalism have extended meaning in a fully oscillatory case. Therefore the coasting beam Floquet ring is worth studying even for electron rings. This adiabatic extension of the coasting ring is at the root of some useful analytical theories such as the Sands theory. These theories are based on concepts such as the dispersion which are strictly speaking meaningless in a three-dimensional oscillator.

In many traditional treatments it is customary to treat the energy z_5 and the path length z_6 separately from the transverse variables. The constant energy is treated like a control parameter of the system. The path length is deprived of its canonical nature and also treated using a special formalism. For example, using the Pythagoras theorem and the definition of the local curvilinear frame, one obtains formulas such as

$$\zeta_6(\vec{z};C) = z_6 + \int_0^C ds \left\{ \left[1 + \frac{\zeta_1(\vec{z};s)}{\rho_s}\right]^2 + \zeta_2^2(\vec{z};s) + \zeta_4^2(\vec{z};s) \right\}^{1/2} . \quad (7.33)$$

This formula defines the path length after one turn in terms of the motion of the transverse plane and relies on a special local coordinate frame. It is certainly useful to have such formulas but it is terrible to elevate them to the status of definitions. In particular, such formulas emphasize a distorted view of the temporal coordinate. Instead of viewing it as a phase space variable, one starts to view it as a separate entity which happens to depend on the transverse coordinates. As problems become more complex as in isochronous rings, it is important to go back to first principles and separate our local and global quantities once more. With these things in mind, we will now revisit the coasting beam normalization and the resulting Floquet ring.

7.3.2.1 Getting to the fixed point in δ

In the oscillatory normalization we insisted on making sure that the canonical transformation be a function of the one-turn map only. By starting with the linear problem, we will see that in the coasting beam case it pays to insist on a little bit more. The problem starts with a matrix around the fixed point which has the following form:

$$L = \begin{pmatrix} N_{11} & \cdots & \cdots & N_{14} & b_1 & 0 \\ \vdots & \ddots & \ddots & \vdots & \vdots & 0 \\ \vdots & \ddots & \ddots & \vdots & \vdots & 0 \\ N_{41} & \cdots & \cdots & N_{44} & b_4 & 0 \\ 0 & 0 & 0 & 0 & 1 & 0 \\ a_1 & \cdots & \cdots & a_4 & \alpha & 1 \end{pmatrix} . \quad (7.34)$$

The first step consists in looking for the δ (or z_5) dependent fixed point of L:

$$L \begin{pmatrix} \eta_1 \\ \vdots \\ \vdots \\ \eta_4 \\ z_5 \\ 0 \end{pmatrix} = \begin{pmatrix} \eta_1 \\ \vdots \\ \vdots \\ \eta_4 \\ z_5 \\ \overline{z}_6 \end{pmatrix}$$

$$\Downarrow$$

$$\vec{\eta} = N\vec{\eta} + z_5\vec{b}. \tag{7.35}$$

This equation has an obvious solution

$$\vec{\eta} = z_5 [1 - N]^{-1}\vec{b} = z_5\vec{\eta}'. \tag{7.36}$$

The linear dispersion $\vec{\eta}'$ at this particular chart is normally defined as the first derivative of the fixed point with respect to δ. One should notice that the uniqueness of $\vec{\eta}'$ is ensured away from integer tunes even for nonlinear maps by the implicit function theorem. The next transformation involves the temporal variable z_6. Our goal is to provide an algorithm which works when the map is not symplectic but reduces to a symplectic transformation in the Hamiltonian case. This is ensured in the nonlinear perturbation theory by the use of Lie operators. In the linear case our methods are "exact" and do not typically involve Lie operators, hence we have to exercise some care. We proceed as follows: first we move to the fixed point in the transverse variables $\vec{x} = (z_1, z_2, z_3, z_4)$. This defines the transverse part of the canonical transformation

$$\vec{t}(\vec{x}) = \vec{x} + \vec{\eta} = (z_1 + \eta_1, z_2 + \eta_2, z_3 + \eta_3, z_4 + \eta_4). \tag{7.37}$$

One notices that we make no mention of linearity here. Indeed if instead of a linear map represented by a matrix such as L we have a nonlinear map $\vec{\zeta}$, it is possible in the k-jet sense to solve for

$$\vec{\zeta}(\vec{\eta}, \delta, 0)\Big|_{\perp} = \vec{\eta} \rightsquigarrow \vec{\eta}(\delta) \ , \tag{7.38}$$

provided that the linear transverse map associated with $\vec{\zeta}$, i.e., the matrix N, does not have a unit eigenvalue. The transformation of Equation (7.37) is not symplectic. This can be seen easily

$$[\vec{x} + \vec{\eta}, z_6] = \frac{\partial \vec{\eta}}{\partial \delta} \neq 0. \tag{7.39}$$

Therefore the time of flight (sixth variable) must be appropriately changed to take this problem into account.

Exercise 7-4

1. *Show that the compositional map* $T = \exp(:J\vec{\eta}\cdot\vec{x} + A(\delta):)$, *where* A *is an arbitrary function of* δ, *is a symplectic representation of the map* $\vec{t}$.

2. *Show that the compositional map $\mathcal{T}$ must remove completely the linear transverse dependence of the time to all orders in δ if η is the correct fixed point to all orders. Hint: you can't prove this without the symplectic condition on the full map $\vec{\zeta}$.*

To devise an algorithm for the nonsymplectic case which reduces to the symplectic algorithm of Exercise 7-4, we need only to remove the linear transverse dependence of the time variable as explained in Exercise 7-4.2. If we assume that the original map in the time variable is

$$\zeta_6(\vec{z}) = z_6 + \Delta(\vec{x}, \delta) \tag{7.40}$$

and that the extension of the map $\vec{t}$ in the sixth variable is

$$t_6(\vec{z}) = z_6 + \vec{x} \cdot \vec{g}(\delta), \tag{7.41}$$

then we can derive an equation for the four-component function $\vec{g}$. It is given by the equation

$$0 = g_i - \sum_{j=1,4} \frac{\partial \zeta_j}{\partial x_i} g_j + \frac{\partial}{\partial x_i} \Delta \Bigg|_{\vec{x}=\vec{\eta}}$$
$$\vec{g} = (D\zeta_{\perp}^{T} - 1)^{-1} \vec{\nabla} \Delta. \tag{7.42}$$

Here all the functions, including the Jacobian $D\zeta_{\perp}^{T}$ of the transverse map, are evaluated at δ and the corresponding $\vec{\eta}(\delta)$. Therefore in conclusion, the full transformation to the fixed point is given by

$$\vec{t}(\vec{z}) = (\vec{x} + \vec{\eta}, z_5, z_6 + \vec{x} \cdot \vec{g} + \frac{\partial A}{\partial \delta}), \tag{7.43}$$

where the function A is the same function as in Exercise 7-4.

7.3.2.2 Full normalization

The full normalization proceeds normally as if the fifth parameter δ is a constant of the motion. Although such a thing makes little sense in the presence of radiation, it is useful to write computer algorithms for k-jets which handle the general case. This is precisely why we described in the previous section the general fixed point transformation. As a byproduct it is worth realizing that we are not limited to canonical variables even if the underlying map is symplectic.

Let us complete this discussion with the complete symplectic algorithm. Given a nonlinear compositional map $\mathcal{M} = \mathcal{LP}$ where $\mathcal{L}$ and $\mathcal{P}$ are the linear and nonlinear part respectively, we transform it using $\mathcal{T} = \mathcal{T}_k \cdots \mathcal{T}_2 \mathcal{T}_1 = \mathcal{T}_{>} \mathcal{T}_1$:

$$
\begin{aligned}
\mathcal{TMT}^{-1} &= \mathcal{T}_{>} \underbrace{\mathcal{N} \exp\left(: \frac{C\alpha_c}{2}\delta^2 :\right)}_{\mathcal{T}_1 \mathcal{L} \mathcal{T}_1^{-1}} \mathcal{T}_1 \mathcal{P} \mathcal{T}^{-1} \\
&= \mathcal{N} \exp\left(: \frac{C\alpha_c}{2}\delta^2 :\right) \underbrace{\overbrace{\mathcal{N}^{-1}\mathcal{T}_{>}\mathcal{N}}^{\text{Transformed } \mathcal{T}_{>}} \quad \overbrace{\mathcal{T}_1 \mathcal{P} \mathcal{T}_1^{-1}}^{\text{Transformed } \mathcal{P}} \quad \mathcal{T}_{>}^{-1}}_{\text{Entirely nonlinear map}} \\
&= \mathcal{N} \exp\left(: \frac{C\alpha_c}{2}\delta^2 :\right) \mathcal{P}^{\dagger}.
\end{aligned} \tag{7.44}
$$

The map $\mathcal{N}$ is associated with the transverse linear part of $\mathcal{L}$, i.e., the submatrix N of Equation (7.34). To proceed further we normalize the map $\mathcal{N}$ into a rotation

$$
\begin{aligned}
R_{\perp} &= \mathcal{A}_1 \mathcal{N} \mathcal{A}_1^{-1} \\
&= \exp\ (: -\mu_1 J_1 - \mu_2 J_2 :),
\end{aligned} \tag{7.45}
$$

which leads to

$$
\mathcal{A}_1 \mathcal{TMT}^{-1} \mathcal{A}_1^{-1} = \underbrace{\mathcal{R}_{\perp} \exp\left(: \frac{C\alpha_c}{2}\delta^2 :\right)}_{\mathcal{R}} \mathcal{A}_1 \mathcal{P}^{\dagger} \mathcal{A}_1^{-1}. \tag{7.46}
$$

The process continues until $\mathcal{A}_1 \mathcal{P}^{\dagger} \mathcal{A}_1^{-1}$ is transformed into a generator of rotations. The tunes are the measurable quantities of Sect. 7.3.1.1 given by

$$
\mu_k = -\frac{\partial h}{\partial J_k}, \tag{7.47}
$$

where the map is given by $\exp\ (: h :)$. In addition, we obtain the amplitude dependent path length as was already mentioned in Sect. 5.1.2:

$$
\Delta z_6 = \frac{\partial h}{\partial \delta}. \tag{7.48}
$$

The result of Equation (7.48) is given in the Floquet frame and thus is not obviously a "physical result." Therefore we must connect it to a gedanken experiment in the original frame of the map $\mathcal{M}$. The answer is quite predictable: the change in the sixth variable which is a constant in the Floquet frame is the turn average change of the sixth variable in the original frame.

Exercise 7-5
Show that $\langle \Delta z_6 \rangle_{\text{turn}}$ *is given by*

$$\langle \Delta z_6 \rangle_{\text{turn}} = \lim_{N\to\infty} \frac{1}{N} \sum_{n=0}^{N} z_6(n+1) - z_6(n)$$
$$= \frac{\partial h}{\partial \delta}(\mathcal{A}^{-1}\vec{J}, \delta) \tag{7.49}$$
$$\text{where } \mathcal{A} = A_k \cdots A_1 T$$
$$\text{and } \mathcal{M} = \mathcal{A}^{-1} \exp(:h:) \mathcal{A}.$$

7.3.2.3 Remarks on the Floquet ring

In the case of oscillatory planes, we indicated in Sect. 7.3.1.2 that if the canonical transformation is a function of the one-turn map only, then the phase advance between matched locations is universal. The same result applies here. The average time of flight is the same between matched locations.

In the case of a coasting beam, there is another quantity which is not an average but ought to be universal. It is the time of flight on the closed orbit. Let us assume that we computed the closed orbits at locations i and j as functions of δ. These orbits are related to each other by the map $\vec{\zeta}_{i\to j}$:

$$(\vec{\eta}^{\,j}, \delta, z_6^j) = \vec{\zeta}_{i\to j}(\vec{\eta}^{\,i}, \delta, z_6^i). \tag{7.50}$$

The difference

$$\Delta z_6 = z_6^j - z_6^i = \zeta_{i\to j;6}(\vec{\eta}^{\,i}, \delta, 0) \tag{7.51}$$

must be a universal number. It is simply the time of flight for a particle on the closed orbit between locations i and j. Therefore we must ask:

is it given automatically by the Floquet map between these locations? Mathematically, we know that the phase advance is just

$$\begin{aligned}\mathcal{R}_{i\to j} &= \mathcal{A}_i\mathcal{M}\mathcal{A}_j^{-1}\\ &= \exp\left(:-\omega_{i\to j}(\vec{J},\delta):\right)\end{aligned} \tag{7.52}$$

and, therefore what can we say about

$$\Delta z_6(\delta) \stackrel{?}{=} -\left.\frac{\partial\omega_{i\to j}}{\partial\delta}\right|_{J=0} ? \tag{7.53}$$

It turns out that nothing further can be said about $\Delta z_6(\delta)$ unless we restrict further the definition of the fixed point transformation $\mathcal{T}$. To be more precise, we must set the free integration term $A(\delta)$ in Equation (7.43) to zero and insist that $\mathcal{T}$ brings us to the correct k-jet fixed point. This may be slightly more subtle than at first glance. Indeed, in a k-jet algorithm the initial transformation $\mathcal{T}$ need only be linear. That is to say that at the very least the linear dispersion of Equation (7.36) must be computed as well as the ensuing correction in the temporal direction given by Equation (7.42) where all quantities are linear jets. Of course, the resulting map $\mathcal{T}\mathcal{M}\mathcal{T}^{-1}$ will contain Lie exponents of the form

$$\begin{pmatrix}\vec{F}_\perp(\delta)\\ 0\\ \vec{x}\cdot\vec{F}^\dagger\end{pmatrix}\cdot\vec{\nabla}$$

where the function $\vec{F}_\perp$ is at least quadratic in δ. This implies that the transformations $\mathcal{A}_k$ will have to remove terms of that form so as to ensure that the final map is on the δ dependent fixed point. The cross-terms between two such transformations will produce, in general, terms of the forbidden form as indicated by the commutator

$$\begin{aligned}\left\{\begin{pmatrix}\vec{F}_\perp(\delta)\\ 0\\ \vec{x}\cdot\vec{F}^\dagger(\delta)\end{pmatrix}\cdot\vec{\nabla},\begin{pmatrix}\vec{G}_\perp(\delta)\\ 0\\ \vec{x}\cdot\vec{G}^\dagger(\delta)\end{pmatrix}\cdot\vec{\nabla}\right\} &= \left\langle\vec{F},\vec{G}\right\rangle\cdot\vec{\nabla}\\ \Rightarrow \left\langle\vec{F},\vec{G}\right\rangle_6 &= \vec{F}_\perp\cdot\vec{G}^\dagger-\vec{G}_\perp\cdot\vec{F}^\dagger.\end{aligned}$$

This is not so surprising since the fixed point of the map $\mathcal{N}_j$ for $0 < j < k$ is not a universal entity in the middle of the k-jet normalization. (Recall the definition $\mathcal{N}_j = \mathcal{A}_j\cdots\mathcal{A}_1\mathcal{T}\mathcal{M}\mathcal{T}^{-1}\mathcal{A}_1^{-1}\cdots\mathcal{A}_j^{-1}$.)

7.3.3 Coasting Beam Concepts as Adiabatic Limits

Anyone who has put his nose into accelerator design knows that it is common to design a ring using concepts which are strictly speaking meaningless in the presence of an RF-cavity. One often sets goals for quantities which are perfectly well-defined for coasting beams but on shaky ground once the beam oscillates in its three planes. More dubious, but still practiced, are simulations on the basis of models which do not even permit the proper inclusion of the longitudinal (temporal) plane. We have said in this book that this practice is not acceptable. In electron rings one must start with a fully radiative code in three planes. Approximations must be consciously and methodically introduced; they cannot be part of an approximate dogma which is constantly patched up to serve one's need. As the reader knows the dichotomous global/local approach of this book owes its genesis to the desire to remedy this situation.

Having reiterated our strong belief in a consistent picture, it remains that for analysis and design purposes (not for honest simulations) it is always useful to take advantage of approximations which simplify analytical formulas. Even though we have not yet defined clearly the difference between a k-jet formula and an analytical one, let us say vaguely that a k-jet formula is a polynomial extracted automatically on the computer by a proper analysis of the map: it can be carried to arbitrary order in any ring your tracking code can simulate. When one asks questions such as "what are the dampings," it should be clear that the answer to this question is buried in a 1-jet. To try to use Sands analytical formalism to answer this question is absolutely misguided.

On the other hand, if one asks general questions related to the design of a ring and the amount of damping, it may be appropriate to use an analytical formalism, such as Sands, which is as simple as possible. Typically analytical formulas may involve nonpolynomial quantities and therefore require in complex cases the use of an algebraic manipulator. One example of such formulas is given by Equation (7.10) or (7.11). These types of formulas are handy: they tell us how certain global observables vary as functions of some parameters. More important they are not specific to a particular ring. In Equation (7.10) only mid-plane symmetry at the location of the perturbation is assumed. The price one pays is two-fold: the order is low (first in this case) and the unperturbed ring usually has some symmetry requirements. Of course since these formulas are most useful in a design stage, these restrictions are not lethal.

Having reiterated our "credo" let us look at some simple limiting concepts which are constantly used and allow us to give formulas for the longitudinally focussed ring in terms of the coasting ring. Let us start by assuming that the third tune (longitudinal tune) is very small. In fact, let us assume that it is vanishingly small.

7.3.3.1 Dispersion

We suppose that an adiabatic process changes the value of δ around a closed circuit. Mathematically, what correct approximate picture can be concocted out of this? First of all, if we start with a particle infinitesimally near its fixed point, then the fixed point itself can be viewed as an average over the transverse motion:

$$\vec{\eta}(\delta) = \langle \vec{x} \rangle = \frac{1}{N} \sum_{n=1}^{N} \vec{x}(n) + O(|\delta(n+1) - \delta(n)|)$$
$$\text{where} \quad N \ll \left| \frac{1}{\nu} \right| . \tag{7.54}$$

This is a crazy definition but it matches our concept of dispersion during adiabatic synchrotron oscillations. The average position of the beam slowly moves across phase space while rotating at a fast rate around an approximate location given by $\vec{\eta}$.

Mathematically one can ask if it is possible to extract this dispersion out of the fully coupled formalism using the definition of Equation (7.54). The answer is yes. One can also ask the reverse question: is there a physically measurable quantity of the three-dimensional oscillator which reduces to the dispersion in the adiabatic limit? The answer is again yes and it is obviously of practical experimental value.

Let us look at the first issue and restrict our calculation to a linear map. With synchrotron oscillations, the linear part of the map is a six by six matrix that we can diagonalized

$$\mathcal{L} = \mathcal{A}^{-1} \mathcal{R} \mathcal{A}$$
$$\text{where } \mathcal{R} = \exp\left(: -\vec{\mu}_\perp \cdot \vec{J}_\perp - \mu_t J_t : \right) . \tag{7.55}$$

Because the map is linear the issue of "small oscillations" does not enter in the discussion. So, let us start with the initial ray $\vec{z}(1) = (0,0,0,0,\delta,0)$ and transform it to Floquet variables using the matrix A^{-1} of Equation (7.55). The answer is simply

$$A^{-1}\vec{z}(1) = \delta(A^{-1}_{15}, \cdots, A^{-1}_{65}). \tag{7.56}$$

We can perform the transverse average of the vector $A^{-1}\vec{z}(1)$ and get

$$\langle A^{-1}\vec{z}(1)\rangle_{\perp} = \delta(0,0,0,0,A^{-1}_{55},A^{-1}_{65}), \tag{7.57}$$

using the fact that rays in Floquet space move on circles with fast frequencies in the two transverse planes. Then we transform this average back to the laboratory frame using A^{-1}

$$\begin{aligned}\vec{\eta}'\delta &= A\langle A^{-1}\vec{z}(1)\rangle_{\perp}|_{\perp} \\ &= \begin{pmatrix} A_{15}A^{-1}_{55} + A_{16}A^{-1}_{65} \\ A_{25}A^{-1}_{55} + A_{26}A^{-1}_{65} \\ A_{35}A^{-1}_{55} + A_{36}A^{-1}_{65} \\ A_{45}A^{-1}_{55} + A_{46}A^{-1}_{65} \end{pmatrix}\delta. \end{aligned}\tag{7.58}$$

This last expression can be expressed entirely and simply in terms of A or A^{-1} in the case of symplectic matrices using the symplectic condition (2.28).

Let us answer the second question: what truly measurable entity in the presence of longitudinal focusing is the dispersion? The answer is somewhat interesting because it involves a system parameter and thus forces us outside the normal six-dimensional phase space. It is clear that longitudinal focusing must be achieved with one or more radio-frequency cavities. Thus we can construct the parameter dependent map $\vec{\zeta}(\vec{z},\omega)$ where ω is the frequency of the RF system. In the case of a single thin cavity in the ring, it is rigorously true that the dispersion can be obtained exactly as the transverse part of the synchronous orbit

$$(\vec{\eta}(\omega), \delta(\omega), t)$$

because the function $\delta(\omega)$ is a constant around the ring. Hence, instead of talking of the average energy of the synchronous orbit around the ring, one can talk of the actual energy in the single cavity case. Of course, because ω is related to δ itself one has in effect a measure of the dispersion. In the case of many cavities, one can only speak of the average of δ around the ring. Only in the case of a small longitudinal tune will the synchronous orbit be related to the dispersion. Indeed one can understand that if big changes in the energy are permitted within a turn, then the connection to coasting beam dispersion ceases to exist.

7.3.3.2 Chromaticities

I will not say much about the chromaticities. Again from a single particle dynamics point of view the chromaticity is an important limiting concept for the three-dimensional oscillator. In the coasting beam problem, chromaticities are well defined quantities—the first order chromaticities are the derivatives with respect to δ of μ_k in Equation (7.47) (see also Equation (5.7)).

One could derive adiabatic expressions for the chromaticities in terms of the transformation $\mathcal{A}$ of the three-dimensional oscillator. The converse problem is more interesting; that is to say what measurable quantity of the real three-dimensional oscillator is the chromaticity in a given plane? One could be tempted to say that since all three tunes are functions of the three actions, then the chromaticities are the derivatives of the transverse tunes with respect to the longitudinal action. Actually, this is completely false!

Inspiring ourselves with the discussion of Sect. 7.3.3.1 we conclude that the chromaticities are in fact the derivatives of the transverse tunes with respect to the RF frequency ω. Of course, this holds only in the adiabatic limits[7.6].

7.3.3.3 Synchrotron tune: the momentum compaction

The synchrotron tune is obviously a measurable quantity in the fully oscillatory system. The issue is why it should be related to the momentum compaction in the adiabatic limit. First, if we imagine a single thin lens cavity at a dispersion-free location, one can write the following exact one-turn longitudinal linear map:

$$\begin{aligned} z_6^f &= z_6 + C\alpha_c\delta \\ \delta^f &= \delta + RF(z_6^f). \end{aligned} \tag{7.59}$$

In Equation (7.59) a matrix of the type of Equation (7.34) is followed by an energy kick expressed by the function RF evaluated at the final time z_6^f. A dispersion-free location implies that both vectors $\vec{a}$ and $\vec{b}$ in (7.34) are null, and thus the temporal plane is decoupled from the transverse as in Equation (7.59). The longitudinal tune is thus related directly to the function RF and the momentum compaction

[7.6] I am grateful to K. Oide of KEK for pointing this out and inspiring this entire discussion on actual measurable quantities in the adiabatic limit.

α_c. For example, suppose that

$$RF(z_6) = -\frac{Vq}{p_0 c} \sin\left(\frac{\omega}{c} z_6\right) \tag{7.60}$$

where the time z_6 is measured in length units as usual, then the linear tune is easily found to be (by taking the trace!):

$$\begin{aligned} |\mu_t| &= 2 \sin^{-1}\left(\sqrt{\frac{Vq\omega C\alpha_c}{4p_0c^2}}\right) \\ &\overset{\text{adiabatic}}{\approx} \sqrt{\frac{Vq\omega C\alpha_c}{p_0c^2}}. \end{aligned} \tag{7.61}$$

For consistency I put an absolute value in the formula for the tune. Indeed, below transition the synchrotron tune is slightly positive and above transition it is slightly negative.

Exercise 7-6

Below transition the momentum compaction is negative (normal behavior) and above transition it is positive. Using the fact that the map of Equation (7.59) *is generated by the Hamiltonian*

$$H(z_5, z_6) = -\frac{\alpha_c}{2} z_5^2 + \delta_p(s) \int_0^{z_6} RF(z')dz, \tag{7.62}$$

show that the phase advance of the ring, Courant-Snyder or (including the integer part) is actually always near zero. This implies that above transition the phase advance goes backward with s.

In the presence of dispersion, it is not a priori evident that the momentum compaction is at all relevant to the synchrotron tune calculation. Assuming once more a single cavity, we can write in the (transverse) linear régime the one-turn map as follows:

$$\begin{aligned} i \neq 5 \quad z_i^{fin} &= \sum_j L_{ij} z_j^0 \\ z_5^{fin} &= z_5^0 + RF(z_6^{fin}). \end{aligned} \tag{7.63}$$

Using the expression for the matrix L in Equation (7.34), we write an expression for the longitudinal plane

$$z_5^{fin} = z_5^0 + RF(z_6^{fin})$$

$$
\begin{aligned}
z_6^{fin} &= z_6^0 + \vec{a}\cdot\vec{x}^0 + \alpha z_5^0 \\
&\Downarrow \text{ linearize } RF(z_6) \approx -\frac{Vq\omega}{p_0c^2} z_6 \\
z_5^{fin} &= z_5^0 - \frac{Vq\omega}{p_0c^2}(z_6^0 + \vec{a}\cdot\vec{x}^0 + \alpha z_5^0)
\end{aligned}
\tag{7.64}
$$

which can be averaged over the slow variables using $\langle \vec{x}^0 \rangle = \vec{\eta}' z_5$:

$$
\begin{aligned}
z_5^{fin} &= z_5^0 - \frac{Vq\omega}{p_0c^2}\left(z_6^0 + \{\vec{a}\cdot\vec{\eta}' + \alpha\}\, z_5^0\right) \\
z_6^{fin} &= z_6^0 + \{\vec{a}\cdot\vec{\eta}' + \alpha\}\, z_5^0.
\end{aligned}
\tag{7.65}
$$

It is easy to check that $\vec{a}\cdot\vec{\eta}' + \alpha$ is the product of the momentum compaction and the circumference; see footnote 4.3 if this circumference bothers you. Since the synchrotron tune is a measurable entity of the three plane oscillator, in the adiabatic limit it can only depend on an observable of the coasting beam system. We have already shown that for a dispersionless cavity location, the synchrotron tune must involve the momentum compaction. The only surprise here is that it involves *only* the momentum compaction in the general case.

Exercise 7-7

Using a simple planar linear model for the transverse plane, show that in the adiabatic limit, for a nonlinear function $RF(t)$, the amplitude dependent synchrotron tune will depend on other observables of the transverse motion such as the Twiss parameters.

CHAPTER 8
POWER SERIES AND ANALYTIC/SYMBOLIC CALCULATIONS

With the introduction of automatic differentiation in the field of accelerator physics, [8.1] it now possible to compute, in perturbation theory, the numerical values of any global (or local) quantities produced by a tracking code. While the global/local point of view had already been established before the arrival of "TPSA-AD" in our field, I consider "TPSA" to be the nail that seals coffin of any other approach. In this chapter, we will see that "TPSA" or jet calculations from tracking codes are not "analytic" calculations in the traditional sense. However they facilitate the evaluation of traditional analytical formulas using the dichotomous global/local theory if one needs to do so. They actually provide the input to analytical formulas because such formulas depend on the global properties of the unperturbed machine. Now, how does one get these global properties in a realistic setting? Answer: "TPSA."

Maybe of even greater importance, it frees us from doing useless analytical calculations. "TPSA" calculations are always self-consistent with the underlying tracking code if the proper programming environment is used or, in FORTRAN, if minimal care is exercised. One could say that in the presence of infinite speed computers, analytical formulas such as the synchrotron integrals of Sands should be used only for guidance in the design process. (Analytical formulas are painful to derive and check, but they may be faster to evaluate. So,

---

[8.1]There is no doubt in my mind that Martin Berz is the individual most responsible for this introduction. However Alex Dragt more than anyone else is responsible for noticing that standard global quantities are obtainable from the Taylor series map. Indeed, this is why I put the concatenation and normal form routines of the code MARYLIE[5] in kick codes[41] before the arrival of Berz on the scene.

I introduce "infinite speed" computer as a rhetorical trick to simplify the discussion.)

To this day theorists try to extend all sorts of formalisms for computational purposes (such as the Teng-Edwards coupling lattice functions [42]); it is a gigantic mistake when the output of such a formalism can be computed, bug-free, automatically, from a tracking code equipped with "TPSA" and an accelerator physics library which can normalize maps of the type described for example in Chapter 5.

In this chapter we will describe the differences between what I term analytical and TPSA calculations. The word analytical could be replaced by symbolic. The two words are complementary. Indeed, a symbolic formula and the rules to manipulate it are purely syntactical. The meaning is imparted by the user of the formula using his "analytical" brain. In a TPSA calculation, we will see that the computer "knows" how to evaluate the symbolic formulas; in fact, the formulas cannot be re-interpreted at will by the user. The tracking program aided by TPSA has already some "semantical" knowledge which is not present in the purely symbolic description.

8.1 WHAT IS A POWER SERIES CALCULATION?

8.1.1 Definition of the Tracking Code

First it is important to define a tracking code. The central purpose of a tracking code is to provide a map $\zeta_{i\to j}$ from $\mathbf{R}^6$ into $\mathbf{R}^6$ which describes the motion from chart i to chart j. But let us be a bit closer to the actual process which generates this map, i.e., the actual piece of software or the actual physical ring. Consider the following map

$$\begin{aligned} &\mathbf{R}^6 \times \mathbf{Z}/\aleph\mathbf{Z} \times \mathbf{Z}/\aleph\mathbf{Z} \times \mathbf{R}^{N_c} \times \mathbf{R}^{N_p} \times \mathbb{B} \longrightarrow \mathbf{R}^6 \\ &\vec{\Omega} = \left(\vec{z}, i, j, \vec{\Delta}, \vec{D}, \mathrm{B}\right) \longmapsto \vec{\zeta}(\vec{\Omega}). \end{aligned} \tag{8.1}$$

For our purpose in this discussion, I will view this function as the tracking code. It allows us to push a ray $\vec{z} \in \mathbf{R}^6$ from position i to position j. From our definition of a ring the positions i and j are modulo $\aleph$ numbers. Furthermore, the tracking code itself, when viewed as a gigantic function, will depend on some N_c control parameters which are locally differentiable. These N_c parameters include magnet strengths, for example. In addition, it will depend on N_p other parameters in either a non-differentiable way or in a physically immutable way. For example, the speed of light may be part of the list of immutable parameters. The actual layout parameters which we will discuss in Part III may be fixed: this is implicitly obvious by having chosen the cardinality $\aleph$ of the ring. Finally the set $\mathbf{B}$ contains all the Boolean parameters that are part of the creation of a ring: the activation of the "bells and whistles" of any tracking code depends ultimately on decisions made by the user.

The set of parameters $\left(\vec{\Delta}, \vec{D}, \mathbf{B}\right)$ describes the "state" of the ring and will correspond to a particular realization of a ring. This has enough generality to represent an actual ring conceptually. This is important if our tracking code is supposed to mimic the operation of an accelerator to which it might even be coupled through its control system.

Exercise 8-1

Write a code (in your favorite language) which simulates the map of Equation (3.49). *Identify the first line as a drift, the second one as a multipole kick (quadrupole and sextupole) and the last one as a drift again. Your code should receive the number of turns and the location of the tracking as inputs. Describe the state of the ring. The partial description of this state and in particular the partial description of the set* $\mathbf{B}$ *is called the manual of the code.*

In a versatile code, we can increase almost infinitum the number of charts; we can also describe different rings. To the extent that one may even be fitting the layout parameters and thus changing the very definition of the "state," it implies that the code is capable of operating, even differentiably, on a "superstate." Passage to the "superstate" is necessary in the early design stage as well as in the description of phenomena outside single particle dynamics (background radiation, Monte-Carlo detector simulations, comparative cost estimates of various designs, mechanical engineering of the ring, etc...). This is why a superclass ("Lagrangian Density"[43, 44]) must exist in a well-designed software environment; in a sense the particular function $\vec{\zeta}$ and the state vector $\vec{\Omega}$ exist in that class as parts of a particular realization of a ring.

Now we are ready for "TPSA" maps. Let us assume that a subset of $N_{\vec{\delta}}$ control parameter $\vec{\delta}$ is isolated from the full set:

$$\delta_k = \Delta_k \text{ for } 1 < k \leq N_{\vec{\delta}} \leq N_c. \tag{8.2}$$

We can now define a map in our more usual way from location i to location j:

$$\vec{z} \longmapsto \vec{\zeta}_{i \to j}(\vec{z}, \vec{\delta}). \tag{8.3}$$

This map can be viewed as a parameter dependent map of phase space.

Because of the assumption of differentiability in a local region of phase space and parameter space, we can state the existence of the k-jet

$$\vec{\zeta}^{[k]}_{i\to j}(\vec{z},\vec{\delta}) = \vec{\zeta}_{i\to j} + d\vec{\zeta}^{[1]}_{i\to j} + \cdots + d\vec{\zeta}^{[k]}_{i\to j}$$
$$\text{where } d\vec{\zeta}^{[m]}_{i\to j} = \sum_{|\vec{n}|=m} A_{\vec{n}} P^{[m]}_{\vec{n}}. \tag{8.4}$$

The functions (or jets) $P^{[m]}_{\vec{n}}$ are monomials of degree m in the variables $(\vec{z},\vec{\delta})$. The coefficients of this k-jet expansion are functions of the full state of the ring. The mapping $\vec{\zeta}$ of Equation (8.1) produces six real numbers (floating points on the computer); the coefficients $A_{\vec{n}}$ are also real numbers.

8.1.2 TPSA: Automatic Differentiation

The coefficients $A_{\vec{n}}$ are in general hard to compute by hand. In fact, they are hard to compute even if the underlying tracking code uses a trivial model. Nevertheless it is possible, at least at the lowest order, to differentiate the functions $\vec{\zeta}$ around some restricted value of the state $\vec{\Omega}$. For example, some codes compute the "feed-down" effects around the design orbit to leading order (i.e., the contribution of every magnet to $d\vec{\zeta}^{[1]}$).

On simple models it is possible to compute higher order effects around an arbitrary orbit of the code with a very restricted set of parameter dependence (usually none). One can certainly get maps from the usual "Taylor Series Codes" or "Transport Codes," which contain a library of low order k-jets ($k \leq 5$) defined around an ideal orbit. Thus one can access the various global observables of an ideal ring through the k-jet. In the early 1980's, it became clear to Dragt that this point of view was ideally suited for complex s-dependent systems such as circular accelerators. However, questions remained: how to use these ideas in arbitrarily complex integrators? How to get the symplectic integrator to generate a Taylor series map around an arbitrary orbit? Clearly the pay-off of such an algorithm would be tremendous: gaining access to all the usual global observables in a self-consistent manner. Synchrotron integrals, tunes shifts, "distortion functions," etc, all computed for the actual function $\vec{\zeta}$ produced by the code!

Remarkably, it is possible to teach a tracking code how to compute derivatives automatically. The general idea is to extend the operations performed on real numbers by the computer to an algebra acting on truncated power series, as reflected by the name "TPSA." The person most responsible for the introduction of this algebra to beam dynamics is Martin Berz. His original code, the TPSA package later known as the "DA package," was the first automatic differentiation package tailored for beam dynamicists.

Mathematically, the real numbers form a field under addition and multiplication: $(\mathbb{R}, +, *)$ is a field. Computers, in floating point mode, approximate this algebra. They do not succeed perfectly as reflected by the inevitable round-off error. But, modulo this round-off error, we will now assume that we do have the field of real numbers [8.2] on the computers. We will come back to the inability to represent real numbers later.

To get the truncated power series algebra one considers an arbitrary k-jet in the restricted variables $(\vec{z}, \vec{\delta})$. Such a k-jet is equivalent to a polynomial of degree k in $(\vec{z}, \vec{\delta})$. Multiplication of two such jets is totally equivalent to polynomial multiplication with an automatic truncation at degrees higher than k, i.e., a truncated power series algebra. The only "tricky" thing to define is division. Consider two jets $f^{[k]}$ and $g^{[k]}$ and their ratio $\frac{f^{[k]}}{g^{[k]}}$. Since the original tracking code performs the calculation of zeroth order jets, it is not permissible at any stage of the calculation to divide by a k-jet $g^{[k]}$ such that $g^{[0]} = 0$. This would produce an overflow on the computer and, in any event, either the map $\vec{\zeta}$ is not differentiable or its implementation on the computer is very poor (for example, $\frac{\sin(x)}{x}$ can present problems around $x = 0$). With this caveat in mind, one rewrites the jets, isolating their zeroth order part

$$
\begin{aligned}
\frac{f^{[k]}}{g^{[k]}} &= \frac{f^{[0]} + Df^{[k]}}{g^{[0]} + Dg^{[k]}} \\
&= \frac{f^{[0]} + Df^{[k]}}{g^{[0]}\left(1 + \frac{Dg^{[k]}}{g^{[0]}}\right)}
\end{aligned}
$$

8.2 In practice the field of rational numbers $(\mathbb{Q}, +, *)$ would be sufficient, but even this field cannot be put on the computer in a satisfactory manner for our needs.

$$= \underbrace{\left(f^{[0]} + Df^{[k]}\right) \left(g^{[0]}\right)^{-1} \sum_{n=0}^{k} \left(-\frac{Dg^{[k]}}{g^{[0]}}\right)^n}_{\left(g^{[k]}\right)^{-1}} . \tag{8.5}$$

Equation (8.5) can be viewed as a search for the inverse function of a jet. The general technique consists in isolating the zeroth order part (i.e., the regular real field term) and expanding the rest as a Taylor series. Because the sum and products of k-jets are k-jets, one gets the correct expression. Let us do another simple example: the exponential function.

$$\begin{aligned}
\exp\left(g^{[k]}\right) &= \exp\left(g^{[0]} + Dg^{[k]}\right) \\
&= \exp\left(g^{[0]}\right) \exp\left(Dg^{[k]}\right) \\
&= \exp\left(g^{[0]}\right) \sum_{n=0}^{k} \frac{\left(Dg^{[k]}\right)^n}{n!}.
\end{aligned} \tag{8.6}$$

Hence, by breaking up functions into a real part and a purely differential part (whenever possible), one can readily compute the k-jet expansion of such a function by knowing its Taylor series expansion around zero and applying to it the algebra of truncated power series. For our purposes it suffices to assume that automatic power series algebra has been installed in our code to perform automatic differentiation.

8.1.3 Redefinition of the Tracking Code

The new tracking code is gotten through a redefinition of Equation (8.1)

$$\begin{gathered}
\mathbf{R}^{\widetilde{N}} \times \mathbf{Z}/\aleph\mathbf{Z} \times \mathbf{Z}/\aleph\mathbf{Z} \times \mathbf{R}^{N_c} \times \mathbf{R}^{N_p} \times \mathbb{B} \times d\mathbb{B} \longrightarrow \mathbf{R}^{\widetilde{N}} \\
\vec{\Omega}^{[k]} = \left(\vec{z}^{[k]}, i, j, \vec{\Delta}, \vec{D}, \mathbf{B}, \mathbf{dB}\right) \longmapsto \vec{\zeta}^{[k]}(\vec{\Omega}^{[k]}).
\end{gathered} \tag{8.7}$$

Two things have changed:

1. We have upgraded the phase space from $\mathbf{R}^6$ to $\mathbf{R}^{\widetilde{N}}$ where

$$\widetilde{N} = \frac{(6 + N_{\vec{\delta}} + k)!}{(6 + N_{\vec{\delta}})!k!}.$$

$\widetilde{N}$ is just the number of monomials in a k-jet and $N_{\vec{\delta}}$ is the number of control parameters.

2. We have symbolically extended the set of Boolean parameters to include **dB** whose purpose is to control the order of the k-jet (i.e., k itself) and the nature of the subset $\vec{\delta}$.

The inclusion of **dB** might appear to the reader as purely symbolic, but actually it has *tremendous* implication on the implementation level. Indeed it implies that one requires from the tracking code the ability to change *at run time* the nature of the k-jet computed. This particular requirement will restrict greatly the type of implementation acceptable to the user.

More important, the physics of rings almost dictates the polymorphic (i.e., at run time metamorphosis) of a real number into a k-jet.

8.1.4 Power Series Analysis

In rings, as we already know, we deduce from the tracking data (i.e., the iteration of 0-jets) the type of calculation one might perform on a k-jet. One-resonance normalizations, tune shift calculations, Fourier transformation of an observable, etc. are all calculations which require a normalization of a k-jet. The k-jet map most useful for such calculations is the k-jet around the fixed point of the one-turn map.

But before going any further it is instructive to see how a tracking code equipped with TPSA can readily allow us to produce a fixed point finder which uses a Newton search algorithm.

INTERLUDE XI
TPSA IMPLEMENTATION OF A NEWTON SEARCH

Consider the following algorithm

$$
\begin{array}{lccc}
\Delta\vec{z}^{[0]} = 0 \longrightarrow & \Rsh & \longrightarrow & \vec{z}^{[k]} = \vec{z}^{[0]} + \Delta\vec{z}^{[0]} + \vec{I} \\
 & & & \downarrow \\
 & & & \vec{\zeta}^{[k]}\left(\vec{z}^{[0]} + \vec{I}\right) = \vec{\zeta}^{[0]} + D\vec{\zeta}^{[k]} \\
 & \uparrow & & \downarrow \\
 & & & \vec{\zeta}^{[0]} + D\vec{\zeta}^{[k]}(\Delta\vec{z}^{[0]}) = \vec{z}^{[0]} \\
 & & & \downarrow \\
 & \llcorner & \longleftarrow & \Delta\vec{z}^{[0]} = \left(D\vec{\zeta}^{[k]}\right)^{-1}\left(\vec{z}^{[0]} - \vec{\zeta}^{[0]}\right) \\
 & & & \downarrow \\
 & \because\ \Delta\vec{z}^{[0]} = 0 & \rightsquigarrow & \vec{z}^{[0]}
\end{array}
$$

whose purpose is to find the fixed point using the redefined tracking code. On the first line one inputs the identity k-jet around the ray $\vec{z}^{[0]}$ which is our initial guess. The second line represents our tracking code—I dropped reference to the location. It produces the real part of the jet as well as the differential part. The third line is not really part of the algorithm because it is mathematically the same as the fourth line. On the fourth line the differential k-jet map is inverted (k-jet wise) and thus can be used to approximate the inversion of the real map $\vec{\zeta}$.

This algorithm is not exact unless the actual tracking code is linear. However it provides a contracting map which will converge to the actual fixed points of the real map within a region around each fixed point. This region is called the basin of convergence. The reader will notice that the traditional Newton search is obtained by specifying that one should use $k = 1$ in the definition of the jets. As explained in Sect. 3.4.3, the k-jet map expressed around the fixed point belongs to a special class of approximations of the one-turn map. It sends the origin of phase space and parameter space back to itself. Hence its Lie operators form a differential algebra that ensures the consistency of all our calculations of global quantities on this k-jet map.

This example shows that without polymorphism it is not possible to use TPSA automatically in a compiled tracking code. Indeed, if one needs to recompile the code each time the definition of a k-jet is changed, it becomes a very awkward tool to use. The user must keep in mind that we will use TPSA for the computation of every usual global quantity—from the lattice functions, the equilibrium beams sizes, all the way to the nonlinear distortion functions.

END OF INTERLUDE

In Sect. 3.4.3 we discussed the "differential algebraic" nature of k-jets around a fixed point. In particular, all the properties of the compositional map extend to k-jets self-consistently if and only if the Lie operators of the k-jets form a differential algebra. For this to be true it is imperative that these Lie operators contain no constant parts:

$$\begin{aligned} &\text{if } \vec{\zeta}^{[k]} = \exp\left(\vec{F}^{[0]} \cdot \vec{\nabla}\right) \exp\left(\vec{F}^{[1]} \cdot \vec{\nabla}\right) \cdots \exp\left(\vec{F}^{[k]} \cdot \vec{\nabla}\right) \\ &\Longrightarrow \vec{F}^{[0]} = 0 \\ &\text{or } \vec{\zeta}^{[k]} \equiv \vec{\zeta}^{[k]} - \vec{\zeta}^{[0]}. \end{aligned} \tag{8.8}$$

Therefore we will normally look at jets around a fixed point of the one-turn map (or even multi-turn maps on occasion). These jets can be represented by Lie operators which are true differential operators when acting on infinitely differentiable functions as well as when acting on k-jets with no constant terms. A good example of this is given by the simple map of Exercise 2-1. Suppose we are interested in the linear tune. As seen in Exercise 2-2, these quantities are very important in determining the topology of the phase space (islands). How do we get these? It is only after expressing the map around its fixed point (in Exercise 2-2), that the actual linear tunes are given by the first order jet (matrix).

Exercise 8-2

1. . *First read or do Exercises* 2-1 *and* 2-2.

2. *Suppose you define the k-jet around the fixed point* $(x, p, d) = (0, 0, 0)$ *in Equation* (2.1) *and furthermore you introduce the variable d in the definition of the k-jet. Question: if you send the resulting expanded map* $\vec{\zeta}^{[k]}(x, p, d)$ *to a normalization algorithm which manipulates "differential algebraic" k-jets (as defined in Equation* (8.8)), *what will be the result for the tune?*

   Answer: The arc cosine of the expression in Equation (2.5) *expanded to degree* $k - 1$ *in d with floating real coefficients.*

As we did with the tracking code, it is useful to conceptualize the code which analyses the map. Since k-jets can be analyzed independent of the code that generated them, we can imagine and in fact design a software package whose purpose is to analyze k-jets. If the reader does not appreciate this fact, then I am doing a very poor job in describing the Hamiltonian-free theory of map analysis which is central to this book. But, just as a reminder it goes like this: if I have a two by two matrix from which the tune is gotten by "taking the trace"—it does not matter where this matrix came from. This is in a nut shell the essence of the global/local approach. Therefore let us imagine the existence of a library (such as the author's LIELIB)[35], based on TPSA, but handling k-jet maps with no constant part, that is to say a differential/Lie algebraic library.

Using this library, one can imagine writing a program which computes a global quantity $\vec{\Gamma}$. This program will be a function similar to the tracking code:

$$\mathbf{R}^{\widehat{N}} \longrightarrow \underbrace{\mathbf{R}^{\widehat{N}} \times \cdots \times \mathbf{R}^{\widehat{N}}}_{\dim \vec{\Gamma} \text{ times}}$$
$$\vec{\zeta}^{[k]} \longmapsto \vec{\Gamma}(\vec{\zeta}^{[k]} - \vec{\zeta}^{[0]}). \tag{8.9}$$

In Equation (8.9) I have eliminated references to Boolean variables. I have explicitly indicated that k-jet maps coming from tracking codes lose their constant part before they are processed by the mapping $\vec{\Gamma}$. The result of the program is the production of $\dim\vec{\Gamma}$ k-jets.

Let us write a program for fitting the tunes with two quadrupole families. What are the steps to be taken?

1. We must decide how many maps are needed in this computation. *Answer*: one map only because the tune is position independent. Therefore our Floquet ring will contain only one map. So we drop reference to the position and call this one-turn map $\vec{\zeta}$ as usual.

2. What is the dimension of the parameter space $N_{\vec{\delta}}$? *Answer*: $N_{\vec{\delta}} = 2$ as one would expect if two quadrupoles are used.

3. What is the minimal order k of the jets computed in this case? *Answer*: $k = 2$. The tune can be extracted out of matrix ($k = 1$), thus the linear dependence of the tunes on quadrupoles must come from a second order map.

Here is the algorithm:

$$\begin{array}{l} \Delta\vec{\delta}^{[0]} = \begin{pmatrix} \delta_1 \\ \delta_2 \end{pmatrix} = 0 \rightarrow \vec{\Gamma} \rightarrow \left\{ \underbrace{\vec{\delta}^{[0]} \leftarrow \vec{\delta}^{[0]} + \vec{\delta}}_{\text{Upgrade}} \right\} \rightarrow \text{Find fixed point } \vec{z}^{[0]} \\ \downarrow \\ \vec{\zeta}^{[2]}\left(\vec{z}^{[0]} + \vec{I}\right) \\ \uparrow \qquad \downarrow \\ \vec{\Gamma}(\vec{\zeta}^{[2]} - \vec{z}^{[0]}) = \begin{pmatrix} \mu_1^{[0]} + \frac{d\mu_1}{d\delta_1} d\delta_1 + \frac{d\mu_1}{d\delta_2} d\delta_2 \\ \mu_2^{[0]} + \frac{d\mu_2}{d\delta_1} d\delta_1 + \frac{d\mu_2}{d\delta_2} d\delta_2 \end{pmatrix} \\ \downarrow \\ \llcorner \longleftarrow \vec{\delta} = \left(D\vec{\Gamma}^{[1]}\right)^{-1} \left(\vec{\mu}^{new} - \vec{\mu}^{[0]}\right) \\ \downarrow \\ \because \vec{\delta} = 0 \rightsquigarrow \text{tunes fitted} \end{array} \tag{8.10}$$

What is absolutely essential to understand is the generality of the algorithm of (8.10). In a true polymorphic implementation of the TPSA in a tracking code, the inclusion of quadrupole parameters in the definition of the jet is trivially done at run time. The first box of the algorithm is the fixed point searcher which is already described in Interlude XI; it is essential if one fits a non-ideal ring or a ring with strange elements—in general the orbit changes with any change of parameters. Again the "at run time polymorphism" is essential because it would be inefficient to run the fixed point searcher with the quadrupole parameters and the order $k = 2$. The situation is even more ridiculous if one fits higher order observables like the amplitude dependent tune shift. Clearly polymorphism (or an approximation of it) becomes a must!

The function $\vec{\Gamma}$ is simply a normalization of the transverse phase space. Again, the library which normalizes the jet is completely general and can handle any k-jet with any number of parameters. In fact, in a proper implementation, the function $\vec{\Gamma}$ can be moved from one code to the next and perform the same task.

Finally, the inversion of the k-jet $D\vec{\Gamma}$ can also be written so as to accept many parameters. In general, one may have different schemes, some for over-determined systems and some for under-determined systems. But all that matters is to realize that whatever routine is called, it is written in a general way, independent of the meaning, degree, or dimensionality of the k-jet sent to it.

Exercise 8-3
The algorithm described in (8.10) *is general. However, in a normalization process one must also get the $\vec{\delta}$ dependent fixed point for free. The canonical transformation will now be $\vec{\delta}$ dependent. Modify the function $\vec{\Gamma}$ and the algorithm so as to give the fixed point searcher a better idea of the new fixed point at each iteration.*

As an example of the generality of this algorithm, I remember being asked by colleagues to fit the momentum compaction as well as the tunes. It should be obvious how to do this with algorithm (8.10). Add one knob $N_{\vec{\delta}} \geq 3$ and add the momentum compaction to $\vec{\Gamma}$. The same normalization routine is called! In fact, if one wants to include the chromaticities and the second order momentum compaction, the algorithm is still the same but both $\vec{\Gamma}$ and $\vec{\delta}$ have three additional entries. In addition the order of the jet goes from $k = 2$ to $k = 3$ because of the sextupoles added to $\vec{\delta}$.

Fitting islands, fixed points, separatrices at a septum, etc. all amount to a map computation with the correct k, the right number of knobs $N_{\vec{\delta}}$, and the right algorithm $\vec{\Gamma}$, which is usually a normalization algorithm since normalizations attempt to compute observables of rings.

8.2 FROM TPSA TO ANALYTIC

In this book we have insisted that the global/local point of view is a paradigmatic shift in the way scientists approach the issue of beam dynamics in a ring. Our point of view is derived in a loose sense from our understanding of modern computing techniques and our wish to exploit them to a maximum. While it is mathematically equivalent to write down a Hamiltonian in some Frenet-Serrat system around some orbit and to use the same Hamiltonian for both tracking and analysis, in practice this approach is not equivalent to the global/local approach. The answer to this is rooted in our computational tools. Whether it is our own brain alone, or our brain with pencil and paper, or our brain with the help of modern computers, or a combination of all these things, the optimization of the strategy one employs to investigate the dynamics of rings depends tremendously on the tools at our disposal.

The advance in computers, algebraic manipulators, Lie methods, TPSA, etc. can certainly help the standard approach because the more tools we have the better is our position[45]. This is almost a tautology. It is not so obvious a priori to decide that a new approach (or paradigm) could make use of the new tools. On a very theoretical level the necessity to change one's way of thinking became obvious to Dragt in the early 1980's. He communicated this belief to the author. The only piece of the puzzle missing was TPSA. Its introduction to our field by Berz just re-enforced this belief of the "Maryland school" and simplified greatly its realization on computers.

Once the reader realizes how important and central to our view is the ability to compute quantities with clarity and precision, we hope that we will be forgiven our excursions into the world of computers, languages and linguistics. In this chapter, these concepts are used to reach a proper understanding of the difference between the world of single-pass systems and rings, between analytical and TPSA calculations. A lot of that language is due to the years of interactions between myself and Johan Bengtsson at LBL. I apologize for the overly philosophical tone.

8.2.1 Turing and Gödel in Brief

Most amongst us have heard of Gödel's theorem and Turing's machine. Gödel proved that sufficiently complex axiomatic systems, such as the arithmetic of natural numbers, must contain well formed propositions which cannot be decided. In other words, he proved that there must exist statements [8.3] in number theory which cannot be proven by the theory. One cannot prove or disprove these statements. The British

[8.3] Fermat's theorem, $x^n + y^n = z^n$ has no solution for x, y, z integers and n an integer greater or equal to three, which may have been proven recently,

mathematician Turing reduced the most general computer to a simple device known as the "Universal Turing machine." He then proved that one could not predict in general if an algorithm on a Turing machine, and thus on a computer, will ever come to a halt. It turns out that Turing's theorem and Gödel's theorem are related. These triumphs in logic had a great impact on modern mathematics, and in particular, they demonstrated that one could not write a complete self-consistent axiomatic theory of mathematics as was attempted by Whitehead and Russell. Of course, once faced with an undecidable proposition, one could enlarge our axiomatic system to include this proposition. But, the resulting system, if not inconsistent, would still contain undecidable propositions. This is a never ending problem.

Let us go back to computers or Turing machines. On them, one can implement a subset of the natural numbers. [8.4] But what about real numbers? The real numbers are not enumerable. This means that no one-to-one mapping between the reals and the natural numbers can be constructed. This is true even though the rationals $\mathbb{Q}$ are dense in $\mathbb{R}$ and are enumerable. Since computer algorithms can only deal with a subset of the natural numbers, it is theoretically impossible for them to simulate exactly a theory based on real numbers. This is a fundamental restriction which shows itself most apparently to the average user through the phenomenon of "round-off error." However up to a point computers can handle correctly rational numbers, and this is at the root of symbolic algebraic manipulators.

The reader is probably thinking about π, e, roots and other transcendental functions which one can manipulate with algebraic manipulators. Is that a contradiction? The answer lies in the concept of "Linguistic Complementarity." If one were to look at the hardware, it is precisely because a rational is exactly a ratio of two integers that it can be represented on the computer. The real number π in an algebraic manipulator has no intrinsic meaning unless the user, using his brain, assigns a meaning to the symbol π with all the theoretical baggage associated with it.

To sum it up, the symbolic manipulator has a well-defined grammar to manipulate symbols (ultimately themselves a string of natural numbers in binary). This grammar is an axiomatic system compatible with the rules of addition and multiplication of the real numbers. Similarly, the manipulator knows rules involving some functions such

---

is an example of a simple proposition exceedingly hard to prove.

[8.4] In fact Gödel's theorem was based on a clever scheme that assigns a number to all possible propositions of number theory.

as the trigonometric functions. But, without the brain of the user, it is not capable of assigning a meaning to the objects being manipulated by the rules.

8.2.2 Modeling Rings versus Single-Pass Systems

The reasons I have chosen to talk about linguistic complementarity and axiomatic systems is related to some philosophical issues about ring dynamics and single-pass systems. Moreover it is related to the information content of jets in rings versus the content in single-pass machines. In Sect. 4.4.3, it was said somewhat loosely and with exaggeration that the information content of a jet in a ring is "zero." In this section we want to use linguistic complementarity and our understanding of computers to sharpen this statement. One corollary of this will be a discussion on the topic of "the-code-that-does-everything." In particular, I am interested in cultural aspects of the field of single-pass systems versus the field of this book—repetitive or periodic systems.

One needs to answer the following questions:

1. Why did the single pass community invest early in the Taylor series (jets) representation of their system? One could name several computer codes which represent the map as a Taylor series around some design orbit.

2. Why isn't it obvious to ring physicists that perturbation theory on the jet in a ring is equivalent to perturbation theory on the Hamiltonian?

Of course, I cannot answer these questions with certainty. What I can do is to contrast both rings and single pass systems from the point of view of an axiomatic system. I hope this will clear up the enormous differences between the two fields.

8.2.2.1 The single pass system

In a single pass system one never iterates the map. From a computer standpoint, it means that the input data (initial conditions) cannot be modified by the program itself (we temporarily neglect fitting procedures). In a ring the data comes out and is fed back into the function $\vec{\zeta}$. It is known that this iterative procedure can lead to very complex behavior even when applied to trivial maps. The reader must have seen Mandelbrot sets and other fractal figures. They are extremely rich and can only appear as one iterates a map.

Let us contrast this with the single pass map and eliminate temporarily all parametric dependence. We first define an aperture $A \subset \mathbb{R}^6$ which is the set of all acceptable initial conditions. A camera or a spectrometer will have such a region of operation: we do not expect these instruments to function for initial rays outside a certain aperture. Secondly, let us consider the k-jet $\vec{\zeta}^{[k]}$ expressed around an orbit somewhere near the center of the aperture A. In a camera or an electron microscope it is natural to select an orbit close to the axis of symmetry of the cylindrical lenses of the ideal instrument. Of course, $\vec{\zeta}^{[k]}$ might include construction errors and thus this axis of "symmetry" is not in general an orbit.

We can ask the following question: what is the maximum discrepancy between the $\vec{\zeta}^{[k]}$ approximation of the map and the map $\vec{\zeta}$ for a ray in the aperture A? Let us assume that we have found this number

$$\exists \widehat{\Delta} \in \mathbb{R} \text{ such that } \forall \vec{z} \in A \quad \left\| \vec{\zeta}(\vec{z}) - \vec{\zeta}^{[k]}(\vec{z}) \right\| \leq \widehat{\Delta}. \tag{8.11}$$

Furthermore, let us assume that we determined by other means that attempting to resolve a ray beyond an accuracy of $\widehat{D}$ is pointless. That is to say, if two machines represented by $\vec{\zeta}$ and $\vec{\xi}$ respectively are such that

$$\forall \vec{z} \in A \quad \left\| \vec{\zeta}(\vec{z}) - \vec{\xi}(\vec{z}) \right\| \leq \widehat{D}. \tag{8.12}$$

then they are experimentally identical. If those conditions are met then there are two cases of interest:

$$\begin{aligned} &\text{Case 1} \quad \widehat{\Delta} \leq \widehat{D} \\ &\qquad\quad \text{or} \\ &\text{Case 2} \quad \widehat{D} \leq \widehat{\Delta}. \end{aligned} \tag{8.13}$$

If Case 1 is realized this means that the k-jet contains all the physics by assumption. We may as well replace the exact map by its Taylor series. If, on the contrary, Case 2 is satisfied we must either use the original code $\vec{\zeta}$ or increase the order k of the jet. With TPSA it is conceivable to increase the order k of the jet. In fact, this process even converges, hopefully in a domain bigger than A, since the underlying Hamiltonian is analytic around the "design orbit."

All our arguments can be extended to include the parameter space $\vec{\Delta}$. The aperture includes physical settings of $\vec{\Delta}$ that can be determined by external conditions. That is to say magnets cannot have infinite strengths or infinite prices:

$$\forall(\vec{z},\vec{\Delta}) \in A \quad \left\|\vec{\zeta}(\vec{z},\vec{\Delta}) - \vec{\zeta}^{[k]}(\vec{z},\vec{\Delta})\right\| \leq \widehat{\Delta}. \tag{8.14}$$

Under Case 1 we can leap over from the real numbers to the rationals. Indeed if one can say that

$$\widehat{\Delta} \leq \widehat{D} \Longleftrightarrow \{\text{The jet has all the Physics}\} \tag{8.15}$$

then we can replace the aperture A by its intersection with the rationals. In fact, we can replace A with its intersection with the rationals represented on the computer.

Let us recapitulate:

1. The set of distinguishable initial conditions can be taken out of the set $A \cap \mathbf{Q}^{\widehat{N}}$ where $\mathbf{Q}$ is the set[8.5] of (representable) rationals.
2. All measurable entities are simple functions of $\vec{\zeta}\left(A \cap \mathbf{Q}^{\widehat{N}}\right)$. In any case we can extend the definition of the map (i.e., the code) to contain all the observables. Of course, they can be projected onto the representable rationals again under Case 1.
3. It follows from our argument that the total number of single pass machines and conditions represented by our code is very large but *finite*.
4. Finally, the total number of relevant rationals is extremely large but finite. The individual machines being represented correctly by k-jets are also completely described by a humongous but finite set.

What have we achieved with this construction? Under our assumptions, which may be totally wrong, the total set of observables,

---

[8.5]The aperture A is a subset of $\mathbf{R}^{\widehat{N}}$ where $\widehat{N}$ is potentially a very large number. So the intersection with the rationals is $A \cap \mathbf{Q}^{\widehat{N}}$. Here I assume that the distance between the closest rationals represented on the computer is much smaller than $\widehat{D}$ and thus $\mathbf{Q}$ is *de facto* equivalent to the set of *representable* rationals.

machines and output is finite. The entire process is like a gigantic "soda" machine where one puts in the initial data (the entire aperture) and produces the set of all possible machines in the neighborhood of some design. Now, if one include in this description the list of all the possible magnets known to mankind, one can conceptually reduce the entire design process to running a trivial mechanical automaton and thus produce a finite but humongous set of output.

The analysis of such a process, which does **not** allow reprocessing of the output data, is not as general as the analysis of a universal Turing machine. We know all the inputs and we know all the outputs. Just as if we had a zillion soda machines running at different prices and providing different flavors and sizes. Theoretically one can predict when the automaton stops and what it will produce. Of course, it may take longer than the age of the universe to taste all the beverages. It might also take longer than the age of the universe to try all the possible electron microscope designs. The lack of self-referencing makes this system finite.

It is interesting to think that the production *and analysis* of a single-pass system is a finite process. One can then dream, as some have done, of the "code-that-does-everything." At least, just as the human brain selects the size of $\widehat{D}$ using physics outside the code (metaphysics of the code?), one can try to trim the parameter space and produce the "code-that-does-almost-everything."

The reader should think of a simple single-pass system like a magnifier. If we assume that our purpose is to start fires with the sun on a clear day at the 45th parallel, it is clear that one can define quite uniquely the initial conditions for the rays (parallel), the precision of the design needed to start a fire, etc. Not surprisingly, it is in the field of light optics that extremely powerful codes are sold commercially. This is consistent with our little analysis, because it is in light optics that the physics of the individual element is simplest. In beam dynamics one can argue that it is more complicated due to the field equations (Maxwell) and to the large range of applications. In addition, collective effects (space charge, beam-beam) come into play so as to negate the sole use of a single particle code. Actually it is often permissible to put the collective effects in the metaphysics and to revisit them later after an initial design is reached. If this statement was false this book would be useless. Or some would say, to be more precise, this book would be even more useless and so would be all the books on single particle dynamics.

8.2.2.2 The repetitive system: rings

Naïvely we might assume that the above analysis, although already utopic for single pass systems, could be carried over to the rings with the same degree of believability. Actually, it is not true.

To understand this we must first go back to the mathematical level. For single pass, we have postulated the existence of the quantity $\widehat{D}$. Typically it is related to the smallest chunk of phase space which our experimental tools can measure. In single pass systems, we can relate this to our ability to measure magnetic fields and predict orbits. At best, we can relate this quantity to our ability to measure differential changes at the output. Let us see in an example why in rings the analysis of Sect. 8.2 must fail. Consider a system which is symplectic and, for simplicity, linear. Although no such object exists in real rings, it would be a great system from the single particle point of view. Let us further assume that the system stays linear under changes of the parameter $\vec{\Delta}$. Using the properties of symplectic matrices, we conclude that for all acceptable changes of $\vec{\Delta}$ our system is still stable (eigenvalues on the unit circle).

Now let us put radiation in the system as in electron rings. Let us assume that the amount of radiation per magnet never produces a relative error in excess of $\epsilon \left\|\widehat{D}\right\|$ where $\epsilon \lll 1$. Naïvely one may ask (and many single-pass specialists ask the question): why do you care about radiation if it is tiny compared to the error in the symplectic system due to measurement uncertainties? The answer to this is at the root of the complexity of a system which reprocesses the data. Indeed, if one computes the norm of the difference between the symplectic and the radiative map, it will be of order ϵ, and thus negligible if one adopts the "camera" model of the previous section. However, it is well-known that a tiny nonsymplectic effect will push the eigenvalues off the unit circle by an amount ϵ. This is not allowed in a symplectic system. Therefore after a number of turns of order ϵ^{-1} the particle, and thus the entire beam, will have shrunk by a factor of order unity. This is a measurable effect in a ring and, in fact, radiation damping is an ingredient of the central mechanism which controls the beam size in electron machines.

We can produce unstable symplectic maps by adding microscopic perturbations to the map. For example, near a third integer, one can destabilize a linear map by adding a tiny sextupole effect. Indeed, the unstable fixed points will be located at a radius of order $\frac{k_s}{\nu-1/3}$. Hence for any finite perturbation, one can change the topology of the map by closing in on the third integer. Given an aperture A, I can choose the tune and the sextupole strength so as to make the sextupole irrelevant according to the discussion of Sect. 8.2, but fundamental in the sense that most of the aperture A is unstable under iteration.

Exercise 8-4

1. *Create a linear map near a third integer. Add a sextupole kick and produce the situation discussed in this section. Define all the concepts needed for the discussion:* $A,\widehat{D},\widehat{\Delta},k_s$, *etc.*

2. *Take a single pass point of view and analyze the map by finding the maximum effect of the sextupole throughout* A.

3. *Subsequently show by tracking or otherwise that the map is unstable throughout most of* A.

Here we are now ready to state the differences between both systems:

1. The number of phenomena which can be observed by iterating a map belonging to a ball A of $\mathbf{R}^{\widehat{N}}$ is infinite. One can just list a few things: strange attractors, resonances, limit cycles, curves of fractal dimension such as Julia sets, etc.

2. Hence, it is not possible to define the meaning of "small" by a trivial error analysis, since small effects can modify drastically the topology of the phase space.

3. Hence, if we try to design "the-code-that-does-everything," we are facing a huge problem. Even though the description of the map is finite (run the same code as in the single pass case), how can we decide what is small and what is big? This must be decided in terms of the actual topological feature one is trying to analyze, but as we just said, the list of potential features is infinite.

In deciding what is the "complete" physics of the code in the single pass, it is conceivable that "metaphysics" is used at the onset and the code can try all the possible cases. Thus "the-code-that-does-everything" is based on the hope that a metalanguage (mathematics and physics outside the actual code) is used once and implemented algorithmically in the code.

In rings, the language of mathematics teaches us that it is impossible to cover all the ground in a system which contains only a subset of the rationals. Hence "the-code-that-does-everything" is a mathematical impossibility. Instead it has to be replaced by the "the-code-that-can-do-everything-with-a-smart-user." One utopia is dropped and is replaced by another one. If one simulates and sees the approximate formation of a topological feature such as a resonance or a strange attractor, the smart user, using the language of mathematics, interprets the results. This is done outside the computer. Then, the "smart-user" changes the code (i.e., the full function $\vec{\zeta}$ including observables and analysis) so as to produce results which illuminate the object whose existence was hinted by the simulation code.

We are now back to the issue originally discussed in Sect. 4.4.3 which was the information content in a k-jet. The jet itself, if restricted to six variables and a small number of parameters, will depend on a finite number of variables. It cannot in any way be representative of the rich physics generated by a real map. Hence, it is the user of "the-code-that-can-do-everything-with-a-smart-user" which imparts to the k-jet a meaning consistent with the actual tracking data, experimental data, and with the meta-language of mathematics and physics. He then implements the various normalization algorithms as well as other simplified models whose purpose is to interpret the imperfect data produced by the code or/and the actual ring. I should add at this point that there exist complex theories to quantify chaos and to try to compute the approximate shapes of fractal curves relevant to chaotic systems. All these calculations are justified by mathematics and are performed to explain actual data. *The user must use his brain to teach the code what to do with the data.*

In rings the k-jet associated with the code may have a large information content (i.e., all the coefficients of the Taylor series), but this information is essentially nil compared to the relevant physics the original code can generate. Hence, one may say that the information is zero in a measure sense. As we said in Sect. 4.4.3, it is the duty of the practitioners of this field to give to the k-jet map the appropriate meaning. This is the essence of linguistic complementarity: the tracking code, when viewed as an axiomatic system acting on a subset of the integers, cannot encompass the entire theory of ring dynamics and design: it is a theoretical impossibility. It is for these reasons that Bengtsson and myself at LBL, and the CLASSIC[44] collaboration, have insisted on codes which are theoretically clean. In such codes, the smart user can easily understand and access the physics. The study of complex phenomena does not allow the code designer to prepare for all eventualities. He must always assume that the user of his code is more competent than he is. In any event, it is always possible to program for "dumb" users standard functions and algorithms.

Now, we are ready to jump into another application of linguistic complementarity: analytic calculations. They also belong to the metaphysics of the tracking code.

8.3 ANALYTIC OR SYMBOLIC CALCULATIONS

In this section we will define the difference between analytic and TPSA calculations. The reader will notice that there is an intersection and thus it is not always clear which calculation one must perform. Let us start with a possible definition of the TPSA calculation which we will then contrast with the analytic case. Then we will reinterpret these definitions in light of the linguistic complementarity discussed in the previous sections.

8.3.1 TPSA: Evaluation of a Functional

Conceptually the tracking code represented by $\vec{\zeta}_{i\to j}$ can be viewed as a particular representation of the ring. The quantities $\vec{\Gamma}$ which we want to compute can be viewed mathematically as functionals of the underlying Hamiltonian. [8.6] In this book, we prefer to say a functional of some map, often the one-turn map. However, just in order to keep the discussion general, I will say the "Hamiltonian." Since the Hamiltonian generates all the maps of interest we do not lose generality.

Hence, given a ring, observables such as the tunes or other quantities can be viewed as functionals of the ring, or worse of the Hamiltonian

$$\vec{\Gamma}\,(H)$$

where H is a function of phase space and parameter space.

In a TPSA calculation, using a map-based theory, we evaluate the functional $\vec{\Gamma}$ for the particular ring represented in our computer program. Furthermore, mathematically we control the parameter definition of the function H at the onset. If the k-jet includes, let us say, five parameters of interest (quadrupole strength, currents, etc.) then the functional is evaluated as a power series in these parameters. All

[8.6] I am grateful to Michelotti for this interpretation.

the coefficients of this expansion are floating points numbers. Let us do a simple example. Consider a rotation followed by a quadrupole kick for which I now write as a computer program:

```
x=0.0
p=0.0
a=0.0
mu=0.3
x1=cos(mu)*x + sin(mu)*p
p=cos(mu)*p - sin(mu)*x
x=x1
p=p-a*x
print x,p
```

The compositional map corresponding to the map of this program is simply $\mathcal{M} = \exp\left(: -\mu\ J :\right) \exp\left(: -\frac{a}{2}\ x^2 :\right)$. Let us imagine that we add to the computation of (x, p) the linear tune using our normalization package. Thus, we declare (x, p) to be 1-jets with no parameter dependence, and we ask what is then the new result for x, p and the tune?

Here is the answer in an obvious format

```
 x= 0.955336489 x_1 + 0.295520207 x_2
 p=-0.295520207 x_1 + 0.955336489 x_2
tune= 0.477464829E-01 revolution
```

One recognizes the one-turn map for the values $\mu = 0.3$ and $a = 0$. Now, let us declare the variable "a" as a TPSA variable in our code and upgrade to a 2-jet calculation. The resulting output should be:

```
 x= 0.955336489 x_1 + 0.295520207 x_2
 p=-0.295520207 x_1 + 0.955336489 x_2
    -0.955336489 x_1 x_3 - 0.295520207 x_2 x_3
tune= 0.477464829E-01 + 0.795774715E-01 x_3  revolution
```

And if the program is instructed to go to 3-jets the output is simply

```
   x= 0.955336489 x_1 + 0.295520207 x_2
   p=-0.295520207 x_1 + 0.955336489 x_2
     -0.955336489 x_1 x_3 - 0.295520207 x_2 x_3
tune= 0.477464829E-01 + 0.795774715E-01 x_3
      - 0.643130829E-01 x_3^2 revolution
```

Hence, the tune, however complex the algorithm which computes it (see (8.10)), can be computed as a function of the parameters of the code, using TPSA and a Lie algebraic library. As we said, all differentiable parameters of the code ($\vec{\Delta}$ of Equation (8.7)) are polymorphic, thus changing definition at run time according to the user's wishes. Here, we instructed the program to treat the quadrupole strength "a" as the third variable of TPSA ($\vec{\delta} = (a)$). Automatically the tune is an expansion in "a" consistent to the order selected for the jets.

The parameter dependence could be interpreted in two different ways:

1. We still evaluate a functional. But the input function H is redefined to be a function of phase space and restricted parameter space. This is none other than the redefinition of the tracking code in Sect. 8.1.3. The new function H is now taking its argument in $\mathbb{R}^{6+N_{\vec{\delta}}}$.

2. We compute and evaluate the derivative of a functional with respect to the chosen $N_{\vec{\delta}}$ parameters.

The two definitions are mathematically equivalent but, algorithmically, our techniques are better reflected by definition (1). Indeed, the map is first computed with the parameters and then sent to a normalization algorithm which is capable of handling parameter dependence. The library which does this normalization has no knowledge of the meaning of the parameter $\vec{\delta}$ chosen for the k-jet. As a matter of fact, for many years I used this library solely as a post-processor. Therefore, the parameter dependent one-turn map (i.e., integrated H) is sent and analyzed, thus reflecting definition (1).

Definition (2) misleads us to believe that one has a formula from which derivatives are taken. This is completely antithetical to the TPSA approach and for that matter to the use of Taylor series in single as well as multi-pass systems. However, it is more in tune with analytical calculations which we now describe.

8.3.2 Analytic: Functional Derivative

Let us consider a formula often quoted in accelerator circles: the leading order tune shift due to a beam-beam force. Assuming that there is no beam-beam force in our code (quite often the case), then there is no way for the code to know about beam-beam forces. Nevertheless, one can actually derive a formula for the leading order tune shift due to beam-beam forces and use it in conjunction with our code. Let us add an arbitrary kick to the map of Sect. 8.3.1:

$$\mathcal{M} = \exp\left(: -\mu\, J :\right) \exp\left(: -\frac{a}{2}\, x^2 :\right) \exp\left(: -V :\right). \qquad (8.16)$$

Since the Lie exponent V is an arbitrary function of the phase space, one cannot use TPSA to evaluate the map or any of its functionals! We could list all its beautiful mathematical properties including differentiability but a fact remains: unless a differentiable functional form is chosen (including parameters) and implemented, TPSA cannot help us compute the effect of V on the tune!

Nevertheless we can immediately provide an abstract expression for the leading order tune shift (setting $a = 0$):

$$\Delta\nu(J) = \frac{1}{4\pi^2}\frac{\partial}{\partial J}\int_0^{2\pi} V\left(\sqrt{2J}\cos(\varphi), -\sqrt{2J}\sin(\varphi)\right) d\varphi. \tag{8.17}$$

Exercise 8-5

Apply the result of Equation (8.17) *to* $V = \frac{a}{2}x^2$ *and regain the leading order tune shift produced by TPSA in the previous section. In fact you will see the coefficient 0.7957747715E-01 is actually* $\frac{1}{4\pi}$. *Derive Equation* (7.10) *trying to justify the steps and then take a look at Sect.* 8.4.

In the analytic formula, the meaning of V and the meaning of the symbol π are not intrinsic to the code. In fact, even if we write a formula for the case $V = \frac{a}{2}x^2$,

$$\nu(J) = \nu_0 + \frac{a}{4\pi}, \tag{8.18}$$

the quality of the knowledge is different. In the TPSA formula, we have an exact derivative for the actual quadrupole placement in the code. By contrast, in the formula of Equation (8.18) we not only get the same information, but also the knowledge that the coefficient in front of "a" or x_3 does not depend on the tune. This result is far from obvious at first glance (ask a single pass expert in passing). There are two ways to describe what has happened.

1. Mathematically we view analytical calculations as functional derivatives of global quantities which depend in an abstract way, on the Hamiltonian on the computer. In practice, they will be functional derivatives of global quantities of an appropriately chosen Floquet ring. That is the art of Hamiltonian-free perturbation theory.

2. Semantically it is clear that all analytic calculations appeal to information which does not exist on the computer. Symbols like

integral signs, π, e or abstract operations like the commutator or Poisson bracket must be re-interpreted by the user using the metalanguage of his physical theory.

This brings us to an interesting historical point: for years it was not obvious that algebraic manipulators could be programmed. Such programs are dependent on linguistic complementarity. Indeed when one sees expressions like

```
[[a,b],c]=abc-bac-cab+cba
```

being generated by a manipulator, the human brain interprets the symbol to be commutators of a noncommutative algebra. The computer has no knowledge of the actual objects `a`,`b` and `c`. These things could be matrices, abstract operators in some Hilbert space, or something else. In fact, the algebraic manipulator knows only how to set up trees corresponding to the product, sum and subtraction operations. The "operators" are mere addresses in computer memory of a character to be printed.

Analytical calculation is therefore an art in symbolic manipulations and abstract grammar. One must define an alphabet of symbols like {a,b,c} and the words which can be made with the alphabet such as "abc" or "bca." These words carry no meaning unless a human with the metalanguage of physics imparts a meaning. The interested reader may consult reference [29]. The author uses the language of alphabets, words and lexicographical order to describe abstract free Lie algebras and to set up algorithms for normalizations and the production of symplectic integrators.

Because there might be an overlap of information between a TPSA and an analytical calculation, it is sometimes difficult to decide what type of calculations one should do. Because the connection between TPSA and Hamiltonian perturbation theory is at the map level and because map-based theories are not fully understood in the ring community as of 1996, it has been my experience that most practitioners jump into analytical calculations when they could instruct a code equipped with TPSA/Lie to produce the desired numerical results. It is our hope that this book will modify this trend in the right direction.

8.3.3 Super-Analytic

Maybe the reader has already noticed that the tracking code can be viewed as a symbolic string, let us be facetious about the whole thing!

The little program which we wrote in Sect. 8.3.1 to explain TPSA calculations is itself a symbolic expression using the syntactical rules of a computer language (FORTRAN for example). It is quite possible in theory to imagine this program, tied to a symbolic system, grinding out an analytic result. Removing the real number assignments at the top, a symbolic manipulator would produce the result

```
print x,p

---> cos(mu) x + sin(mu) p ,
    -(sin(mu)+a*cos(mu)) x + (cos(mu)-a sin(mu)) p
```

In this particular example, the symbolic manipulator was the author's brain for whatever that is worth! What is interesting to realize is that I needed only syntactic rules of FORTRAN to convert the computer program into the syntactic rules of mathematics. The reader appreciates that in the line

```
x=x-a*x
```

the = symbol has a syntactic function quite different from its mathematical homonym.

The purpose of this discussion is simply to emphasize that the boundary between "symbolic/syntactic" and "semantic" is very hazy. The computer code itself, when viewed line by line, is a symbolic representation of the ring and could be fed to a symbolic manipulator such as a human or electronic brain. This manipulator could then generate an analytic one-turn map as we just did. In TPSA calculations, we do not take this point of view: the symbolic lines of codes are compiled in order to generate the map $\zeta^{[k]}_{i\to j}$. Moreover, due to the size of a ring, the complete symbolic representation of the map in a useful format is out of the question. This problem is furthermore compounded by the need to normalize maps to extract useful information. Hence, we do not see any use in complete symbolic representation of maps for realistic systems. Only simple FODO cells and other small modules are represented symbolically for pedagogical purposes.

Analytic calculations are useful when they are highly discriminating on what is a symbolic variable and what is not. Hence we advocate the use of a tracking code with TPSA to generate the input to a symbolic manipulator (or formula) which handles only a tiny subset of critically important parameters of a design. In other words, a function $\vec{\Gamma}$ is computed, producing information at selected locations and the result is fed to the output of a symbolic manipulator, i.e., to a formula.

In the global/local approach the tracking code is as close as possible to the real ring, and thus, the lines of coding which symbolically represent the maps between sections of the ring, are not suited for symbolic computation. In the global/local approach it really does not matter because the analytical theory is based on a judiciously selected set of maps which the code can either compute (tracking and/or TPSA) or help compute by producing the necessary input to an algebraic formula.

Hence the lines of coding have no semantical meaning as far as analytic calculations are concerned. This meaning must be imparted to the functions $\vec{\Gamma}$ which the user/programmer attaches to the code.

8.4 RULES OF ANALYTIC CALCULATIONS

At this stage the reader should take another look at Chapter 7 and in particular at Sect. 7.3. The concept of a Floquet ring is essential to the rules of analytic calculations. Observables in a ring are usually properties of the one-turn map. So, we will concentrate at this stage on quantities defined at one location such as the invariants or the tunes.

8.4.1 Perturbing the One-turn Map

Our goal is to modify the zeroth order ring by a perturbation belonging to a large class of functions. For example when one derives an analytical formula for the second order tune shifts with sextupoles, mathematically we take the functional derivative of the tunes with respect to a restricted class of functions: vector potentials of sextupoles. The resulting formulas will depend on global quantities of the unperturbed rings. The formula for a first order quadrupole tune shift depends on the unperturbed beta function at the point of insertion in the case of a midplane symmetric zeroth order ring. Hence, once the formula has been derived using pencil and paper or sophisticated symbolic manipulators, it is clear that its evaluation on an actual ring will depend *critically* on our ability to compute zeroth order observables such as the beta function.

The practitioner of the global/local theory has no fear at this stage because we know exactly where, who, what and how the zeroth order observables are obtained. They come from the underlying tracking code through TPSA and Lie algebraic libraries. Moreover the tracking code itself may be used to check the analytical formulas if the class of functions in the functional derivatives is programmed in the code.

It is typical to hear horror stories about complex "linear coupled formalism" being analytically grinded out for the sole purpose of producing floating point numbers which are to be fed into other analytic formulas. This is a practice which reveals the conceptual problems which still plague the field of rings. When you need numbers, your first instinct should be to run a code equipped with TPSA/Lie tools. Only computer limitations should lead you away for this "primordial reflex."

8.4.1.1 Plain computation of the perturbed one-turn map

We have said before that physics dictates how many maps the mathematical ring must contain. Let us concentrate on observables which require only a one-turn map. For example we might ask what the map is. This is kind of dumb but we will see that it is a first step in the rules of analytical calculations out of which will come other observables such as the tunes.

Clearly, if we are interested only in the map at one chart $i = 0$, the TPSA calculation (regular phase space plus parameters) requires only one map. The ring is simply:

$$Ring_1 = \{\mathcal{M}_0\}, \tag{8.19}$$

and the tracking code is instructed to approximate the compositional map $\mathcal{M}_0$ by the function (code) $\vec{\zeta}_{0\to\aleph}$ where $\aleph$ is the number of maps in the actual code. The reader can set this to infinity later and regain Hamiltonian theory from the map method.[8.7]

Next we ask the following question: if a perturbation to the Hamiltonian ϵV_s is introduced how does one compute the ϵ^κ correction to $\mathcal{M}_0$ analytically and what definition of the ring allows us to perform this calculation?

Our starting point is similar to perturbation theory and Feynman's propagator methods. Imagine a localized perturbation at "s"

$$\Longrightarrow \epsilon\, \delta(s - s')\, ds V_s .$$

The length ds was added to keep the units consistent and to ease the link with the continuous (Hamiltonian) limit. In the discontinuous

[8.7]The reverse is hard when dealing with the product of Dirac delta functions as an accurate representation of the theory requires. See Sect. 6.3.1 which compares Guignard's traditional work with ours.

approximation which is good if V_s is truly a very peaked function around s', like a box of width Δ and height Δ^{-1}, we can show easily that the one-turn map at $s = 0$ is given by:

$$\mathcal{M}(\epsilon) = \mathcal{M}_{0\rightarrow s'} \exp\ (: -\epsilon\ ds\ V_{s'} :) \mathcal{M}_{s'\rightarrow N} + O(\Delta\epsilon). \quad (8.20)$$

These results become exact as Δ approaches zero. Indeed this is the case of a localized perturbation: the usual kick. If we return to a smooth function V_s, we can imagine replacing it by a sum over N_V such kicks. That is the essence of integration. The new function is

$$V_s = \sum_k^{N_V} \delta(s - s_k)\ ds\ V_{s_k} \quad (8.21)$$

which is in effect what is usually done in tracking codes (sextupole kicks, beam-beam kicks, etc.). The map corresponding to (8.21) is given exactly by

$$\begin{aligned} \mathcal{M}(\epsilon) = \mathcal{M}_{0\rightarrow 1} \exp\ (: -\epsilon\ ds\ V_{s_1} :) \mathcal{M}_{1\rightarrow 2} \cdots \quad (8.22) \\ \times \mathcal{M}_{k-1\rightarrow k} \exp\ (: -\epsilon\ ds\ V_{s_k} :) \mathcal{M}_{k\rightarrow N} + O(\Delta\epsilon). \end{aligned}$$

Of course, this is not a very useful form of the map for analytical calculations because the unperturbed propagator is split in $k+1$ pieces. In the Hamiltonian limit, we would have an infinite number of splits. Hence, we need to rewrite this map in a form where the perturbations are added not multiplied. First, we introduce a series of identities and move the unperturbed propagator in the exponent of each perturbation

$$\begin{aligned} \mathcal{M}(\epsilon) &= \exp(: -\epsilon \mathcal{M}_{0\rightarrow 1} ds\ V_{s_1} :) \cdots \exp(: -\epsilon \mathcal{M}_{0\rightarrow k} ds\ V_{s_k} :) \mathcal{M}_{0\rightarrow N} \\ & \quad (8.23) \\ &\stackrel{\text{or}}{=} \mathcal{M}_{0\rightarrow N} \exp\left(: -\epsilon \mathcal{M}_{1\rightarrow N}^{-1} ds\ V_{s_1} :\right) \cdots \exp\left(: -\epsilon \mathcal{M}_{k\rightarrow N}^{-1} ds\ V_{s_k} :\right). \end{aligned}$$

These expressions have simple (and exact) interpretations. The first factorization corresponds to all the perturbations at the beginning of the ring and the unperturbed map $\mathcal{M}_{0\rightarrow N}$ follows. Each exponent is advanced by the appropriate unperturbed propagator, i.e., phasors, and the perturbations operate sequentially to all orders correctly for $\Delta = 0$. In the last line the perturbations operate after the compositional map $\mathcal{M}_{0\rightarrow N}$. Perturbations must be reversed in phase to reflect the new position. Again, the expressions are exact.

Now comes the fundamental approximation. We are in a position to replace this product by a sum using the BCH formula:

$$\mathcal{M}(\epsilon) = \exp\{: -\epsilon \sum_{n=1}^{N_V} \mathcal{M}_{0\to n} ds\ V_n$$
$$+ \frac{\epsilon^2 ds^2}{2} \sum_{n<n'}^{N_V} [\mathcal{M}_{0\to n} V_n, \mathcal{M}_{0\to n'} V_{n'}] + O(\epsilon^3) :\} \mathcal{M}_{0\to \aleph}$$
$$\stackrel{\text{or}}{=} \mathcal{M}_{0\to \aleph} \exp\{: -\epsilon \sum_{n=1}^{N_V} \mathcal{M}_{n\to \aleph}^{-1} ds\ V_n +$$
$$\frac{\epsilon^2 ds^2}{2} \sum_{n<n'}^{N_V} [\mathcal{M}_{n\to \aleph}^{-1} V_n, \mathcal{M}_{n'\to \aleph}^{-1} V_{n'}] + O(\epsilon^3) :\} . \quad (8.24)$$

For those who would like a continuous result (Hamiltonian theory), one needs to replace all the sums by integrals:

$$\mathcal{M}(\epsilon) = \exp\{: -\epsilon \int_0^C ds\ \mathcal{M}_{0\to s}\ V_s$$
$$+ \frac{\epsilon^2}{2} \int_0^C ds \int_s^C ds' [\mathcal{M}_{0\to s} V_s, \mathcal{M}_{0\to s'} V_{s'}] + O(\epsilon^3) :\} \mathcal{M}_{0\to \aleph}$$
$$\stackrel{\text{or}}{=} \mathcal{M}_{0\to C} \exp\{: -\epsilon \int_0^C ds\ \mathcal{M}_{s\to C}^{-1}\ V_s +$$
$$\frac{\epsilon^2}{2} \int_0^C ds \int_s^C ds' [\mathcal{M}_{s\to C}^{-1} V_s, \mathcal{M}_{s'\to C}^{-1} V_{s'}] + O(\epsilon^3) :\}. \quad (8.25)$$

Finally, similar formulas can be derived for general vector fields where one replaces the Poisson bracket by the vector field bracket. There is no additional complexity.

At this stage it is time to answer the initial question. While the TPSA calculation of a one-turn map requires a ring with one element (i.e., that map), what kind of rings must we consider conceptually (i.e., in the metalanguage outside the tracking code) in order to perform an analytical calculation? The answer is simple: if one performs a κth order calculation in the perturbation, we need to consider a ring with $\kappa + 1$ unperturbed propagators and κ perturbations. As in quantum electrodynamics, a first order calculation can be viewed as the sudden (local in time or s) appearance of a perturbation which is then summed over all possible times or positions. In quantum mechanics, a probability amplitude controls the probability of an interaction between the unperturbed field and the perturbation. In classical mechanics, the actual distribution is man-made unless random distributions are included in a calculation.

Let us list the rings which enter in a first order calculation:

$$Ring_3 = \{\mathcal{M}_{0\to 1}\ ,\ \exp\,(:-\epsilon\ ds\ V_1\ :)\ ,\ \mathcal{M}_{1\to N}\}\,. \tag{8.26}$$

This ring with 3 maps is made of two unperturbed propagators and the perturbation at $j = 1$. Then we factorize the map (let us pick the reverse Dragt-Finn factorization):

$$Ring_2 = \{\exp\,(:-\epsilon\ ds\ \mathcal{M}_{0\to 1}\ V_1\ :)\ ,\ \mathcal{M}_{0\to N}\}\,. \tag{8.27}$$

This ring has two maps: the original unperturbed map and the perturbation expressed in terms of an exponent. We then sum/integrate over all occurrences of the perturbation. Hence the final result is also a ring with two maps:

$$Ring_2 = \left\{\exp\left(:-\epsilon \sum_{n\,=\,1}^{N_V} ds\ \mathcal{M}_{0\to k}\ V_k\ :\right)\ ,\ \mathcal{M}_{0\to N}\right\}. \tag{8.28}$$

Just to get the trend, let us consider a second order analytic calculation. So without more ado, here are the rings:

$$Ring_5 = \left\{\mathcal{M}_{0\to 1}, e^{:-\epsilon\ ds\ V_1:}, \mathcal{M}_{1\to 2}, e^{:-\epsilon\ ds\ V_2:}, \mathcal{M}_{2\to N}\right\}, \tag{8.29}$$

followed by

$$Ring_2 = \left\{\exp(:-\epsilon\ ds\ V_1^{\dagger} - \epsilon\ ds\ V_2^{\dagger} + \frac{\epsilon^2\ ds^2}{2}\left[V_1^{\dagger}, V_2^{\dagger}\right]:), \mathcal{M}_{0\to N}\right\}$$

$$\text{where}\quad V_n^{\dagger} = \mathcal{M}_{0\to k} V_n. \tag{8.30}$$

We then sum over the perturbations

$$Ring_2 = \left\{\exp(:-\epsilon \sum_{n=1}^{k} ds\ V_n^{\dagger} + \frac{\epsilon^2}{2} \sum_{n<n'}^{N_V} ds^2 \left[V_n^{\dagger}, V_{n'}^{\dagger}\right]:), \mathcal{M}_{0\to N}\right\}. \tag{8.31}$$

8.4.1.2 General rules for the analytical one-turn map

Now we can easily guess the general perturbation rules for an ϵ^{κ} calculation in a perturbing potential.

Rules for an ϵ^κ perturbation of the One-Turn Map

1. Insert κ distinct perturbations in the ring numbered from one to κ in order of appearance. This ring has by definition $\aleph = 2{*}\kappa{+}1$.
2. Compute the one-turn map by "phase advancing" all the perturbations at the beginning or end of the ring using the zeroth order propagator.
3. Lump all the perturbations into one exponent using the BCH theorem.
4. Sum/integrate over the actual distribution of the perturbations using the index j labeling the perturbations $(1 \geq j \leq \kappa)$ as a time-ordering label.

The end result of this calculation is the one-turn map expressed in terms of the propagator of the unperturbed problem. In effect, the various powers of ϵ will give us the functional derivatives of the one-turn map. Moreover, if we let this map act on some function, it will give us the functional derivative of the output function. The generality of this derivative depends on the class of potentials represented by V. Typically we restrict ourselves to simple functions such as multipoles or beam-beam forces in the weak-strong régime.

In rings we must ask how to obtain the observables found from normalization algorithms. In a map based theory we have carried out the first step: the computation of the one-turn map.

8.4.1.3 Rules in Floquet variables

Our goal is not to produce one-turn maps for their own sake but to extract from them the global quantities of interest. The first step in this must start with our ability to perform such a calculation on the TPSA k-jet produced by the tracking code. Secondly, since normalization produces simpler maps and thus simpler rings (Floquet ring), we must try to express the rules of Sect. 8.4.1.2 in the k-jet Floquet ring produced by the tracking code and the Lie/DA/TPSA normalization library.

We remind the reader of the definition of a transformed ring (see Equation (7.9)). Once a ring has been defined for a certain purpose, one can transform each location i by a map $\mathcal{A}_i$. The propagation in the new system is given by the ring

$$\mathcal{A}\left[Ring_{\aleph}\right] = \left\{ \mathcal{R}_{i\to i+1} \;\middle|\; \begin{array}{c} i = 0, \aleph - 1 \\ \mathcal{R}_{i\to i+1} = \mathcal{A}_i \mathcal{M}_{i\to i+1} \mathcal{A}_{i+1}^{-1} \end{array} \right\}$$

where the maps $\mathcal{R}_{i \to i+1}$ are rotations in a complete normalization of a symplectic map. In a coasting beam normalization, these maps are rotations in the transverse planes and drift-like in the longitudinal plane.

We must ask ourselves how our rules are modified by working in the Floquet ring instead of the original ring. To answer this one needs to transform the perturbed ring by the maps $\mathcal{A}_i$'s. We get for the perturbation:

$$\mathcal{A}_n e^{:-\epsilon\, ds\, V_n:} \mathcal{A}_n^{-1} = e^{:-\epsilon\, ds\, \mathcal{A}_n V_n:},$$

which implies that the potentials $V_n^\dagger$ in Equation (8.31) are redefined to be

$$\begin{aligned} V_n^\dagger &\equiv \mathcal{A}_0 \mathcal{M}_{0\to n} \mathcal{A}_n^{-1} \mathcal{A}_n V_n = \mathcal{R}_{0\to n} \mathcal{A}_n V_n \\ &\overset{\text{or}}{\equiv} \mathcal{A}_0 \mathcal{M}_{n\to N}^{-1} \mathcal{A}_n^{-1} \mathcal{A}_n V_n = \mathcal{R}_{n\to N}^{-1} \mathcal{A}_n V_n . \end{aligned} \tag{8.32}$$

The first and second line of Equation (8.32) represent respectively the perturbations at the start and the end of the ring.

Hence it follows that the rules in Floquet variables are the same but the perturbation is transformed according to Equation (8.32).

Let us look at one example through an exercise. Here our purpose is to illustrate the difference between an analytical Floquet map and the map without Floquet variables.

Exercise 8-6

As an example of a periodic system, we will look at a simple "FODO" cell with one degree of freedom.

Super Analytical

Here we compute the full map as one single analytical expression. This is a highly unlikely possibility in a real system. It corresponds to the fully symbolic representation of the tracking code as discussed in Sect. 8.3.3.

1. *Consider the following ring with six maps:*

$$\begin{aligned} Cell_6 &= \{\mathcal{Q}_k, \mathcal{D}_L, \mathcal{Q}_{-k}, \mathcal{Q}_{-k}, \mathcal{D}_L, \mathcal{Q}_k\} \\ \text{where} \quad & \begin{array}{l} \mathcal{Q}_k = \exp\left(: -\frac{k}{2}\, x^2 :\right) \\ \mathcal{D}_L = \exp\left(: -\frac{L}{2}\, p^2 :\right) \end{array} . \end{aligned} \tag{8.33}$$

Compute the one-turn map from at $i = 0$. Answer:

$$M_0 = \begin{pmatrix} 1 - 2k^2L^2 & 2kL^2 + 2L \\ 2k^3L^2 - 2k^2L & 1 - 2k^2L^2 \end{pmatrix}. \qquad (8.34)$$

2. *Normalize M_0 using a Courant-Snyder transformation (Equation (2.75)). Answer:*

$$\mu = 2\sin^{-1}(kL) \qquad (8.35.a)$$

$$\beta_0 = \gamma_0^{-1} = \sqrt{\frac{1 + kL}{k^2(1 - kL)}} \text{ for } 0 < kL < 1 \qquad (8.35.b)$$

$$\alpha_0 = 0. \qquad (8.35.c)$$

3. *The one-turn map at a distance s such that $0 < s < L$ is given by $\mathcal{M}_s = \mathcal{D}_s^{-1}\mathcal{Q}_k^{-1}\mathcal{M}_0\mathcal{Q}_k\mathcal{D}_s$. Using this map compute the lattice functions at s. (Hint: propagate the invariant or look at Sect. 2.6.2).*

$$\beta_s = \beta_0\left(1 + \frac{s^2}{\beta_0^2} - 2ks + k^2s^2\right)$$

$$\gamma_s = \frac{1}{\beta_0}\left(1 + \beta_0^2k^2\right)$$

$$\alpha_s = \beta_0 k - \frac{s}{\beta_0} - \beta_0 k^2 s. \qquad (8.36)$$

TPSA

Here we display the result of a tracking code which simulates this FODO cell.

4. *Suppose that a tracking code (with 5 significant digits) equipped with TPSA is capable of simulating the FODO cell of this exercise. Let us set the parameters as follows:*

$$L = 1.0000$$
$$k = 0.67301$$

Compute the 1-jet in (x, p). Answer:

$$\begin{pmatrix} \overline{x} \\ \overline{p} \end{pmatrix} = \begin{pmatrix} 0.094115\ x + 3.3460\ p \\ -0.29622\ x + 0.094115\ p \end{pmatrix}.$$

5. *What is the tune? Answer:*

$$\nu = \frac{\mu}{2\pi} = 0.2350$$

6. *Suppose the code is instructed to extract 2-jets; will the answer change? Answer: no, because the system is linear.*

7. *Thus, keeping the jet order at 2, declare the parameter L as the third TPSA parameter. What is now the tune? Answer:*

$$\nu = 0.2350 \; + \; 0.28963 \underbrace{x_3}_{dL}$$

First Order Analytic Calculation

We do a really bizarre calculation where we extend the length of the half cell.

8. *What is the perturbation mapping for an infinitesimal length added to a ring at some arbitrary location in Floquet variables (Hint: Use the Anti-Courant-Snyder choice Equation* (2.102))*? Answer:*

$$\mathcal{P} = \exp\left(: -\frac{dL_s}{2} \gamma_s p^2 : \right).$$

9. *Apply the rules of perturbation theory to this Floquet map and get the perturbed mapping. Answer:*

$$\mathcal{N} = \exp\left(: -\frac{1}{2} \sum_s dL_s \gamma_s \mathcal{R}_{0 \to s} p^2 : \right) \mathcal{R}_0.$$

10. *Refine the result to take advantage of the absence of magnets in the half-cell. Answer:*

$$\mathcal{N} = \exp\left(: -\frac{1}{2} \left\{ dL_1 \gamma_{0_+} \mathcal{R}_{0 \to 0_+} + dL_2 \gamma_{L/2_+} \mathcal{R}_{0 \to L/2_+} \right\} p^2 : \right) \mathcal{R}_0,$$

where 0_+ and $L/2_+$ refer to the beginning of the first and second drift sections respectively while dL_1 and dL_2 are the added lengths.

11. *Show that*

$$\Delta\nu = \sum_n \frac{1}{4\pi}\Delta L_n\ \gamma_n.$$

12. *Apply this analytic formula to our lattice. Clearly you need to compute the TPSA γ function. At how many places? Answer: γ is a constant in this cell because the quadrupoles are thin lenses. The final answer for the tune shift as a function of the half-cell length is*

$$\Delta\nu = \frac{1}{4\pi}\Delta L\left(\gamma_{L/2}+\gamma_{3L/2}\right) = \frac{\gamma_{L/2}}{2\pi}\Delta L. \qquad (8.37)$$

From TPSA, one deduces that the gamma function in the unperturbed half-cell is 1.8199 *and thus the analytical result is*

$$\Delta\nu = 0.28963\ \Delta L$$

just as before.

13. *The reader notices that the analytical formula for the tune shift is a function of the perturbation itself* (ΔL *in our case*) *and the unperturbed global observables* (γ_s). *This formula has a lot of generality since it only depends on global properties of the system and the perturbation. In fact, the only thing that came into play is the shape of the invariant in the half-cell. One could have a fully coupled ring elsewhere or a gutter. It does not matter! Of course, the result should agree with "super analytical" results. Hence show that the expansion of the tune* (8.35.a) *and the substitution of the "super analytical" result for gamma* (8.36) *into the analytical perturbation theory result of Equation* (8.37) *gives the same answer.*

8.4.1.4 Rules for the normalization

Once the map is computed to the appropriate order, we must extract an observable. Usually this observable requires a normalization of the map. In analytical calculations of the perturbed Floquet map, we obtain a result of the form

$$\mathcal{N}_0 = \exp\left(:\sum_{k=1}^{\kappa}\epsilon^k f_k:\right)\mathcal{R}_0. \qquad (8.38)$$

The purpose of the analytical normalization is to reduce the entire map into a normalized form (say a rotation) to order κ. This is a recursive process as it is also in a TPSA calculation. Hence let us assume that we have performed this process to order ι

$$\mathcal{N}_\iota = \exp\left(:\sum_{k=1}^{\kappa} \epsilon^k f_k^\iota :\right) \mathcal{R}_0. \tag{8.39}$$

The Lie transforms $: f_k^\iota :$ for $k \leq \iota$ will not change in the subsequent iterations. Presumably the Lie transforms are expressed at any stage in terms of the global quantities of the unperturbed motion.

To proceed one order higher in the perturbation we need (bare minimum) to remove the terms of order $\iota+1$ using a canonical transformation. Hence, we write the transformation as $\mathcal{A}_{\iota+1} = \exp\left(: \epsilon^{\iota+1} F_{\iota+1} :\right)$ and act on the map $\mathcal{N}_\iota$:

$$\begin{aligned}
\mathcal{N}_{\iota+1} &= \mathcal{A}_{\iota+1}\mathcal{N}_\iota\mathcal{A}_{\iota+1}^{-1} \\
&= \exp\left(: \epsilon^{\iota+1} F_{\iota+1} :\right) \exp\left(:\sum_{k=1}^{\kappa} \epsilon^k f_k^\iota :\right) \mathcal{R}_0 \exp\left(: -\epsilon^{\iota+1} F_{\iota+1} :\right) \\
&= \mathcal{R}_0 \exp\left(: \epsilon^{\iota+1} \mathcal{R}_0^{-1} F_{\iota+1} :\right) \exp\left(:\sum_{k=1}^{\kappa} \epsilon^k \mathcal{R}_0^{-1} f_k^\iota :\right) \\
&\quad \times \exp\left(: -\epsilon^{\iota+1} F_{\iota+1} :\right) \\
\Longrightarrow \quad & \mathcal{R}_0^{-1} f_{\iota+1}^\iota + \mathcal{R}_0^{-1} F_{\iota+1} - F_{\iota+1} \in \left\{\text{Generators of } \mathcal{R}_0\right\}.
\end{aligned} \tag{8.40}$$

The last line ensures that the map to order $\iota+1$ is normalized correctly. For example, in a complete normalization we insist that the Lie operator be a function of the actions only, and of the energy also, if the beam is coasting longitudinally. In a one-resonance normalization we insist that the Lie operator be invariant under a discrete subgroup of the planar rotations. This subgroup acts in the plane of the resonance. In the other plane we have full invariance as we do in a complete normalization.

At any order the BCH theorem is applied to order κ for the computation of $\mathcal{N}_{\iota+1}$. The process is recursive and can be summarized as follows:

$$\begin{array}{ccc}
\mathcal{N}_0 = \exp\left(: \sum_{k=1}^{\kappa} \epsilon^k f_k :\right) \mathcal{R}_0 & \rightarrow & F_{\iota+1} = -\left\{\mathcal{I} - \mathcal{R}_0^{-1}\right\}^{-1} \prod f_k^{\iota} \\
 & \uparrow & \downarrow \\
 & \llcorner & \mathcal{N}_{\iota+1} = \mathcal{A}_{\iota+1} \mathcal{N}_\iota \mathcal{A}_{\iota+1}^{-1} \\
 & & \downarrow \\
 & & \mathcal{N}_\kappa
\end{array} \tag{8.41}$$

In Equation (8.41), Π is a projection operator onto the set of Lie operators which must be removed by the normalization.

The observable can then be extracted from the canonical transformation or, if it is a tune, from the normalized map itself. We have said earlier that the quantities of interest defined in perturbation theory can be viewed as functionals of the k-jet map. A TPSA calculation is mathematically equivalent to the evaluation of the said functional on the actual map (i.e., Hamiltonian) implemented in one's computer program. The rules described here are more difficult to implement since they are functional derivatives to order κ. These functional derivatives are usually carried out with respect to a subset of all functions (quadrupole or sextupoles, for example), thus allowing with some pain the computations of the higher terms. This will be illustrated to second order in Chapter 9.

8.4.1.5 Conclusion: comparison with Hamiltonian normalization

The emphasis on map-based theory is not a blanket seal of approval for all calculations performed in physics or applied mathematics. First of all for s-independent systems the map and the Hamiltonian can be confused without introducing approximations. In such cases, the map for an arbitrary time s is the exponential of the operator $: H :$ itself (Equation (6.1))

$$\forall s \in \mathbb{R} \quad \mathcal{M}_{S \rightarrow S+s} = \exp\left(: -sH :\right), \tag{8.42}$$

and thus the effect of a canonical transformation can be moved directly onto the Hamiltonian

$$\begin{aligned} \forall s \in \mathbb{R} \quad \mathcal{A}\mathcal{M}_{S \rightarrow S+s}\mathcal{A}^{-1} &= \mathcal{A} \exp\left(: -sH :\right) \mathcal{A}^{-1} \\ &= \exp\left(: -s\mathcal{A}H :\right). \end{aligned} \tag{8.43}$$

In Equation (8.43) the one-turn map is an arbitrary map, why not take the map for an infinitesimal time? This map is of course no other than the Hamiltonian operator. We see in Equation (8.43) that working on the Hamiltonian has the advantage of avoiding the BCH theorem. Everything is explicitly written in terms of brackets.

The above discussion is trivial. Deprit would not have extended Hori's algorithm to time dependent Hamiltonians for no reason. When is the Hamiltonian theory more effective computationally than the map approach when applied to a periodic time dependent H?

The answer to this question is rooted in the form of Equation (6.42). As in the time independent case, the new Hamiltonian K_s is given in explicit form in terms of Poisson brackets

$$K_s = \exp(: w_s :) H_s + \frac{\exp(: w_s :) - 1}{: w_s :} \partial_s : w_s : . \qquad (8.44)$$

Hence, if the s-dependence is trivial so that a few Fourier modes in s can be kept, this equation can be used to normalize H_s to arbitrary order without the help of the BCH formula. In theory this is a great advantage since we can combine the computation of the perturbed map and the normalization in one recursive algorithm. However, in practice this is only feasible on simple Hamiltonians. Thus, for accelerators, where the one-turn map is more that an abstract Poincaré surface of section, there are conceptual and computational reasons to discard this procedure. This was illustrated in Sect. 6.3.1 where we showed how the standard accelerator formalism boils down to a Fourier expansion of delta functions.

Exercise 8-7

Using the map procedure as well as Equation (8.44), completely normalize the Hamiltonian

$$H_s = \nu J + \epsilon \sin(s)\, x^2 \qquad (8.45)$$

to second order in ϵ.

CHAPTER 9
EXAMPLES OF THE ANALYTICAL NORMALIZATION

In this chapter, we derive some formulas by applying the rules of Chapter 8. The calculations are done to first and second order. We also apply the formalism to a problem in quantum mechanics.

9.1 EXAMPLES OF FIRST ORDER NORMALIZATION

I start with a computation of the effect of a sextupole to leading order on a coasting lattice without coupling and then with coupling. The surprising result of this calculation to the standard accelerator physicist is the appearance of longitudinal effects for free!

9.1.1 Effect of Sextupoles on Chromaticities and Momentum Compaction

The first step is the definition of the perturbation:

$$\mathcal{P}_s = \exp\left(: -ds \, \frac{k_3(s)}{3}(x^3 - 3xy^2) : \right). \tag{9.1}$$

Under the assumption of mid-plane symmetry, we can compute the effect of a Courant-Snyder transformation on this sextupole insertion:

$$\begin{aligned} \mathcal{A}_s \mathcal{P}_s \mathcal{A}_s^{-1} &= \exp\left(: -ds \, \frac{k_3(s)}{3} \mathcal{A}_s (x^3 - 3xy^2) : \right) \\ &= \exp\left(: -ds \, \frac{k_3(s)}{3} (x^{\dagger\,3} - 3x^\dagger y^{\dagger\,2}) : \right) \end{aligned}$$

$$\text{where} \quad \left\{ \begin{array}{lcl} x^\dagger &=& \sqrt{\beta_x} x + \eta_x \delta \\ y^\dagger &=& \sqrt{\beta_y} y \end{array} \right. \tag{9.2}$$

We can immediately compute the first order one-turn map using our rules:

$$\mathcal{N}_0 = \exp\left(: -ds\ \frac{k_3(s)}{3} \mathcal{R}_{0\to s}(x^{\dagger\,3} - 3x^{\dagger} y^{\dagger\,2}) : \right) \mathcal{R}_0. \quad (9.3)$$

To perform the normalization one rewrites the Lie exponent in the resonance basis

$$\begin{aligned} x^{\dagger\,3} &\Longrightarrow \left(\sqrt{\beta_x} \frac{r_{1+} + r_{1-}}{2} + \eta_x \delta \right)^3 \\ x^{\dagger} y^{\dagger\,2} &\Longrightarrow \left(\sqrt{\beta_x} \frac{r_{1+} + r_{1-}}{2} + \eta_x \delta \right) \beta_y \left(\frac{r_{2+} + r_{2-}}{2} \right)^2. \end{aligned} \quad (9.4)$$

The terms which are left in the map belong to the Lie algebra of the coasting normalization which is made of operators depending only on $J_1 = \frac{r_{1+} r_{1-}}{2}$, $J_2 = \frac{r_{2+} r_{2-}}{2}$ and δ. Thus the result of $\mathcal{A}_1 \mathcal{N}_0 \mathcal{A}_1^{-1}$ is just

$$\mathcal{N}_1 = e^{:\oint ds\ \frac{k_3(s)}{3} (\eta_x^3 \delta^3 + 3\eta_x \delta \{\beta_x J_1 - \beta_y J_2\}):} \mathcal{R}_0. \quad (9.5)$$

Defining the angles in the usual way (see Equation (2.54)), one gets for the effect of the perturbation on the angle variables:

$$\frac{d\nu_i}{d\delta} = \frac{(-1)^{i+1}}{2\pi} \oint ds\ k_3 \eta_i \beta_i. \quad (9.6)$$

These are the two chromaticities, which in the case of a mid-plane symmetric system, are identified with the horizontal and vertical planes.

The same map, if applied to the longitudinal plane, will give us the various momentum compactions:

$$\Delta \ell = -2\pi \frac{d\vec{\nu}}{d\delta} \cdot \vec{J} - \delta^2 \oint ds\ k_3\ \eta_x^3. \quad (9.7)$$

We can say a few things about the form of Equation (9.7). First, the amplitude dependence of the path length, which is of some relevance in isochronous rings, is to leading order, the chromaticities. This is a result of the symplectic condition and of the normal form itself. It does not depend on the original ring and thus it is rigorously true for all rings which are represented by a Hamiltonian (see Equation (5.8)). This fact often escapes accelerator physicists as one can see in references [22, 23]. In fact, the author in [23] claims erroneously that only the chromaticity of the quadrupoles produces amplitude dependent tune shifts. A coasting beam normal form makes this result totally implausible. The reader can guess why such a mistake is made: standard accelerator theory treats the longitudinal plane separately. Indeed, in [23] the author tries to derive the amplitude dependence of the momentum compaction by averaging Equation (7.33) which is a special formula of restricted applicability. Of course, a mistake must still be made if one is to conclude incorrectly that the chromaticity originating in bends produces no amplitude dependent momentum compaction. In [23], the author assumes that the turn average $\left\langle \int_0^C ds \frac{\zeta_1(\vec{z};s)}{\rho_s} \right\rangle$ is zero to second order. This is false because it neglects the second order normalization map $\mathcal{A}_3 = \exp\ (: F_3 :)$ which is needed to bring the expression to be averaged in normalized variables to quadratic order in the transverse variables. Of course, in a quadrupole channel such terms do not exist but in small bends there are second order geometric terms which **must** conspire to produce the bend chromaticity.

Fortunately all this nonsensical discussion is rendered useless by our full 6-d treatment of the normal form for coasting beams.

Exercise 9-1

A colleague of mine used Equation (9.7) *in a paper [46] and received comments of disbelief from the reviewer. Using dimensional analysis the reviewer pointed out that one would expect terms proportional to the beta functions to appear in the expression for the momentum compaction. Again one can guess that the reviewer was not aware of Jordan normal forms for coasting beams and was probably struggling with Equation* (7.33). *The reviewer's dimensional analysis was correct but did not take into account the symplectic condition! The following exercise demonstrates that the second order momentum compaction depends only on the dispersion vector (including momenta) if the unperturbed problem is symplectic.*

1. *Show that even in the case of a coupled unperturbed ring the second order momentum compaction is only a function of the linear dispersion vector*

$$\Delta\ell = -2\pi \frac{d\vec{\nu}}{d\delta} \cdot \vec{J} - \delta^2 \oint ds\ k_3 \left\{\ \eta_x^3 - 3\eta_x\eta_y^2 \right\}. \qquad (9.8)$$

The solution can be found in Sect. 9.1.2.

2. *Consider the nonsymplectic "sextupole kick"*

$$\Delta\vec{z}_\perp = ds\ \vec{F}_\perp \tag{9.9}$$
$$\textit{where } F_{\perp\, i} = -k_3 \sum_{l,m=1}^{4} F_i^{lm} z_{\perp l} z_{\perp m}$$
$$\textit{and } \vec{z}_\perp = (x, p_x, y, p_y).$$

Compute the second order momentum compaction.

Hints:

(a) Show that the full coasting beam Courant-Snyder transformation must include

$$\bar{\ell} = \ell + \left\{ \sqrt{\beta_x}\eta_x' + \frac{\alpha_x}{\sqrt{\beta_x}}\eta_x \right\} x - \frac{\eta_x}{\sqrt{\beta_x}} p_x, \tag{9.10}$$

or see Sect. 7.3.2 if you are confused.

(b) Using Equation (4.29), show that the Lie operator of the nonsymplectic kick in Floquet variables in given by

$$\begin{aligned} \ell - \textit{component} &= -ds\ \mathrm{J}\vec{\eta} \cdot \mathcal{A}\vec{F}\ \partial_\ell \\ \textit{where } \mathrm{J}\vec{\eta} &= \begin{pmatrix} \eta_x' \\ -\eta_x \\ \eta_y' \\ -\eta_y \end{pmatrix}, \end{aligned} \tag{9.11}$$

in the longitudinal direction.

(c) Average over the transverse directions (i.e., assume the existence of $\mathcal{A}_2 = \exp(\vec{A}_2 \cdot \vec{\nabla})$) and regain the result of Equation (9.7) for the real sextupole.

9.1.2 Same thing with Coupling

Of course the same rules apply. Concentrating on the symplectic case, we can assume the existence of an unperturbed Floquet transformation $\mathcal{A}$ which depends on s.

$$\begin{aligned} \bar{x} &= A_{11}x + A_{12}p_x + A_{13}y + A_{14}p_y + \eta_x\delta \\ &\vdots \\ \bar{p}_y &= A_{41}x + A_{42}p_x + A_{43}y + A_{44}p_y + \eta_y'\delta. \end{aligned} \tag{9.12}$$

We substitute this into the Lie operator of a sextupole

$$\mathcal{A}_s V = -ds\, k_3 \frac{\begin{pmatrix} A_{11}x \\ +A_{12}p_x \\ +A_{13}y \\ +A_{14}p_y \\ +\eta_x\delta \end{pmatrix}^3 - 3 \begin{pmatrix} A_{11}x \\ +A_{12}p_x \\ +A_{13}y \\ +A_{14}p_y \\ +\eta_x\delta \end{pmatrix} \begin{pmatrix} A_{31}x \\ +A_{32}p_x \\ +A_{33}y \\ +A_{34}p_y \\ +\eta_y\delta \end{pmatrix}^2}{3}. \tag{9.13}$$

The Lie operator after normalization is

$$\begin{aligned} h(\vec{J};\delta) = -k_3 \{ & \eta_x\delta \{ (\beta_{xx} - \beta_{yx}) J_x + (\beta_{xy} - \beta_{yy}) J_y \} \\ - & \{ C_x J_x + C_y J_y \} \eta_y \delta - \frac{1}{3} \{ \eta_x^3 - 3\eta_x \eta_y^2 \} \delta^3 \Big\}. \end{aligned} \tag{9.14}$$

The quantities appearing in the analytical formula for the normalized Lie operator are all observables of the unperturbed motion. For example, the four beta functions are extensions of the usual two beta functions of the uncoupled problem. One can show that

$$\begin{pmatrix} \langle x^2 \rangle \\ \langle y^2 \rangle \end{pmatrix} = \frac{1}{2} \begin{pmatrix} \hat{x}^2 \\ \hat{y}^2 \end{pmatrix} = \begin{pmatrix} \beta_{xx} & \beta_{xy} \\ \beta_{yx} & \beta_{yy} \end{pmatrix} \begin{pmatrix} J_x \\ J_y \end{pmatrix}$$

and

$$\langle xy \rangle = C_x J_x + C_y J_y. \tag{9.15}$$

As we said earlier, the purpose of analytical perturbation theory in rings is to express the observables of the perturbed system in terms of the observables of the unperturbed system. Of course, one can express the same formulas in terms of the unperturbed canonical transformation $\mathcal{A}$. For example the beta functions are

$$\begin{aligned} \beta_{xx} &= A_{11}^2 + A_{12}^2 \\ \beta_{xy} &= A_{13}^2 + A_{14}^2 \\ \beta_{yx} &= A_{31}^2 + A_{32}^2 \\ \beta_{yy} &= A_{33}^2 + A_{34}^2 \\ C_x &= A_{11}A_{31} + A_{12}A_{32} \\ C_y &= A_{13}A_{33} + A_{14}A_{34}. \end{aligned} \tag{9.16}$$

One notices that all these quantities are invariant under the group $SO(2) \times SO(2)$. This group is precisely that of the phase advances and we know that canonical transformations are known up to a phase only.

Exercise 9-2
As a trivial exercise, take the anti-Courant-Snyder transformation of Equation (2.101) *and show that* $A_{11}^2 + A_{12}^2$ *is indeed the beta function. Conversely show that* $A_{21}^2 + A_{22}^2$ *in the case of the Courant-Snyder transformation must be the Twiss parameter gamma. What is the alpha function? Answer: it is just* $-A_{21}A_{11} - A_{22}A_{12}$ *and is proportional to the average of* $\langle xp \rangle$. *For a review see Sect.* 2.4.5.

9.1.3 The Quantum Harmonic Oscillator

Map methods, Lie methods, operators—these are words which are often identified with quantum mechanics. Historically it is not fair since Sophus Lie himself introduced Lie operators in the context of differential equations. In this book, we certainly go back to Lie. In all fairness, Hori in his seminal paper [15] was the first scientist to use Lie methods in classical mechanics and quite fairly jumped over Poincaré and cited solely Lie in his 1966 article. In this section we make the explicit connection between classical and quantum methods on the perturbed harmonic oscillator.

The quantum harmonic oscillator is described by the Hamiltonian

$$H = \frac{p^2}{2m} + \frac{1}{2}m\omega^2 x^2 \tag{9.17}$$

where the symbols x and p are operators satisfying the commutation relation

$$\{x, p\} = i\hbar. \tag{9.18}$$

If we redefine the Lie bracket to include one power of $i\hbar$

$$[\![x, p]\!] = \frac{1}{i\hbar}\{x, p\} = 1, \tag{9.19}$$

then the quantum operators x, p have the same Lie algebra as the symplectic operators $:x:$ and $:p:$. Remarkably, the same applies to the quadratic operators. Consider the following identification between the symplectic and quantum operators:

$$\begin{array}{llcl} \text{Name} & \text{Quantum} & & \text{Symplectic} \\ O_{11} & x^2 & \rightarrow & :x^2: \\ O_{12} & \frac{1}{2}(xp+px) & \rightarrow & :xp: \\ O_{22} & p^2 & \rightarrow & :p^2: \end{array} \tag{9.20}$$

then the Lie algebra of these operators is the same:

$$\begin{aligned}\{O_{11}, O_{12}\} &= 2O_{11}\\ \{O_{22}, O_{12}\} &= -2O_{22}\\ \{O_{11}, O_{22}\} &= 4O_{12}.\end{aligned} \tag{9.21}$$

The commutator symbol in Equation (9.21) is to be interpreted depending on the context.

The Lie algebra describes infinitesimal elements of the Lie group hence it is not obvious that the Lie group should be the same and indeed it turns out to be different in the quantum and classical case. In the classical case, it is possible to find a finite dimensional representation of the Lie group which is faithful, namely the matrices of $Sp(2)$. In the quantum mechanical case the full group has only infinite-dimensional representations.

9.1.3.1 Schrödinger's equation

Wave dynamics starts with the Schrödinger equation which takes the form

$$\begin{aligned}\frac{d}{dt}\Psi_t &= \frac{1}{i\hbar}H\Psi_t\\ \text{where } p &= \frac{\hbar}{i}\frac{d}{dx}.\end{aligned} \tag{9.22}$$

This equation describes the time evolution of a square-integrable function of a variable x. We can turn this equation into a map equation as we did in the symplectic case:

$$\begin{aligned}&\text{if } \Psi_t = \mathcal{U}_{0\to t}\Psi_0\\ &\Downarrow\\ \frac{d}{dt}\mathcal{U}_{0\to t} &= \frac{1}{i\hbar}H\mathcal{U}_{0\to t}\\ &= \widehat{H}\mathcal{U}_{0\to t} \text{ where } \widehat{H} = \frac{1}{i\hbar}H.\end{aligned} \tag{9.23}$$

Equation (9.23) is identical to the map equation of classical mechanics for the *inverse* symplectic map (See Exercise 2-24). Indeed the inverse map $\mathcal{M}_{0\to t}^{-1}$ obeys the equation

$$\frac{d}{dt}\mathcal{M}_{0\to t}^{-1} = :H:\mathcal{M}_{0\to t}^{-1}, \tag{9.24}$$

which implies that if one knows the Lie representation of the inverse classical map and if the Lie algebras are identical one can actually construct the quantum map $\mathcal{U}_{0\to t}$ from the classical map $\mathcal{M}_{0\to t}^{-1}$. From our discussion of quadratic operators, one sees that this is true for classical linear maps. This is the basis of Fourier optics which connects the symplectic linear group $Sp(2)$ to the quantum metaplectic group. However, for nonlinear Hamiltonians, the Lie algebras will differ and it is not possible to construct the exact quantum propagator from the classical compositional map.

Let us now look at perturbation theory applied to the harmonic oscillator. We will see that, in the operator formalism, it is formally very similar to the classical normal form theory presented in this book.

9.1.3.2 Creation and annihilation operators

Let us first look at the time-independent Hamiltonian of Equation (9.17). We know that the classical Hamiltonian can be put into resonance basis using the transformation

$$\overline{h}_{\pm} = \frac{1}{\sqrt{\beta}}x \pm i\sqrt{\beta}p$$

$$\text{where} \quad \begin{cases} \beta &= \dfrac{1}{m\omega} \\ \mu &= \omega \end{cases}$$

$$\Downarrow$$

$$H = \mu\overline{h}_{+}\overline{h}_{-} = \omega\overline{h}_{+}\overline{h}_{-}. \tag{9.25}$$

This particular factorization of the classical Hamiltonian also works for the quantum case provided that we take into account the lack of commutativity between the x and p operators. In this case, one gets quantum mechanically the result

$$H = \frac{\omega}{2}\left\{\hbar + \overline{h}_{-}\overline{h}_{+}\right\} \equiv \frac{\omega}{2}\left\{\hbar + a^{\dagger}a\right\}. \tag{9.26}$$

One recognizes in Equation (9.26) the lowering and raising operators. It is easy to show that

$$\begin{aligned} \left\{a, a^{\dagger}\right\} &= 2\hbar \\ \left\{H, a\right\} &= -\hbar\omega a \\ \left\{H, a^{\dagger}\right\} &= \hbar\omega a^{\dagger}. \end{aligned} \tag{9.27}$$

Exercise 9-3

Consult a book on quantum mechanics if you have difficulties. The solutions to these exercises can be found in reference [47].

1. *Show that the expectation value of the energy is positive.*
2. *Using the properties of the operator a, show that there must exist a ground state Ψ_0 such that $a\Psi_0 = 0$. Thus the eigenvalue of the ground state is $\frac{\hbar\omega}{2}$.*
3. *Show that*

$$Ha^{\dagger^n}\Psi_0 = (n + \frac{1}{2})\hbar\omega a^{\dagger^n}\Psi_0. \tag{9.28}$$

4. *By deriving a relation between the norm of $\langle a^{\dagger^n}\Psi_0 \mid a^{\dagger^n}\Psi_0 \rangle$ and $\langle a^{\dagger^{n-1}}\Psi_0 \mid a^{\dagger^{n-1}}\Psi_0 \rangle$, show that if $\langle \Psi_0 \mid \Psi_0 \rangle = 1$ then the set of eigenvectors*

$$\Psi_n = \frac{1}{\sqrt{n!}}\left(\frac{a^\dagger}{\sqrt{2\hbar}}\right)^n \Psi_0 \tag{9.29}$$

is orthonormal.

9.1.3.3 Quantum perturbation theory

Consider a nonlinear perturbation added to the Hamiltonian H:

$$H(\epsilon) = H_0 + \epsilon 16x^4. \tag{9.30}$$

Following our classical rules, we express this Hamiltonian in terms of "Floquet" variables:

$$H(\epsilon) = \frac{\omega}{2}\left\{\hbar + a^\dagger a\right\} + \epsilon\beta^2\left(a + a^\dagger\right)^4. \tag{9.31}$$

As in the classical case, let us consider a canonical transformation of order ϵ on the Hamiltonian

$$\begin{aligned} H_1(\epsilon) &= \exp(\epsilon F_1)\, H(\epsilon) \exp(-\epsilon F_1) \\ &= \exp(\epsilon\ \mathrm{ad}F_1)\, H(\epsilon) \\ &= H_0 + \frac{\epsilon\omega}{2}\mathrm{ad}F_1 a^\dagger a + \epsilon\beta^2\left(a + a^\dagger\right)^4 + O(\epsilon^2)\cdots \end{aligned} \tag{9.32}$$

The operator $\mathrm{ad}F_1$ is the adjoint of the operator F_1 and simply takes the commutator of F_1 with the operator on which it acts. Hence, we rewrite Equation (9.32) as

$$\frac{\omega}{2}\left\{F_1, a^\dagger a\right\} + \beta^2\left(a + a^\dagger\right)^4 = \lambda_0 + \lambda_1 a^\dagger a + \lambda_2\left(a^\dagger a\right)^2$$
$$\lambda_0, \lambda_1, \lambda_2 \in \mathbf{R}. \tag{9.33}$$

Our goal is to select the canonical transformation so as to make the Hamiltonian a function of the "number" operator $a^\dagger a$ where $a^\dagger a \Psi_n = 2n\Psi_n$. Using the commutation relation $\left\{a, a^\dagger\right\} = 2\hbar$, we obtain the result

$$\begin{aligned} \lambda_0 &= \beta^2 12\hbar^2 \\ \lambda_1 &= \beta^2 12\hbar \\ \lambda_2 &= \beta^2 6 \end{aligned} \tag{9.34}$$

and thus the normalized Hamiltonian is

$$\begin{aligned} H_1(\epsilon) = \frac{\omega\hbar}{2} &+ \epsilon\beta^2 12\hbar^2 + \left\{\epsilon\beta^2 12\hbar\right\} a^\dagger a \\ &+ \epsilon\beta^2 6\left(a^\dagger a\right)^2 + O(\epsilon^2)\cdots \end{aligned} \tag{9.35}$$

The Hamiltonian $H_1(\epsilon)$ is fully diagonal when acting on the original eigenmodes Ψ_n as defined by Equation (9.29). One notices that the insertion of a quartic force causes

1. a shift of the ground state,
2. a modification of the quadratic term also,
3. and finally, as in the classical case, a nonlinear effect similar to the tune shift with amplitude.

The first two effects are purely quantum mechanical: the Lie algebras are different once we include nonlinear terms. A quadratic perturbation of the harmonic oscillator such as x^2 would modify the level of the ground state (a purely quantum effect), however the solution would still be in a one-to-one correspondence with the classical solution.

Exercise 9-4

1. *Show that the normalized* Ψ_0 *is*

$$\Psi_0(x) = \left(\frac{1}{\beta\hbar\pi}\right)^{1/4} \exp\left(-\frac{x^2}{2\beta\hbar}\right). \tag{9.36}$$

2. *Using the recursive formula for* Ψ_n *compute* Ψ_1 *and show that*

$$\begin{aligned}\langle\Psi_0|\epsilon 16x^4|\Psi_0\rangle &= \int_{-\infty}^{\infty} \epsilon 16x^4 \Psi_0(x)^2 dx = \epsilon\beta^2 12\hbar^2 \\ \langle\Psi_1|\epsilon 16x^4|\Psi_1\rangle &= \epsilon\beta^2 60\hbar^2.\end{aligned} \tag{9.37}$$

3. *Check that it agrees with the eigenvalues of* $H_1(\epsilon)$ *given by*

$$E_n = \omega\hbar(n+\frac{1}{2}) + \epsilon\beta^2\hbar^2(12+24n+24n^2). \tag{9.38}$$

9.2 EXAMPLES: SECOND ORDER NORMALIZATION

Power series calculations can be carried out with TPSA to an arbitrary order with an arbitrary number of parameters and are thus very useful in tracking codes. Analytical calculations are more complex and must be avoided whenever possible. We have said at the beginning of this chapter that it is typical to launch oneself into complex and unnecessary analytical calculations while it is possible to use a code and Lie library based on TPSA to perform the same tasks. With this warning in mind, we will look into simple second order analytical calculations to illustrate the rules of analytical perturbation theory. One of the great virtues of map methods in rings resides in the use of a formalism which integrates linear and nonlinear calculations. We will first display this feature explicitly on a tune shift calculation to second order.

Finally, in Sect. 9.2.4, using the rules of analytic perturbation theory learned in Chapter 8, we will upgrade the result of the one-resonance normalization obtained previously to an arbitrary system of more than two sextupole kicks.

We hope that these examples will illustrate the power of map methods for analytic calculations.

9.2.1 Tune Shifts to ϵ^2

We start by assuming that the map is in Floquet variables and all the perturbations are properly phase advanced at the beginning of the ring. Hence, the one-turn map of this system is given by

$$\begin{aligned} Ring &= \left\{\mathcal{R}_{01}, e^{:-\epsilon ds \mathcal{A}_1 V_1:}, \mathcal{R}_{12}, e^{:-\epsilon ds \mathcal{A}_2 V_2:}, \mathcal{R}_{20}\right\} \\ &\Downarrow \\ \mathcal{M}_0 &= e^{:\epsilon V_1^\dagger:} e^{:\epsilon V_2^\dagger:} \mathcal{R}, \end{aligned} \tag{9.39}$$

where

$$V_i^\dagger = -ds\mathcal{R}_{0i}\mathcal{A}_i V_i. \tag{9.40}$$

To second order we lump the perturbations into one Lie exponent

$$\begin{aligned} \mathcal{M}_0 &\overset{O(\epsilon^3)}{=} \exp\left(:\epsilon V^\dagger + \frac{\epsilon^2}{2}\left[V_1^\dagger, V_2^\dagger\right]:\right)\mathcal{R}, \\ \text{where } V^\dagger &= V_1^\dagger + V_2^\dagger. \end{aligned} \tag{9.41}$$

The next step consists in applying a canonical transformation $\mathcal{A}_1 = \exp\ (:\epsilon F:)$ to the map of Equation (9.41). We separate $V^\dagger$ into a sum of elements of the kernel and range of $\mathcal{R} - \mathcal{I}$

$$\begin{aligned} \mathcal{M}_1 &= e^{:\epsilon F:}\mathcal{M}_0 e^{-:\epsilon F:} \\ &\overset{O(\epsilon)}{=} \exp\left(\epsilon : K^\dagger + R^\dagger + F - \mathcal{R}F :\right)\mathcal{R}, \\ \text{where } V^\dagger &= \underbrace{K^\dagger}_{\text{Kernel}} + \underbrace{R^\dagger}_{\text{Range}}. \end{aligned} \tag{9.42}$$

This can be solved as usual to eliminate the operator in the range

$$F = (\mathcal{R} - \mathcal{I})^{-1} R^\dagger. \tag{9.43}$$

Using this result, we carry the calculation of $\mathcal{M}_1$ to second order in ϵ. The result is

$$\mathcal{M}_1 = \exp\left(:\epsilon K^\dagger + \frac{\epsilon^2}{2}\left\{[V_1^\dagger, V_2^\dagger] + [F + K^\dagger, R^\dagger]\right\} + \cdots :\right)\mathcal{R}. \tag{9.44}$$

Our goal is to compute the tune shift with amplitude and thus we must, in theory, remove from Equation (9.44) the terms in ϵ^2 belonging to the range. (See Exercise 5-3 where the same thing was carried out in the one-resonance example). Since we are mostly interested in the tune shift formula, let us first identify the tune shift terms as we did in Exercise 5-3. Rewriting $V_k^\dagger$ as $R_k^\dagger + K_k^\dagger$, the Lie exponent in (9.44) becomes

$$\epsilon K^\dagger + \frac{\epsilon^2}{2}\left\{\left[R_1^\dagger, R_2^\dagger\right] + \left[R_1^\dagger, K_2^\dagger\right] + \left[R_2^\dagger, K_1^\dagger\right] + \left[F + K^\dagger, R^\dagger\right]\right\}$$

and one concludes that tune shift can only originate from

$$\epsilon K^\dagger + \frac{\epsilon^2}{2}\left\{\left[R_1^\dagger, R_2^\dagger\right] + \left[F, R^\dagger\right]\right\}.$$

To proceed further one needs to express these quantities in action-angle variables (i.e., resonance basis). Starting with $R_k^\dagger$, one gets:

$$R_k^\dagger = \sum_{\vec{m}} X_{\vec{m}}^k(\vec{J}) \exp\left(i\vec{m}\cdot\vec{\phi}\right) \tag{9.45}$$

which implies that

$$\begin{aligned} F_k^\dagger &= (\mathcal{I} - \mathcal{R})^{-1} R_k^\dagger \\ &= \sum_{\vec{m}} \frac{1}{\exp\left(i\vec{m}\cdot\vec{\mu}\right) - 1} X_{\vec{m}}^k(\vec{J}) \exp\left(i\vec{m}\cdot\vec{\phi}\right) \\ &= \sum_{\vec{m}} Y_{\vec{m}}^k(\vec{J}) \exp\left(i\vec{m}\cdot\vec{\phi}\right). \end{aligned} \tag{9.46}$$

In many calculations the tune $\vec{\mu}$ is a constant because one uses the linear map as the zeroth order propagator. However here we will not yet assume linearity. Examining the bracket $\frac{\epsilon^2}{2}\left[R_1^\dagger, R_2^\dagger\right]$ in action-angle variables, one gets

$$\frac{\epsilon^2}{2}\left[R_1^\dagger, R_2^\dagger\right] = \frac{\epsilon^2}{2}\sum_{\vec{m},\vec{m}'} i\left\{X_{\vec{m}}^1 \vec{m}\cdot\partial_{\vec{J}} X_{\vec{m}'}^2 - X_{\vec{m}'}^2 \vec{m}'\cdot\partial_{\vec{J}} X_{\vec{m}}^1\right\} e^{i(\vec{m}+\vec{m}')\cdot\vec{\phi}}$$

$$\Downarrow$$

$$\text{tune term} = \frac{\epsilon^2}{2}\sum_{\vec{m}} i\vec{m}\cdot\partial_{\vec{J}}\left\{X_{\vec{m}}^1 X_{-\vec{m}}^2\right\}. \tag{9.47}$$

The entire result can be written as

$$\text{tune term} = \frac{\epsilon^2}{2}\sum_{\vec{m}} i\vec{m}\cdot\partial_{\mathcal{J}}\left\{X^1_{\vec{m}}X^2_{-\vec{m}}\right\}$$

$$+\frac{\epsilon^2}{2}\sum_{\substack{k=1,2\\k'=1,2}}\sum_{\vec{m}} i\vec{m}\cdot\partial_{\mathcal{J}}\left\{Y^k_{\vec{m}}X^{k'}_{-\vec{m}}\right\}. \tag{9.48}$$

Using our rules, we can extend this result to an arbitrary distribution. In the first term, the indices 1 and 2 appear asymmetrically and thus are interpreted as time-ordering labels. The final result is

$$\text{tune term} = \frac{\epsilon^2}{2}\sum_{\substack{\vec{m}\\k<k'}} i\vec{m}\cdot\partial_{\mathcal{J}}\left\{X^k_{\vec{m}}X^{k'}_{-\vec{m}}\right\}$$

$$+\frac{\epsilon^2}{2}\sum_{\substack{k=1,N\\k'=1,N}}\sum_{\vec{m}} i\vec{m}\cdot\partial_{\mathcal{J}}\left\{Y^k_{\vec{m}}X^{k'}_{-\vec{m}}\right\}. \tag{9.49}$$

Of course, the continuous result of Hamiltonian theory can be obtained by replacing sums with integrals.

In this calculation we have not assumed linearity of the unperturbed problem. It is traditional to do so in most calculations. The resulting simplified expression is:

$$\text{tune term} = \frac{\epsilon^2}{2}\sum_{\substack{\vec{m}\\k<k'}} i\vec{m}\cdot\partial_{\mathcal{J}}\left\{X^k_{\vec{m}}X^{k'}_{-\vec{m}}\right\} \tag{9.50}$$

$$+\frac{\epsilon^2}{4}\sum_{\vec{m}}\cot\left(\frac{\vec{m}\cdot\vec{\mu}}{2}\right)\sum_{\substack{k=1,N\\k'=1,N}}\vec{m}\cdot\partial_{\mathcal{J}}\left\{X^k_{\vec{m}}X^{k'}_{-\vec{m}}\right\}.$$

Now is the time for examples!

9.2.2 Tune shifts due to Normal Quadrupoles

In this case the function $V_i^\dagger$ can be written as

$$V_s^\dagger = -ds\mathcal{R}_{0s}\mathcal{A}_s\underbrace{V_s}_{k_s^Q\frac{x^2-y^2}{2}}$$

$$= -dsk_s^Q\mathcal{R}_{0s}\frac{\left(\sqrt{\beta_x^s}x+\eta_x^s\delta\right)^2-\beta_y^s y^2}{2} \tag{9.51}$$

$$= -dsk_s^Q \frac{\left(\sqrt{2\beta_x^s J_x}\frac{e^{i\phi_x^s}+e^{-i\phi_x^s}}{2} + \eta_x^s\delta\right)^2 - 2\beta_y^s J_y\left(\frac{e^{i\phi_y^s}+e^{-i\phi_y^s}}{2}\right)^2}{2}$$

where $\begin{cases} \phi_x^s = \mathcal{R}_{0s}\phi_x = \phi_x + \mu_{0s;x} \\ \phi_y^s = \mathcal{R}_{0s}\phi_y = \phi_y + \mu_{0s;y} \end{cases}$.

TABLE III. Driving Terms $X_{(m,n)}$

(m,n)	$X_{(m,n)}$
$(2,0)$	$-\frac{1}{4}dsk_s^Q\beta_x^s J_x e^{i2\mu_{0s;x}}$
$(0,2)$	$\frac{1}{4}dsk_s^Q\beta_y^s J_y e^{i2\mu_{0s;y}}$
$(1,0)$	$-\frac{1}{2}dsk_s^Q\sqrt{2\beta_x^s J_x}\,\eta_x^s\delta e^{i\mu_{0s;x}}$
$(0,0)$	$-\frac{1}{2}dsk_s^Q\left\{\beta_x^s J_x - \beta_y^s J_y + \eta_x^{s2}\delta^2\right\}$

The first order tune shifts are generated by

$$K^\dagger = \sum_{s=1}^{N} -\frac{1}{2}dsk_i^Q(s)\left\{\beta_x(s)J_x - \beta_y(s)J_y + \eta_x^2(s)\delta^2\right\}, \quad (9.52)$$

where the sum can be replaced by an integral in the continuous (usual) limit. We can write down the first term of Equation (9.50)

$$\begin{aligned} &= \sum_{s<s'}\frac{i}{2}\left\{\frac{1}{4}dsds'k_s^Qk_{s'}^Q\beta_x^s\beta_x^{s'}\exp\left(i2\left\{\mu_{0s;x}-\mu_{0s';x}\right\}\right)J_x\right. \\ &+ \frac{1}{4}dsds'k_s^Qk_{s'}^Q\beta_y^s\beta_y^{s'}\exp\left(i2\left\{\mu_{0s;y}-\mu_{0s';y}\right\}\right)J_y \\ &+ \left.\frac{1}{2}dsds'k_s^Qk_{s'}^Q\sqrt{\beta_x^s\beta_x^{s'}}\,\eta_x^s\eta_x^{s'}\delta^2\exp\left(i\left\{\mu_{0s;x}-\mu_{0s';x}\right\}\right)\right\} \\ &+ \text{Complex Conjugate}\,. \end{aligned} \quad (9.53)$$

This first term can be written as

$$= \sum_{s<s'} \frac{1}{2} \left\{ -\frac{1}{2} ds ds' k_s^Q k_{s'}^Q \beta_x^s \beta_x^{s'} \sin\left(2\left\{\mu_{0s;x} - \mu_{0s';x}\right\}\right) J_x \right.$$

$$- \frac{1}{2} ds ds' k_s^Q k_{s'}^Q \beta_y^s \beta_y^{s'} \sin\left(2\left\{\mu_{0s;y} - \mu_{0s';y}\right\}\right) J_y$$

$$\left. - ds ds' k_s^Q k_{s'}^Q \sqrt{\beta_x^s \beta_x^{s'}}\ \eta_x^s \eta_x^{s'} \delta^2 \sin\left(\left\{\mu_{0s;x} - \mu_{0s';x}\right\}\right) \right\}. \quad (9.54)$$

Similarly for the second term, one can write

$$= \sum_{s,s'} \frac{1}{4} \left\{ \frac{1}{2} ds ds' \cot(\mu_x) k_s^Q k_{s'}^Q \beta_x^s \beta_x^{s'} \cos\left(2\left\{\mu_{0s;x} - \mu_{0s';x}\right\}\right) J_x \right.$$

$$+ \frac{1}{2} ds ds' \cot(\mu_y) k_s^Q k_{s'}^Q \beta_y^s \beta_y^{s'} \cos\left(2\left\{\mu_{0s;y} - \mu_{0s';y}\right\}\right) J_y \quad (9.55)$$

$$\left. + ds ds' k_s^Q k_{s'}^Q \cot\left(\frac{\mu_x}{2}\right) \sqrt{\beta_x^s \beta_x^{s'}}\ \eta_x^s \eta_x^{s'} \delta^2 \cos\left(\left\{\mu_{0s;x} - \mu_{0s';x}\right\}\right) \right\}.$$

Finally one notices that these two summations can be conveniently lumped into one summation

$$\Delta\mu_{x,y} = -\frac{\partial}{\partial J_{x,y}} \sum_{s \le s'} \frac{ds ds' k_s^Q k_{s'}^Q (2 - \delta_{ss'})}{4}$$

$$\times \left\{ \frac{1}{2} \beta_x^s \beta_x^{s'} \left[\frac{1}{\sin(\mu_x)} \cos\left(2\left\{\mu_{0s;x} - \mu_{0s';x}\right\} + \mu_x\right) \right] J_x \right.$$

$$+ \frac{1}{2} \beta_y^s \beta_y^{s'} \left[\frac{1}{\sin(\mu_y)} \cos\left(2\left\{\mu_{0s;y} - \mu_{0s';y}\right\} + \mu_y\right) \right] J_y \quad (9.56)$$

$$\left. + \sqrt{\beta_x^s \beta_x^{s'}}\ \eta_x^s \eta_x^{s'} \delta^2 \frac{2}{\sin\left(\frac{\mu_x}{2}\right)} \cos\left(\left\{\mu_{0s;x} - \mu_{0s';x}\right\} + \frac{\mu_x}{2}\right) \right\}.$$

Exercise 9-5

We revisit the "FODO" cell of Exercise 8-6. We recall that this system was given by the ring

$$\begin{array}{ll} & Cell_6 = \{\mathcal{Q}_k, \mathcal{D}_L, \mathcal{Q}_{-k}, \mathcal{Q}_{-k}, \mathcal{D}_L, \mathcal{Q}_k\} \\ where & \begin{array}{l} \mathcal{Q}_k = \exp\left(: -\frac{k}{2}\ x^2 :\right) \\ \mathcal{D}_L = \exp\left(: -\frac{L}{2}\ p^2 :\right) \end{array} \end{array} . \quad (9.57)$$

In this exercise, we consider a perturbation of this cell at each quadrupole.

1. *Write down the rings which a second order calculation must use. First the conceptual ring of the analytic calculation and then the ring inside the tracking code which a TPSA calculation uses to feed the unperturbed propagators and observables (lattice functions etc.) to the analytical formulas. The answers are for the analytic and TPSA calculation respectively*

$$Cell_{2+2} = \{\mathcal{M}_{0s}, \mathcal{Q}_{k_s}, \mathcal{M}_{ss'}, \mathcal{Q}_{k_{s'}}, \mathcal{M}_{s'2}\},$$

and

$$\begin{aligned} Cell_2 &= \{\mathcal{M}_{01}, \mathcal{M}_{12}\} \\ \text{where } \mathcal{M}_{01} &= \mathcal{Q}_k \mathcal{D}_L \mathcal{Q}_{-k}, \\ \mathcal{M}_{12} &= \mathcal{Q}_{-k} \mathcal{D}_L \mathcal{Q}_k. \end{aligned}$$

In this case, the rings are similar because the analytic formula is applied at two locations only.

2. *Compute numerically (using any method at your disposal) the zeroth order information provided by $Cell_2$. Answer:*

Position	β_x	β_y	$\nu_{0s;x}$	$\nu_{0s;y}$	η_x
1	*0.65690*	*3.3609*	*0.11750*	*0.11750*	*0.00000*
2	*3.3609*	*0.65690*	*0.23500*	*0.23500*	*0.00000*
Total Tunes			*0.23500*	*0.23500*	

3. *Using the above information and Equation* (9.56) *compute the tunes as a function of the changes in the first and second quadrupoles.*

$$\begin{aligned} \Delta\nu_x &= 0.26745 k_1 + 0.052274 k_2 \\ &\quad -0.021244 k_1^2 - 0.088237 k_1 k_2 - 0.81155\ 10^{-3} k_2^2 \\ \Delta\nu_y &= -0.052274 k_1 - 0.26745 k_2 \\ &\quad -0.81155\ 10^{-3} k_1^2 - 0.088237 k_1 k_2 - 0.021244 k_2^2 \\ \Delta\ell &= 0. \end{aligned} \tag{9.58}$$

The answers to Exercise 9-5 where obtained in two different ways by the author: by plugging the values of zeroth order calculation (done with TPSA) into the formulas, and by doing a TPSA calculation of the tunes which includes the quadrupole strengths as additional parameters as was explained for example in the algorithm of (8.10).

The advantages and disadvantages of both approaches should now be clear to the reader. Analytical results are more difficult to derive. The quality of the result is higher but the pain to get it is also far greater. In a TPSA calculation the zeroth order ring (i.e., what is in the tracking code before parameters are introduced) is as complex as the tracking code permits.

Exercise 9-6

For the quadrupoles, we expressed the tune shift formula in one sum. This result is good for any perturbation potential which depends on position only. In this exercise, prove that the tune term can be written as

$$\begin{aligned}
\text{tune term} &= \sum_{\{\vec{m}_>\},s\le s'} \frac{(2-\delta_{ss'})}{2}\vec{m}\cdot\partial_J\left(\chi^s_{\vec{m}}\chi^{s'}_{\vec{m}}\right)\left\{\frac{\cos\left(\vec{m}\cdot\left[\vec{\mu}_{0s}-\vec{\mu}_{0s'}+\frac{1}{2}\vec{\mu}\right]\right)}{\sin\left(\frac{\vec{m}\cdot\vec{\mu}}{2}\right)}\right\} \\
\textit{where} \quad & \{\vec{m}_>\} = \textit{Half the resonances} \\
\textit{and} \quad & X^s_{\vec{m}} = \chi^s_{\vec{m}} \exp(i\vec{m}\cdot\vec{\mu}_{0s}).
\end{aligned} \tag{9.59}$$

9.2.3 Tune shifts due to Normal Sextupoles

Map methods emphasize the formal similarity between a linear and a nonlinear calculation. Just to drive the point home we repeat verbatim the next section in a "fill-in-the-blank" mode.

In this case the function $V_s^\dagger$ can be written as

$$\begin{aligned}
V_s^\dagger &= -ds\mathcal{R}_{0s}\mathcal{A}_s \underbrace{V_s}_{k_s^S\frac{x^3-3xy^2}{3}} \\
&= -dsk_s^S\mathcal{R}_{0s}\frac{\left(\sqrt{\beta_x^s}x+\eta_x^s\delta\right)^3-3\sqrt{\beta_x^s\beta_y^{s2}}xy^2}{3} \\
&= \frac{-dsk_s^S}{3}\left\{\left(\sqrt{2\beta_x^s J_x}\frac{e^{i\phi_x^s}+e^{-i\phi_x^s}}{2}+\eta_x\delta\right)^3\right. \\
&\qquad \left. -3\sqrt{8\beta_x^s\beta_y^{s2}}J_x^{1/2}J_y\left(\frac{e^{i\phi_x^s}+e^{-i\phi_x^s}}{2}\right)\left(\frac{e^{i\phi_y^s}+e^{-i\phi_y^s}}{2}\right)^2\right\}
\end{aligned} \tag{9.60}$$

$$\text{where} \quad \left\{\begin{array}{lllll} \phi_x^s & = & \mathcal{R}_{0s}\phi_x & = & \phi_x+\mu_{0s;x} \\ \phi_y^s & = & \mathcal{R}_{0s}\phi_y & = & \phi_y+\mu_{0s;y} \end{array}\right. .$$

From this equation, we extract the function χ of Exercise 9-6:

TABLE IV. Driving Terms $\chi_{(m,n)}$

(m,n)	$\chi_{(m,n)}$
$(3,0)$	$-dsK_s^S\frac{\sqrt{2}}{12}\beta_x^{s\,3/2}J_x^{3/2}$
$(1,0)$	$-dsK_s^S\frac{\sqrt{2}}{2}\sqrt{\beta_x^s J_x}\left\{\frac{1}{2}\beta_x^s J_x-\beta_y^s J_y+\eta_x^{s\,2}\delta^2\right\}$
$(1,2)$	$dsK_s^S\frac{\sqrt{2}}{4}\sqrt{\beta_x^s J_x}\beta_y^s J_y$
$(-1,2)$	$dsK_s^S\frac{\sqrt{2}}{4}\sqrt{\beta_x^s J_x}\beta_y^s J_y$
$(2,0)$	$-dsK_s^S\frac{1}{2}\beta_x^s J_x\eta_x^s\delta$
$(0,2)$	$dsK_s^S\frac{1}{2}\beta_y^s J_y\eta_x^s\delta$
$(0,0)$	$-dsK_s^S\left(\eta_x^s\delta\left\{\beta_y^s J_y-\beta_x^s J_x\right\}+\frac{1}{3}\eta_x^{s\,3}\delta^3\right)$

Using table IV and the result of Exercise 9-6, one gets for tune term:

$$
\begin{aligned}
=\Bigg\{&\frac{1}{8}F_{3,0}^{s,s'}\beta_x^{s\,3/2}\beta_x^{s'\,3/2}J_x^2\\
&+F_{1,0}^{s,s'}\Bigg\{\frac{3}{8}\beta_x^{s\,3/2}\beta_x^{s'\,3/2}J_x^2+\frac{1}{2}\beta_x^{s\,1/2}\beta_x^{s'\,1/2}\beta_y^s\beta_y^{s'}J_y^2\\
&-\frac{1}{2}\left[\beta_x^{s\,3/2}\beta_x^{s'\,1/2}\beta_y^{s'}+\beta_x^s\beta_x^{s'\,3/2}\beta_y^s\right]J_xJ_y\\
&+\frac{1}{2}\left[\beta_x^{s\,3/2}\beta_x^{s'\,1/2}\eta_x^{s'\,2}+\beta_x^{s\,1/2}\beta_x^{s'\,3/2}\eta_x^{s\,2}\right]\delta^2J_x\\
&-\frac{1}{2}\beta_x^{s\,1/2}\beta_x^{s'\,1/2}\left[\beta_y^s\eta_x^{s'\,2}+\beta_y^{s'}\eta_x^{s\,2}\right]\delta^2J_y+\frac{1}{2}\beta_x^{s\,1/2}\beta_x^{s'\,1/2}\eta_x^{s\,2}\eta_x^{s'\,2}\delta^4\Bigg\}\\
&+F_{2,0}^{s,s'}\beta_x^s\beta_x^{s'}\eta_x^s\eta_x^{s'}\delta^2J_x+F_{1,2}^{s,s'}\beta_x^{s\,1/2}\beta_x^{s'\,1/2}\beta_y^s\beta_y^{s'}\left\{\frac{1}{8}J_y^2+\frac{1}{2}J_xJ_y\right\}\\
&+F_{-1,2}^{s,s'}\beta_x^{s\,1/2}\beta_x^{s'\,1/2}\beta_y^s\beta_y^{s'}\left\{-\frac{1}{8}J_y^2+\frac{1}{2}J_xJ_y\right\}\\
&+F_{0,2}^{s,s'}\beta_y^s\beta_y^{s'}\eta_x^s\eta_x^{s'}\delta^2J_y\Bigg\}\,ds\,ds'K_s^SK_{s'}^S.
\end{aligned}
\tag{9.61}
$$

The form factor $F_{\vec{m}}^{s,s'}$ is given by

$$F_{\vec{m}}^{s,s'} = \frac{(2-\delta_{ss'})}{2} \frac{\cos\left(\vec{m}\cdot\left[\vec{\mu}_{0s}-\vec{\mu}_{0s'}+\frac{1}{2}\vec{\mu}\right]\right)}{\sin\left(\frac{\vec{m}\cdot\vec{\mu}}{2}\right)}. \tag{9.62}$$

Exercise 9-7

The general formula for the tune shifts all have a bizarre asymmetry, namely the summation is carried over the domain $s \leq s'$. Because the tune shift is position independent, one ought to be able to symmetrize Equation (9.59). Show that the tune shift term can be written as

$$\sum_{\{\vec{m}_>\},s,s'} \frac{(2-\delta_{ss'})}{2}\vec{m}\cdot\partial_{\mathcal{J}}\left(\chi_{\vec{m}}^{s}\chi_{\vec{m}}^{s'}\right)\left\{\frac{\cos\left(\vec{m}\cdot\left[\frac{1}{2}\vec{\mu}-\Delta\vec{\mu}_{s\to s'}\right]\right)}{\sin\left(\frac{\vec{m}\cdot\vec{\mu}}{2}\right)}\right\}$$

$$\textit{where} \quad \{\vec{m}_>\} = \textit{Half the resonances.} \tag{9.63}$$

The phase difference $\Delta\vec{\mu}_{s\to s'}$ is always the growth in phase from position s to position s' in the direction of propagation. This phase advance is obviously independent of the reference position chosen (i.e., the $s=0$ location).

9.2.4 Upgrading the 1/3 Integer Resonance

We have said earlier that the "pet" problem which consisted in normalizing Equation (5.31) using a resonant normal form was more general than it looked. Following our rules of second order normalization, it follows that, in one degree of freedom, one must consider the following sextupole map:

$$\begin{aligned}\mathcal{M} = \mathcal{R}_{0s}\exp&\left(:-k_s\beta_s^{3/2}\frac{1}{3}x^3-\frac{\beta_s^2}{4}k_{o;s}x^4:\right)\\ &\times\mathcal{R}_{ss'}\exp\left(:-k_{s'}\beta_{s'}^{3/2}\frac{1}{3}x^3:\right)\mathcal{R}_{s'\aleph}.\end{aligned} \tag{9.64}$$

Our goal is to derive the functional form of the invariant h_c by generalizing Equation (5.47). This can be done by dimensional analysis and redefinition of the various constants. We first notice that

$$\begin{array}{ccc} k_s & \longrightarrow & -\frac{k_{s'}\beta_{s'}^{3/2}}{24} \\ 1 & \xrightarrow{\text{Dimensional analysis}} & \frac{k_s\beta_s^{3/2}}{24} \\ k_o & \longrightarrow & -\frac{k_{o;s}\beta_s^2}{64} \end{array} \tag{9.65}$$

We start with the leading order term in sextupole strengths:

$$h_{c3}^3 = \sum_s \frac{i\frac{1}{12}\pi\epsilon k_s\beta_s^{3/2}e^{i3\Delta\mu_{s\to\aleph}}}{1-\exp(i2\pi\epsilon)}r_+^3 + \frac{-i\frac{1}{12}\pi\epsilon k_s\beta_s^{3/2}e^{-i3\Delta\mu_{s\to\aleph}}}{1-\exp(-i2\pi\epsilon)}r_-^3. \tag{9.66}$$

Then, we must compute the tune shift term h_{c4}^0 which is more complex:

$$\begin{aligned} h_{c4}^0 = \sum_{s\le s'} J^2 \Bigg(& \frac{1}{16}\left\{\cot(\pi\epsilon) - \frac{2\pi\epsilon}{1-\cos(2\pi\epsilon)}\right\} \\ & \times\left\{k_s^2\beta_s^3 + 2k_sk_{s'}\beta_s^{3/2}\beta_{s'}^{3/2}\cos(3\Delta\mu_{s\to s'})\right\} \\ & +\frac{3}{16}\cot\left(\frac{\mu}{2}\right)\left\{k_s^2\beta_s^3 + 2k_sk_{s'}\beta_s^{3/2}\beta_{s'}^{3/2}\cos(\Delta\mu_{s\to s'})\right\} \\ & +\frac{1}{8}\Big\{-3k_{o;s}\beta_s^2 + k_sk_{s'}\beta_s^{3/2}\beta_{s'}^{3/2}\sin(3\Delta\mu_{s\to s'}) \\ & +3k_sk_{s'}\beta_s^{3/2}\beta_{s'}^{3/2}\sin(\Delta\mu_{s\to s'})\Big\}\Bigg). \end{aligned} \tag{9.67}$$

Exercise 9-8 (Other Observables)

The theory does not restrict us to tune shifts. We can also compute observables such as $< x^2 >$. In a midplane symmetric system this is equal to the Courant-Snyder invariant times the beta function. Using Equation (2.63) compute the beta function change due to a sextupole in the leading order, i.e., $\frac{d\beta}{d\delta}$.

1. *Using the map $\mathcal{M}_{0s}\exp\left(:-\frac{k_s}{3}(x^3-3xy^2):\right)\mathcal{M}_{s\aleph}$ as prescribed by our rules, show that it is diagonalized by*

$$\mathcal{A} = \exp(:F:)\exp(:\delta F_\delta:)\exp\left(:\delta^2 F_{\delta^2}:\right)\mathcal{A}_0$$

where (9.68)

$$\mathcal{A}_0 \equiv \textit{Courant-Snyder Transformation}$$

$$F_\delta = k_s\eta_{s;x}\left\{\frac{\beta_{s;x}e^{-i2\mu_{0s;x}}r_{1+}^2}{4(1-e^{-i2\mu_x})} - \frac{\beta_{s;y}\,e^{-i2\mu_{0s;y}}r_{2+}^2}{4(1-e^{-i2\mu_y})}\right\} + \begin{matrix}\textit{Complex}\\ \textit{Conjugate}\end{matrix}$$

$$F_{\delta^2} = \frac{-k_s\eta_{s;x}^2\sqrt{\beta_{s;x}}\,e^{-i\mu_{0s;x}}r_{1+}}{2(1-e^{-i\mu_x})} + \begin{matrix}\textit{Complex}\\ \textit{Conjugate}\end{matrix}$$

$$F = \frac{k_s}{8}\left\{\beta_{s;x}^{3/2}\left[\frac{e^{-i3\mu_{0s;x}}r_{1+}^3}{3(1-e^{-i3\mu_x})} + \frac{e^{-i\mu_{0s;x}}r_{1+}^2r_{1-}}{(1-e^{-i\mu_x})}\right]\right.$$
$$-\beta_{s;x}^{1/2}\beta_{s;y}\left[\frac{e^{-i\mu_{0s;x}}r_{1+}r_{2+}r_{2-}}{(1-e^{-i\mu_x})} - \frac{e^{-i\mu_{0s;x}-i2\mu_{0s;y}}r_{1+}r_{2+}^2}{(1-e^{-i\mu_x-i2\mu_y})}\right.$$
$$\left.\left.-\frac{e^{-i\mu_{0s;x}+i2\mu_{0s;y}}r_{1+}r_{2-}^2}{(1-e^{-i\mu_x+i2\mu_y})}\right] + \begin{matrix}\textit{Complex}\\ \textit{Conjugate}\end{matrix}\right\}.$$

2. *Show that the average of x^2 at position 0 around the δ-dependent fixed point is given by*

$$\langle x^2\rangle = \beta_{0;x}\left\{1 - \sum_s \frac{\delta}{2}\frac{\cos(2\mu_{os;x}-\mu_x)}{\sin(\mu_x)}\beta_{s;x}\eta_{s;x}k_s\right\}\varepsilon_x.$$

A similar expression holds for y^2:

$$\langle y^2\rangle = \beta_{0;y}\left\{1 + \sum_s \frac{\delta}{2}\frac{\cos(2\mu_{os;y}-\mu_y)}{\sin(\mu_y)}\beta_{s;y}\eta_{s;x}k_s\right\}\varepsilon_y.$$

In all the formulas, ε_x and ε_y are the δ dependent invariant of the linear maps which are just

$$\varepsilon_x = \left(\frac{x-\eta_x\delta}{\sqrt{\beta_x}}\right)^2 + \left(\sqrt{\beta_x}\left[p_x - \eta_x'\delta\right] + \frac{\alpha_x\left[x-\eta_x\delta\right]}{\sqrt{\beta_x}}\right)^2$$
$$\varepsilon_y = \left(\frac{y}{\sqrt{\beta_y}}\right)^2 + \left(\sqrt{\beta_y}p_y + \frac{\alpha_y y}{\sqrt{\beta_y}}\right)^2.$$

3. *There are no amplitude dependent contributions to the beam sizes due to sextupoles, however they do move the average position of the beam. Compute the average of x.*

$$\langle x\rangle = \sum_s \sqrt{\beta_{0;x}} k_s \frac{\cos\left(\mu_{0s;x} - \frac{1}{2}\mu_x\right)}{\sin\left(\frac{1}{2}\mu_x\right)} \times \left\{ \frac{\delta^2}{2}\beta_{s;x}^{1/2}\eta_{s;x}^2 - \frac{1}{4}\beta_{s;x}^{3/2}\varepsilon_x + \frac{1}{4}\beta_{s;x}^{1/2}\beta_{s;y}\varepsilon_y \right\}. \quad (9.69)$$

The part which does not depend on amplitude is the regular δ^2 dispersion. It is a non-dynamical effect. It is the change of the fixed point as a function of energy in a coasting beam normalization. The amplitude dependent terms are dynamical. A beam of finite size, on momentum ($\delta = 0$), appears shifted as if the fixed point moved.

9.3 CONCLUSION

The entire theory of rings and quasi-periodic systems has only been glanced at without a serious look at applications. They are several reasons for this lack of completeness. The author's own abilities are certainly restricted and his experience in the field of accelerators quite limited: it would be presumptuous on my part to go into a discourse on the design of accelerators. It is our experience that the methods presented here give increased computational and conceptual power to those willing to make the effort to learn them. The work of Irwin on Lie methods applied to accelerator modules shows how one can make quantum leaps by using the appropriate tools when faced with a complex design problem.

This book does not go into fancy nonlinear dynamics. Again it should be understood that our goal is to provide the language necessary to attack complex problems efficiently within the context of particle accelerators and rings in particular. Sometimes it is better to revert to simplistic models when trying to understand complex nonlinear phenomena. The emphasis of this book is at the other end: the phenomena are assumed to be simple but the system is a horrible s-dependent mess.

So, we are now ready for the "horrible s-dependent mess!" Now, let us dive into the local theory of accelerators.

PART III

Local Description: Means

CHAPTER 10
THE LAYOUT IN THE LABORATORY FRAME

Naïvely the word "local" seems to imply that we should jump immediately into the discussion of individual magnets and, by hook and crook, find a way to glue all these beautiful objects into our computer codes. Actually this is a mistake. Although we will look at the "trees" and even the "branches," first we must ask whether or not these will be used in a forest, a garden or on the side walk!

We must ask how magnets as distinct physical objects, junky pieces of metal, plastic and liquid, fit "organically" or "functionally" in our tracking code. In a tracking code the magnets will have a mathematical function or rôle to play: the clearer the rôle is defined, the cleaner will be the structure of the code, and the more efficient will be the use of modern object-oriented languages. Therefore a full understanding of the mathematical model (looking at the entire forest instead of just the bark) is a prerequisite for a successful modeling and implementation on the computer of both the local and global aspects of our theory.

Consider as an example the authorities of a country creating a database to police its citizens. In the computer sense we can ask if it makes sense to have a class of humans splitting immediately into male and female classes with different attributes? Clearly the answer to this question would have been completely different a few decades ago. Nowadays we can say that functionally both sexes operate similarly as far as politics are concerned. Fifty years ago one could have created two classes: an adult male class and a minor class that includes adult females for the simple reason that our civil codes treated women like children.

We can see that a proper definition of the objects and classes on the computer must reflect our understanding not only of the physical object itself (male, female, magnets,etc.) but also of their complex functional and organic inter-relations. Going back to magnets we say quite naturally that our purpose is to look at particles propagating through them. First let us define a magnet as a physical object producing an electro-magnetic field but surrounded by a region of free field: this is quite reasonable since a group of magnets with interacting fringe fields are single objects even from a magnet design point of view.

10.1 CONDITIONS FOR "MAGNET INDEPENDENCE"

We must ask a set of critical questions about the system we intend to study:

1. Under what conditions can a magnet be an object in a functional sense so that operations on a single magnet are well defined? This will lead to an object oriented implementation of the software.
2. If some phenomena must be "weeded out" from our description to allow magnets to become independent objects, is it fundamental or just the result of our poor understanding of the theory?
3. If it is fundamental, where do we put them back in?

The answer to question 1 is simple and easy to guess by anyone who has read the other parts of this book: single particle dynamics allows the definition of magnet-objects. By contrast collective effects *prevent* a coherent definition based on functionally independent magnets. The answer to question 2 is unfortunately yes. Collective effects present us with a fundamental problem that cannot be resolved. One has to decide at the onset whether the code structure is inherently collective (space charge, impedance effects, etc.) or single particle.

In a magnet-object code, we must be able to define things such as "propagation through a magnet" which is implicitly assumed when maps such as $\vec{\zeta}_{i\to j}$ were used in the previous chapters. Whether it is phase space, spin, or moments, it is important to realize that we take their values in a plane slicing the beam pipe in front of the magnet and produce the final values in a plane on the other side of the magnet. Implicitly we always assume that this knowledge is gotten from information pertaining to this magnet alone.

In collective phenomena, I say that this is a theoretical impossibility.

Consider a bunch made of two interacting particles. Suppose that particle 1 is in magnet a while particle 2 has entered magnet

b. Through the inter-particle force, particle 1 already sees the field of magnet b and vice-versa. If we put a smallness parameter $\varepsilon_{k\text{—}j}$ in front of all the interactions, then particle 1 will start seeing magnet b at the order $\varepsilon_{1\text{—}2}\varepsilon_{2\text{—}b}$. From a software design point of view, this would not be in itself a problem if all the forces were simple. What is a problem is its effect on the magnet-object definition in our code. It is not possible. The *absolutely* correct description of interacting particles in a bunch requires us to use time (of a clock) as the independent variable of propagation since the particles will interact on the basis of their relative positions and velocities at a given time. One cannot define propagation through magnet a without knowing magnet b. Therefore the magnet is no longer an object from a functional point of view. While one can still have object-oriented graphical routines ("a magnet knows how to draw itself"), dynamically a magnet cannot know how to propagate a distribution through itself. What remains is the Lagrangian for the whole ring and the forces between the particles.[10.1]

A similar situation occurs in the discussion of stochastic cooling of a coasting beam which extends longitudinally around the ring. What is an appropriate description of such an object and its interaction with the magnets? In this case it is convenient to describe the magnetic effects of the ring using Fourier transforms involving the periodic variable s. As we know, this approach has limitations and is not ideal for single particle dynamics. But, if the collective effects are important, we must compromise somewhere if one is to compute useful quantities on a computer.

Of course, if collective effects are introduced as perturbations, treating cross-terms between magnets and particle-particle forces in the lowest order, then collective effects can be added as "localized" magnet-like effects. The most famous phenomenon that is modeled in this manner is the beam-beam interaction. Even the strong-strong forces are often modeled by collisions of "pancakes" which are localized lenses based on the actual distribution. Therefore the class structure of single particle dynamics remains useful for the simulation of several important collective effects.

[10.1] There are models of the internal dynamics of magnets which are compatible with both "time" tracking and "s" tracking. However it remains a fact that the magnet-object paradigm must be discarded even in such codes. See reference [48] in which the author implements time tracking within the code "TEAPOT" [6].

The last paragraph partly answered question 3. But the question is still unanswered for "real collective effects." It turns out that the layout we will describe in this chapter can still be used to compute single particle quantities relevant to the collective calculation (like Fourier modes, phase advances), but during the calculation process it can only act as a convenient way to store the magnet information without any dynamical functionality. A different class, which probably creates the entire ring as a single object (decomposable into Fourier modes of some sort), would exist in parallel with its own set of propagators where layout and magnets cannot be decoupled. Of course this type of ring can communicate with a single particle type of ring only at the data level: coordinates and velocities of the individual particles of the distribution in the laboratory frame can be converted into coordinates in the layout frame. A superclass englobing the single particle description and the collective effects (and other things as well as) must exist above everything. We call it the Lagrangian class. The full Poincaré group operates on the members of this class and layouts are objects created in the Lagrangian class.

This applies also to our political analogy. Anyone who has had to deal with a multicultural environment can attest to the failure of generalizations. This can be seen daily in the marketing decision of tobacco companies in the United States and the opposing governmental agencies: they have a very complex representation of the public going beyond the male/female categories. In the end however, the result is the same: the persons buy or do not buy the product. The same thing is true for a particle which, through the course of its life, may be part of an interacting ensemble or a self-contained entity: its has six coordinates. The simplicity of the final states hides the potentially complex interaction mechanisms which are necessary to describe efficiently the evolution of the particle. And, in conclusion, attempting to lump all these mechanisms into one grand description (single, collective, quantum etc...) will lead to a model with restricted computational power. So let us go back to the task of exploiting to a maximum the theoretical framework allowed by single particle dynamics.

10.2 DISCOVERING THE LAYOUT

As Feynman said about the whole of physics and as we did in the discussion of the global theory, we must try to motivate the description of the layout from a small set of quintessential properties. Since it is our intention to look first at the forest, we can literally do so through simple questions: what does anyone see flying above a (transparent) ring?

One sees a tunnel that surrounds a beam pipe. Like a roller coaster, this pipe may or may not be in one plane. One notices a second important thing: there are "aggregates of junk" at various places in the ring surrounded by free sections of pipe. Third crucial point: propagation in the free section must be along straight lines (to a good approximation) because one never sees free section of pipes with kinks. And thus the free propagator must be a translationally invariant operator. (The earth magnetic field could be a problem in very low energy electron rings.)

So, just through a cursory analysis, we see that "aggregates of junk," which we will now call a magnet/block, can be thought to rest on arcs of circles linked smoothly to the entrance and exit beam pipes which are themselves straight. We can conjecture safely that magnets supported by arcs of circle must have the primary function of turning the beam around so as to ensure that it stays in the tunnel. We also notice that some magnets have parallel entrance and exit planes. These are straight blocks whose purpose is at this point mysterious.

This may be a surprise to the reader but we have enough information to start designing our layout including a description of the Euclidean group of translations and rotations as it acts on a magnet/block. In fact, we will describe the action of the Euclidean group in two situations: when the magnet/block is a graphical object (to be drawn by a graphical routine) and when it is a dynamical object, i.e., a map between two charts. In an object-oriented computer implementation, the layout is created in a class in which a generic description of space-time is possible. In fact, we call it the "Lagrangian" class. A complex (and complete) multi-particle description of a ring can coexist in the space time class with our single particle description and its associated propagators. However they can only exchange data at the level of a Lorentz frame such as the laboratory frame. With that in mind, we proceed with the description of the layout and its single particle operators.

10.3 THE ACTUAL LAYOUT

The laboratory sits in a Lorentz frame whose origin can be anywhere (Lagrangian or space time class). On Figure 10.1 we put it near the center of the ring and drew a planar ring for simplicity.

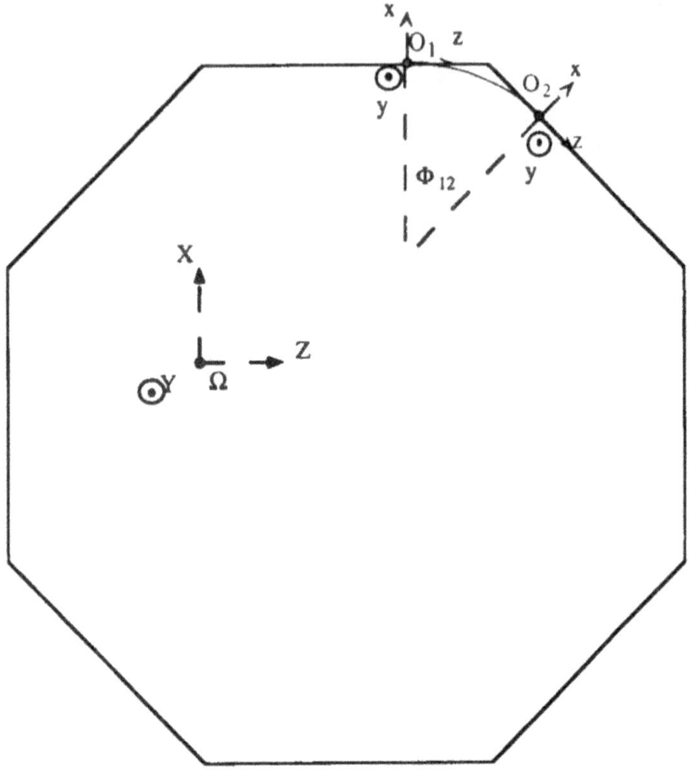

Figure 10.1: Layout of the Ring.

The ring is made out of arcs of circles joined smoothly to each other. In general arcs of circle can be rotated so as to bring the layout into the third dimension. The best visual equivalents are the newest roller coasters which are appearing all over the world. In our case the axis of the roller coaster is made out of straight lines and arcs of circles glued together.

In Figure 10.1 a layout block is displayed. Its entrance is at point O_1 and its exit at point O_2. Local Cartesian frames of reference (charts) are put around O_1 and O_2. In the dynamical implementation, it might be necessary to introduce "torsion." Torsion is a rotation in the local transverse $x - y$ plane which is necessary to connect the exit frame of a block with the entrance frame of the following block.

This is often necessary when dynamical observables are propagated from one chart to the next in a nonplanar layout. This rotation can be placed just before the entrance plane of a block. It is interesting to know that torsion, in rings, is unavoidable in general, and that in some cases involving only 90° horizontal and vertical bends we might still need it (see [13]).

The chart in configuration space, let us say at O_1, is generally defined in term of the basis $\left(O_1, \vec{b}_1^x, \vec{b}_1^y, \vec{b}_1^z, 1\right)$ of the space $(O_1, \mathbf{R}^3 \otimes \mathbb{R})$. The first vectors are related to space and the fourth entry is time. The point O_1 is thus a location in space and time. Quite often we may measure time relative to a fixed clock which, for example, can be the time of an ideal synchronous particle. More in tune with the common parlance, we can say that a particle has three coordinates plus time (x, y, z, t) when expressed in the basis[10.2] around O_1. Typically we will refer to the coordinates (x, y) as the transverse positions. Unfortunately the third coordinate used in accelerators, which is t, is often called the "longitudinal" coordinate instead of the temporal coordinate. In single particle dynamics, the third dependent variable is clearly the time t whether we call it longitudinal or temporal. The confusion arises because of our wish to unify single and collective effects. As I have said earlier, this is not really possible without losing the magnet as a mathematical object. However, it is often done anyway in treatments of collective effects where single particle dynamics is sufficiently simplified for the sake of the complex collective effects.

In the canonical formalism, we add three more vectors to the system: $\left(O_1, \vec{b}_1^x, \vec{b}_1^y, \vec{b}_1^z, 1, \vec{p}_1^x, \vec{p}_1^y, \vec{p}_1^z\right)$. The point O_1 is now a point in extended phase space: three positions, three momenta and time. Thus a particle has seven coordinates usually denoted by $(x, y, z, t, p_x, p_y, p_z)$. We can relate any observable measured in the chart around O_1 to the global laboratory frame $(\Omega, X, Y, Z, T, P_x, P_y, P_z)$ of our Lagrangian class. The same applies to the full extended phase space.

In single particle dynamics we perform a little trick at every chart. First, we include (minus) energy as the eighth variable:
$(x, y, z, t, p_x, p_y, p_z, -E)$. Of course this (dependent) variable is redundant. Then we choose to think of the variable p_z as the locally

[10.2] If this basis was a Frenet-Serrat coordinate system, z would correspond to a distance in the direction of the tangent, x would be the normal and finally y would be the bi-normal. The time t would be simply a parameter describing the curve. Regular time does the job for a trajectory!

redundant variable:

$$(x, y, z, t, p_x, p_y, p_z, -E) \longrightarrow (x, y, t, z, p_x, p_y, -E).$$

Such a thing is possible if, for all particles of interest, the dependence (locally) of z on the parameter t is monotonic. In other words particles should move forward in the beam pipe and never execute u-turns. This was discussed at the mathematical level in Chapter 3. Moreover, it can be shown (see again Chapter 3) that the motion of the remaining variables is symplectic and generated by the Hamiltonian $-p_z$. The time t is now a dependent position with minus the energy as its canonical momentum. In this book, I prefer to use the energy as a canonical position and time as a canonical momentum as explained in footnote 4.2.

As I have said, locally at O_1, there are two transverse Cartesian positions x and y and their Cartesian directions (momenta or the tangents $\frac{dx}{dz}$ and $\frac{dy}{dz}$ where z is the Cartesian variable perpendicular to the plane $x - y$) supplemented by a temporal coordinate and an energy coordinate. It is crucial to realize as I have been clamoring that the other position variable is the new "time-like" variable describing the layout. In effect, it is the discrete variable i defined modulo $\aleph$.

The establishment of the layout is approximately model independent. Curved blocks are normally used in locations where one expects to find elements whose primary purpose is to bend the beam as we hinted above. Finally, the collection of layout charts forms what mathematicians call an atlas. I referred to the collection of charts and the maps between them as a ring (see Chapter 7). The maximal ring is the Hamiltonian because it must contain an infinite number of charts. It is usually an overkill to think of a problem in terms of the Hamiltonian. The desire to deal with smaller rings is at the root of the global/local approach of ring theory, simulation, and experiment.

So, to sum it up, the actual clumps of matter around the beam pipe determine the ring $Ring_{\aleph}$. The layout allows us to connect this ring with the actual Lorentz frame of the laboratory whose use is not tied to single particle dynamics.

10.4 ACTION OF THE EUCLIDEAN GROUP

To understand how the group acts on a graphical or on a dynamical object we must first define these objects! At this point, we have a collection of charts linking planes around a ring without any dynamical or graphical form. The reasons why we explain in parallel the effect of the affine group on a map and a picture is to emphasize the magnet as an "object" in our single particle representation.

Let us define the Euclidean group: it is the usual group of translations and rotations of real objects of our every day world. The world in which we live as three-dimensional creatures is an affine world, not a vectorial world. In a vector space the origin is unique: it is the null vector which in $\mathbb{R}^3$ always has components $(0,0,0)$; it is the identity element of vector addition. If planes are parallel in a vector space they are in fact equal. This is not the physical world in which we live: the choice of origin in Figure 10.1 is arbitrary. Let us defined the affine space $\mathbb{E}$ associated with the vector space[10.3] $\mathbb{R}^3$. If $\mathbb{E}$ is affine there must exist a relation ϑ from the points in $\mathbb{E}$ to the vectors in $\mathbb{R}^3$ such that

$$\begin{aligned} &\forall(A,B) \in \mathbb{E}^2 \ \vartheta(A,B) \in \mathbb{R}^3 \\ &\forall(A,B,C) \in \mathbb{E}^3 \ \vartheta(A,B) + \vartheta(B,C) = \vartheta(A,C). \end{aligned} \tag{10.1}$$

The usual notation for the vector $\vartheta(A,B)$ is simply $\overrightarrow{AB}$; this leads to the popular definition of a vector as an "arrow" connecting two points in affine space. Therefore, it is natural and thus "physical" to define first an affine space because it is our everyday life world corresponding to Euclid's geometry. Mathematically it is more natural to construct an affine space out of a vector space (see Arnold's book for example [12]).

Exercise 10-1
In his book, Arnold claims that pre-Galilean physics lived in a vector space rather than an affine space. What does he mean? See the answer below.

Answer To Exercise 10-1
Of course until general relativity the space that we perceive had always been a three-dimensional affine vector space, and, as we have said above, its origin is arbitrary. In pre-Galilean physics it was believed that the center of the earth played a crucial rôle in the establishment of the laws of physics (Platonic spheres, epicycles, etc.). Hence, even though the ancients believed our world to be embedded in an affine space (Euclidean geometry) they thought that the laws of physics were

[10.3] The mathematically oriented readers may complain that a vector space is only defined when a field of scalars and its action on a vector is defined. Here, we use $\mathbb{R}$ itself as the field of scalar.

not invariant under translation and thus "vectorial" to the extent that the origin was unique: the center of the Earth.

To create a dynamical map we simply assume that someone gives us a mapping in $\mathbf{R}^6 = \mathbf{R}^3 \times \mathbf{R}^3$ connecting the coordinates at i to those at $i+1$; the three additional dimensions are needed because of the "momenta-like" coordinates used to specify the full state of a particle. In a single particle description attached to the layout, the position space and the momentum space do not possess a full vector space structure because one of the "positions" is time and one of the "momenta" is energy. Hence representing the Euclidean group on the maps defined on the layout is more complex as we will soon see. The actual construction of the map is irrelevant at this stage. The six coordinates are defined at the point O_1 and O_2 of Figure 10.2.

10.4.1 Euclidean Group on Graphical Elements

> *Nota Bene: This section is treacherous if one confuses a vector with its coordinates. In this section vectors are called vectors and coordinates are called "coordinates vectors." Remember that vectors like points exist independently of any coordinates. Coordinates need a basis in order to be defined; vectors and points do not. They simply are! In this section there is a frame in the middle of the layout and a frame attached to the body of the magnet. It is correct to think of these two frames as belonging to different affine spaces. Placing the magnet in the layout is actually an isomorphism between these spaces. We denote it by the letter $\mathfrak{P}$.*

We arbitrarily create a frame in the middle of the layout planes whose origin is O_{12} (Figure 10.2) on which the element (graphical and dynamical) is ideally centered. Our goal is to move this element away from this ideal location. Hence, we also imagine a frame attached to the element denoted by the origin O' in Figure 10.2. Before moving the element, the O_{12} and O' frames coincide, that is to say that the isomorphism $\mathfrak{P}$ is an "identity map."

$$\mathfrak{P}(O', \vec{b}'^1, \vec{b}'^2, \vec{b}'^3) = (O_{12}, \vec{b}^1_{12}, \vec{b}^2_{12}, \vec{b}^3_{12}). \tag{10.2}$$

Just as it is done in rigid body motion, we will use the body frame to describe the motion in affine space. The definition of a graphical

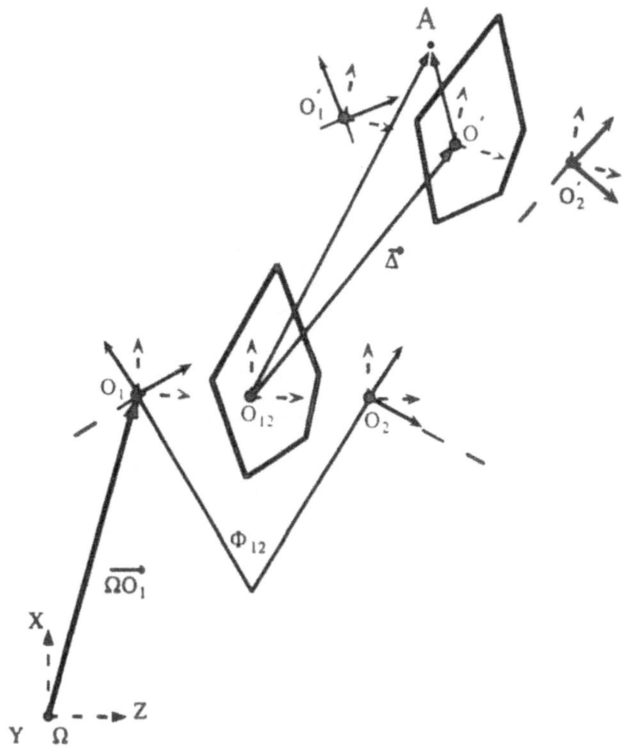

Figure 10.2: Moving an Magnet/Block.

element can be very complex, ascribing colors, shapes and shadows to the actual object presumably sitting somewhere between the planes of the layout; here we will keep things simple. In the O' frame we define the graphical object as a set of points[10.4] whose coordinates are

$$Q = \left\{ A_j \in \mathbb{E}^3 \;\middle|\; \overrightarrow{O'A_j} = \sum_{i=1}^{3} \lambda_i^j \vec{b}'^i \right\}. \tag{10.3}$$

[10.4]Pixels with colors on an actual computer: the author is not an expert, our definition is purely mathematical for illustration purposes. Presumably in a proper object-oriented design the graphical object will "know" how to draw itself.

As we said before, the frames $\left(O_{12}, \vec{b}_{12}^1, \vec{b}_{12}^2, \vec{b}_{12}^3\right)$ and $\left(O', \vec{b}'^1, \vec{b}'^2, \vec{b}'^3\right)$ are initially "identical," hence, in the layout frame, the pictorial magnet is obtained by changing frames through Equation (10.2).

$$\mathfrak{P}(Q) = \left\{ A_j \in \mathbf{E}^3 \;\middle|\; \overrightarrow{O_{12}A_j} = \sum_{i=1}^{3} \lambda_i^j \vec{b}_{12}^i \right\}. \tag{10.4}$$

The set Q is an invariant because it represents a rigid object in its own body frame. As we move it, as we change the definition of $\mathfrak{P}$, we express this set of points with respect to the centered layout frame $(O_{12}, \vec{b}_{12}^1, \vec{b}_{12}^2, \vec{b}_{12}^3)$.

Our first displacement is a translation of the figure by a vector $\vec{\Delta}$. The new translated object is given by the set

$$\mathfrak{P}_{\vec{\Delta}}(Q) = \left\{ A_j \in \mathbf{E}^3 \;\middle|\; \overrightarrow{O_{12}A_j} = \vec{\Delta} + \sum_{i=1}^{3} \lambda_i^j \vec{b}_{12}^i \right\}. \tag{10.5}$$

We then rotate this figure around its own body axes. This is achieved by rotating the body vectors. Initially, we start with $\mathfrak{P}(\vec{b}'^i) = \vec{b}_{12}^i$ indicating a proper design orientation—this frame is now rotated into $\mathfrak{P}(\vec{b}'^i) = R_{ij}\vec{b}_{12}^j$ (summed over j)

$$\begin{aligned} \forall i \;\; \mathfrak{P}(\vec{b}'^i) &= R_{ij}\vec{b}_{12}^j \\ &= \left(Z(\alpha_z)Y(\alpha_y)X(\alpha_x)\right)_{ij}\vec{b}_{12}^j. \end{aligned} \tag{10.6}$$

These rotations are the usual rotations of $SO(3)$. Because of the special nature of the z-direction in the dynamical case, we select the $y-z$, $x-z$, and $x-y$ rotations given by:

$$\begin{aligned} X(\alpha_x) &= \begin{pmatrix} 1 & 0 & 0 \\ 0 & \cos(\alpha_x) & \sin(\alpha_x) \\ 0 & -\sin(\alpha_x) & \cos(\alpha_x) \end{pmatrix} \\ Y(\alpha_y) &= \begin{pmatrix} \cos(\alpha_y) & 0 & \sin(\alpha_y) \\ 0 & 1 & 0 \\ -\sin(\alpha_y) & 0 & \cos(\alpha_y) \end{pmatrix} \\ Z(\alpha_z) &= \begin{pmatrix} \cos(\alpha_z) & \sin(\alpha_z) & 0 \\ -\sin(\alpha_z) & \cos(\alpha_z) & 0 \\ 0 & 0 & 1 \end{pmatrix}. \end{aligned} \tag{10.7}$$

The resulting picture is given by:

$$\mathfrak{P}_{\vec{\Delta},\vec{\alpha}}(Q) = \left\{ A_j \in \mathbb{E}^3 \;\middle|\; \overrightarrow{O_{12}A_j} = \vec{\Delta} + \sum_{i,k=1}^{3} \lambda_i^j R_{ik} \vec{b}_{12}^k \right\}. \tag{10.8}$$

The original set Q was expressed in terms of "coordinate vectors" $\vec{\lambda}^j$ in the body frame $\left(O', \vec{b}'^1, \vec{b}'^2, \vec{b}'^3\right)$. We need the new coordinates of this object in the layout frame:

$$\mathfrak{P}_{\vec{\Delta},\vec{\alpha}}(Q) = \left\{ A_j \in \mathbb{E}^3 \;\middle|\; \overrightarrow{O_{12}A_j} = \sum_{i=1}^{3} \left(\underbrace{\vec{\Delta} \cdot \vec{b}_{12}^i}_{d_i} + R_{ik}^{-1} \lambda_k^j \right) \vec{b}_{12}^i \right\}. \tag{10.9}$$

We have made use of the fact that the inverse of a rotation is its transpose. The coordinate vector $\vec{d}$ with components $d_i = \vec{\Delta} \cdot \vec{b}_{12}^i$ gives us the actual numerical values of the translation in the layout frame. We define the affine transformation on the coordinate vectors of a graphical object to be

$$\mathfrak{p}_{\vec{d},\vec{\alpha}}^{-1}(\vec{\lambda}) = \vec{d} + R^{-1}\vec{\lambda}\,. \tag{10.10}$$

The inverse operation is simply

$$\mathfrak{p}_{\vec{d},\vec{\alpha}}(\vec{\lambda}) = R(\vec{\lambda} - \vec{d})\,. \tag{10.11}$$

The reader will notice that I purposely associated the affine space map $\mathfrak{P}$ with $\mathfrak{p}^{-1}$.

In summary, the first operation is a translation acting in all three planes; the second operation rotates the object. Rotations in $\mathbb{R}^3$ do not commute and therefore it is important to realize that in Equation (10.6) the object is first rotated around its internal x-axis, then around its y-axis and finally around its z-axis. The resulting affine set is given by Equation (10.8). It is clear that this set can be expressed in the global frame around Ω or any other frame if one desires to do so.

10.4.2 Euclidean Group Representation: Time Based

The action of the Euclidean group on a dynamical element of the layout is more complex because the variable measuring actual time is now a dependent variable. However, when a regular time description is used in Cartesian coordinates, the action of the Euclidean group on a transfer map from time t_1 to t_1 is trivially obtained by a similarity transformation. First let us extend the map $\mathfrak{p}$ from the coordinate space of the full Cartesian phase space of time based dynamics. Since the momenta transform like the positions under rotation, the extension is

$$\mathfrak{p}_{\vec{d},\vec{\alpha}}(\vec{q},\vec{p}) = \left(R(\vec{q}-\vec{d}), R\vec{p}\right). \tag{10.12}$$

Suppose we are given a map $\vec{\zeta}_{1\to2}$ that propagates a ray from time t_1 to t_2 in the original configuration. By definition a particle with initial phase space coordinates $\vec{z}_1$ will reach, at time t_2, the value $\vec{z}_2$ given by

$$\vec{z}_2 = \vec{\zeta}_{1\to2}(\vec{z}_1).$$

When we rotate an[10.5] element, we must answer the question: keeping the initial phase space coordinates fixed, what will be the output if the entire device generating the fields is rotated by the map $\mathfrak{P}$ of Sect. 10.4.1? This implies that the coordinate $\vec{Z}_1$ attached to the "magnet" will rotate as the coordinate $\vec{\lambda}$ of the graphical element. In the frame attached to the magnet we know by definition that the propagator is $\vec{\zeta}_{1\to2}$, thus we can combine all these remarks into:

$$\begin{gathered}
\mathfrak{p}^{-1}_{\vec{d},\vec{\alpha}}(\vec{Z}_1) = \vec{z}_1 \\
\text{and } \vec{Z}_2 = \vec{\zeta}_{1\to2}(\vec{Z}_1) \\
\Downarrow \\
\vec{z}_2^{\,new} = \left(\mathfrak{p}^{-1}_{\vec{d},\vec{\alpha}} \circ \vec{\zeta}_{1\to2} \circ \mathfrak{p}_{\vec{d},\vec{\alpha}}\right)\left(\vec{z}_1\right).
\end{gathered} \tag{10.13}$$

The last equation implies that the new compositional map is

$$\mathfrak{p}_{\vec{d},\vec{\alpha}}\left[\mathcal{M}_{1\to2}\right] = \mathcal{P}_{\vec{d},\vec{\alpha}}\mathcal{M}_{1\to2}\mathcal{P}^{-1}_{\vec{d},\vec{\alpha}}, \tag{10.14}$$

where $\mathcal{P}_{\vec{d},\vec{\alpha}}$ is the compositional map associated with $\mathfrak{p}_{\vec{d},\vec{\alpha}}$. If the map is generated by a Hamiltonian H, the new Hamiltonian is

$$\begin{aligned}
H(\vec{z};t) \xrightarrow{\mathfrak{p}_{\vec{d},\vec{\alpha}}} \left(\mathcal{P}_{\vec{d},\vec{\alpha}}H\right)(\vec{z}) &= H\left(\mathfrak{p}_{\vec{d},\vec{\alpha}}(\vec{z});t\right) \\
&= H\left(R(\vec{q}-\vec{d}), R\vec{p}; t\right).
\end{aligned} \tag{10.15}$$

[10.5] We should say "the element" because the entire "universe" is rotated in a time based representation.

The compositional map $\mathcal{P}_{\vec{d},\vec{\alpha}}$ decomposes into a translation and a rotation in the usual manner:

$$\mathcal{P}_{\vec{d},\vec{\alpha}} = \mathcal{T}_{\vec{d}}\mathcal{R}_{\vec{\alpha}}$$
$$\text{where } \mathcal{R}_{\vec{\alpha}} = \mathcal{X}(\alpha_x)\mathcal{Y}(\alpha_y)\mathcal{Z}(\alpha_z)$$
$$\text{and}$$
$$\mathcal{X}(\theta) = \exp\,(\,\theta : \underbrace{yp_z - zp_y}_{L_x} :\,) \tag{10.16.a}$$
$$\mathcal{Y}(\theta) = \exp\,(\,\theta : \underbrace{xp_z - zp_x}_{-L_y} :\,) \tag{10.16.b}$$
$$\mathcal{Z}(\theta) = \exp\,(\,\theta : \underbrace{xp_y - yp_x}_{L_z} :\,) \tag{10.16.c}$$
$$\mathcal{T}_x(d) = \exp\,(d : p_x :) \tag{10.16.d}$$
$$\mathcal{T}_y(d) = \exp\,(d : p_y :) \tag{10.16.e}$$
$$\mathcal{T}_z(d) = \exp\,(d : p_z :)\,. \tag{10.16.f}$$

If we were using the laboratory frame, this would be the end of the dynamical Euclidean group. It is just a subgroup of the full Poincaré group of special relativity (relativistic boosts + $O(3)$ + translations).

It is important to realize that all of space is rotated and thus the word "magnet" must refer to the entire ring in a time dependent description. Even when dealing with a single particle, such a theory is collective by construction. We must be able to rotate each magnet independently. Thus, even though the connection between the graphical and time based dynamical element is transparent, it is of little value in a theory where each magnet must be an independent object.

10.4.3 Two Euclidean Group Representations for Layout Based Dynamics

We now turn our attention to the single particle description in which time is a dependent coordinate. A layout map, before displacement, relates by definition the planes O_1 to O_2. If we apply naïvely Equation (10.14) to the compositional map $\mathcal{M}_{12}$, the resulting map

$$\mathcal{P}\left[\mathcal{M}_{12}\right] \neq \mathcal{P}\mathcal{M}_{12}\mathcal{P}^{-1}$$
$$\text{where } \mathcal{P} = \mathcal{T}\mathcal{R}. \tag{10.17}$$

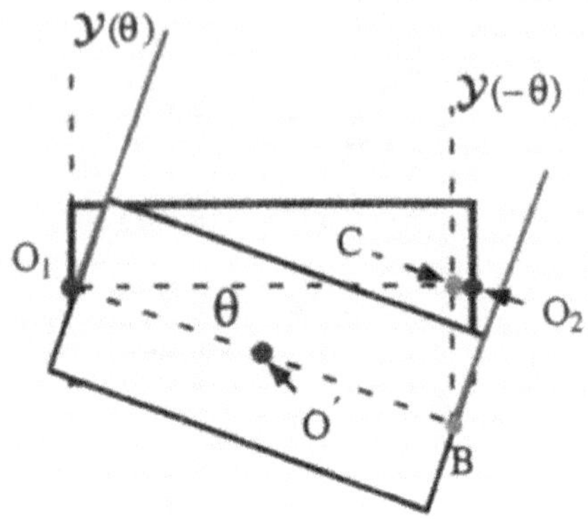

Figure 10.3: Rotation of a Straight Element.

is not the properly displaced map. To grasp how formula (10.14) fails if applied to a map relating the layout of frames at O_1 and O_2, it is best to apply it mindlessly and look at the consequences of our actions! The results are depicted on Figure 10.3. We consider a straight element rotated in the $x - z$ plane by an angle θ around the origin of the entrance. If we follow in Figure 10.3 the operations on a ray, we see that a particle must "rotate" from the original layout plane at O_1 to the rotated plane at O_1 (in red). This is done by the map $\mathcal{Y}(\theta)$. The map $\mathcal{M}_{12}$ of the ideal magnet/block propagates the rays from O_1 to B, and finally we operate with $\mathcal{Y}(-\theta)$ which rotates the ray at B back into the dashed frame parallel to the original layout. This is the effect of the map $\mathcal{Y}(\theta)\mathcal{M}_{12}\mathcal{Y}(-\theta)$ which obeys Equation (10.14). Unfortunately, as one can see in Figure 10.3, the point B and the dashed line do not coincide at all with the layout frame at O_2.

Fortunately, the frame at O_2 can be "reached" easily with the help of two translations. The first translation is a simple redefinition of the x coordinate from the point B to the point C

$$x \longmapsto x - BC' \quad \text{where} \quad BC' = \left\| \overrightarrow{BC'} \right\| . \tag{10.18}$$

Then we must propagate in the forward z-direction using the field-free propagator for a distance $C'O_2$. The conclusion is that the correct expression for $\mathcal{P}[\mathcal{M}_{12}]$ is

$$\mathcal{P}[\mathcal{M}_{12}] = \mathcal{Y}(\theta)\mathcal{M}_{12}\mathcal{Y}(-\theta)\mathcal{T}_x(BC')\mathcal{T}_z(C'O_2) . \tag{10.19}$$

At this stage, it is instructive to express the lengths BC and CO_2 in terms of the distance between the layout planes O_1O_2. These are given by the formulas

$$\begin{aligned} BC &= O_1O_2 \sin(\theta) \\ CO_2 &= O_1O_2 (1 - \cos(\theta)). \end{aligned} \tag{10.20}$$

We notice that the corrections to the usual formula $\mathcal{Y}(\theta)\mathcal{M}_{12}\mathcal{Y}(-\theta)$ involve two translations which become the identity if the length O_1O_2 of the original layout vanishes. This remark is very important because it tells us that, if the magnet/block is thin ($O_1O_2 = 0$), then the usual formula $\mathcal{Y}(\theta)\mathcal{M}_{12}\mathcal{Y}(-\theta)$ would apply exactly. Of course, this is not physically possible since magnets must have a finite extent in the z-direction. The question we must ask is whether or not it is possible to realize the exact dynamics with blocks of zero width and use the usual Euclidean transformations on them as simply as it was done on the graphical element.

The reader should not be surprised: zero width blocks are for the layout the equivalent of events happening at the same time in a regular Lorentz frame where the graphical element sits (in the laboratory frame centered at Ω on Figure 10.1). In the usual time based dynamics the rotations and translations of a element happen at a fixed time. We will see later, in Sects. 10.4.3.2 and 10.4.3.3, that there are two important types of blocks which can be rotated easily: the 0 degree (zero width) block and the 180 degree block. Later, in Sect. 10.4.4, we derive a general formula which is valid for any block.

But before going into the actual representation of the Euclidean group and its connection with the usual graphical group of Sect. 10.4.1, we will describe in more detail the six canonical transformations associated with the six usual Euclidean transformations. This violates our Feynmanesque ideal of keeping irrelevant details off limits and introducing only the quintessential aspects of the theory. We will try to identify what is absolutely needed from what is not.

10.4.3.1 The layout translations and rotations

There are six transformations needed to move an element: 3 translations and 3 rotations. In the regular time based dynamics, as we have just seen, they are simply the graphical transformations augmented with a prescription for the momenta. In the layout representation, two translations and one rotation keep their usual form. These are the transformations that move a 0 degree block in the transverse $x-y$ plane only and thus do not affect the time-like variable z. Their associated compositional maps are given by Equations (10.16.c), (10.16.d) and (10.16.e) which, as a reminder, were

$$\begin{aligned} \mathcal{Z}(\theta) &= \exp\,(\;\theta : \underbrace{xp_y - yp_x}_{L_z} :\;) \\ \mathcal{T}_x(d_x) &= \exp\,(d_x : p_x :) \\ \mathcal{T}_y(d_y) &= \exp\,(d_y : p_y :)\,. \end{aligned} \tag{10.21}$$

The remaining transformations that correspond to $\mathcal{X}$, $\mathcal{Y}$, and $\mathcal{T}_z$ are more complicated and differ substantially from the graphical equivalents of Sect. 10.4.1 because they involve propagation across the layout. Fortunately they have a very simple dynamical interpretation which we can deduce from Figure 10.3. They are field-free propagators. The simplest of the transformations is $\mathcal{T}_z$—a "drift" or free propagator in the z-direction. It has the following Lie representation:

$$\begin{aligned} \mathcal{T}_z(d_z) &= \exp\left(\;: d_z \underbrace{\sqrt{(1+\delta)^2 - p_x^2 - p_y^2}}_{p_z} :\;\right) \\ &= \exp\,(: d_z p_z :) \end{aligned} \tag{10.22}$$

We see that the Lie operator for propagation in the z-direction happens to be the momentum p_z as was explained in Sect. 3.1.2. The associated transfer map is just the usual drift

$$x^{new} = x + d_z \frac{p_x}{\sqrt{(1+\delta)^2 - p_x^2 - p_y^2}} \tag{10.23.a}$$

$$y^{new} = y + d_z \frac{p_y}{\sqrt{(1+\delta)^2 - p_x^2 - p_y^2}} \tag{10.23.b}$$

$$\ell^{new} = \ell + d_z \frac{1+\delta}{\sqrt{(1+\delta)^2 - p_x^2 - p_y^2}}\,. \tag{10.23.c}$$

The time of flight $\tau = ct$ can be gotten easily from Equation (10.23.c). The rotations $\mathcal{X}$ and $\mathcal{Y}$ have a slightly more complex interpretation. Looking at point O_1 in Figure 10.3, we see that $\mathcal{Y}$ involves propagation from the layout plane to the magnet plane (in red). At the same time the frame of reference, the x-axis, must rotate to align itself along the red line. Hence we conclude that the propagator for $\mathcal{Y}$ is the field-free propagator in cylindrical coordinates where the radius coordinate is the x-coordinate. This propagator is given by

$$\mathcal{Y}(\theta) = \exp\left(\theta : x\sqrt{(1+\delta)^2 - p_x^2 - p_y^2} :\right) \tag{10.24}$$

and similarly for $\mathcal{X}(\theta)$

$$\mathcal{X}(\theta) = \exp\left(\theta : y\sqrt{(1+\delta)^2 - p_x^2 - p_y^2} :\right). \tag{10.25}$$

Because Equations (10.24) and (10.25) represent drifts in polar coordinates, their associated transfer maps can be derived geometrically. This was noticed by Dragt in [49] and was also used in reference [6]. The geometrical interpretation implies that the equation of motion parameterized by θ for the Hamiltonian

$$-x\sqrt{(1+\delta)^2 - p_x^2 - p_y^2}$$

can be integrated in terms of the usual trigonometric functions. The answer for $\mathcal{Y}$ is

$$x^{new} = \frac{x}{\cos(\theta)\left(1 - \frac{p_x \tan(\theta)}{p_z}\right)} \tag{10.26.a}$$

$$p_x^{new} = p_x \cos(\theta) + \sin(\theta)\, p_z \tag{10.26.b}$$

$$y^{new} = y + \frac{p_y x \tan(\theta)}{p_z\left(1 - \frac{p_x \tan(\theta)}{p_z}\right)} \tag{10.26.c}$$

$$p_y^{new} = p_y \tag{10.26.d}$$

$$\ell^{new} = \ell + \frac{(1+\delta) x \tan(\theta)}{p_z\left(1 - \frac{p_x \tan(\theta)}{p_z}\right)} \tag{10.26.e}$$

$$\text{where } p_z = \sqrt{(1+\delta)^2 - p_x^2 - p_y^2}\,.$$

For $\mathcal{X}$ the rôles of the x and y planes are interchanged. The reader will notice that the expressions in Equation (10.26) are analytic in a neighborhood around the origin. Although we started with cylindrical coordinates where x played the rôle of the radius, in the final formula it can be reinterpreted correctly as the usual Cartesian coordinate which, unlike the radius is defined for negative values.

Finally the three transformations $\mathcal{X}$, $\mathcal{Y}$, and $\mathcal{T}_z$ must all modify the path length variable ℓ (or temporal τ). This is expected because they are "drifts" from one plane to another.

In summary, we have established a few things which do not depend on details:

1. The arbitrary displacement of a thick element must involve dynamical corrections.
2. These corrections vanish if the element is thin—the entrance and exit planes of the layout coincide.
3. The dynamical corrections are just the dynamical version of $\mathcal{X}$, $\mathcal{Y}$, or $\mathcal{T}_z$, in either cylindrical or Cartesian coordinates. They can all be computed in closed form with trigonometric functions in the case of an element immersed in a field-free region.

We must now exploit these facts for a coherent representation of the Euclidean transformations which unite the graphical and dynamical objects.

Exercise 10-2 (The intuitive way MAD rotates!)
They are two geometrical ways to look at an arbitrary dynamical rotation $\mathcal{R}(\vec{\alpha})$. One can either "rotate" first or "propagate" first. The latter is associated with explicit symplectic integration in sector bends: it was first used in the code TEAPOT [6]. It is displayed in Figure 12.2 *and discussed in Sect.* 12.1.3.

Alternatively one can first rotate and then propagate in the medium. This is depicted on Figure 10.4. *We will now see that this is a very powerful geometrical description. (This is how MAD[50, 51] does it!) We now derive our $\mathcal{Y}(\theta)$ using this technique.*

1. *Remembering that p_z is just the Hamiltonian, rotate a ray from the entrance plane to the final plane using Equation* (10.12). *(The initial value of z is arbitrary and can be set to zero.)*
   Answer:

$$\begin{pmatrix} x \\ y \\ 0 \\ p_x \\ p_y \\ p_z \end{pmatrix} \longrightarrow \begin{pmatrix} \cos(\theta)\, x \\ y \\ -\sin(\theta)\, x \\ \cos(\theta)\, p_x + \sin(\theta)\, p_z \\ p_y \\ \cos(\theta)\, p_z - \sin(\theta)\, p_x \end{pmatrix}. \tag{10.27}$$

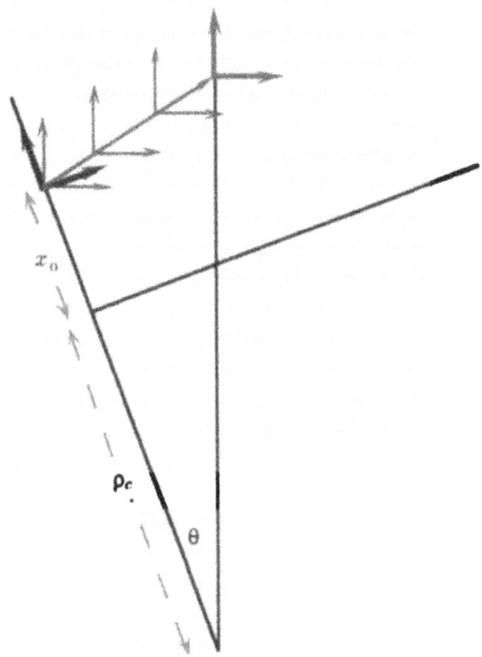

Figure 10.4: The "Rotate First" Geometrical Interpretation of $\mathcal{Y}(\theta)$.

2. *Propagate backwards a ray (including time) by an amount equal to the new value of z and show that the final map is just the same as that of Equation (10.26).*

This geometrical formulation has several advantages. First it decouples neatly the geometrical and dynamical part. It should be pointed that neither operations are symplectic, but their product is. One can see how to extend this procedure to a medium which is not field-free. Furthermore, the geometrical step can be done with **any** *rotation. This is perhaps the greatest advantage of the method.*

10.4.3.2 Making a thick block thin

The most general element is depicted on Figure 10.2: the entrance and exit plane are separated by a distance $L_{12} = O_1 O_2$ and make an angle Φ_{12}. It is always assumed, as is possible even in non-planar rings, that the y_1 and y_2 axes are aligned so that the x_1 and x_2 axes intersect at the angle Φ_{12}. This is the situation in Figures 10.1 and 10.2.

Let us consider the following rewriting of the map $\mathcal{M}_{12}$:

$$\begin{aligned} \mathcal{M}_{12} &= \mathcal{Y}^{1/2}\mathcal{T}_z^{1/2}\mathcal{T}_z^{-1/2}\mathcal{Y}^{-1/2}\mathcal{M}_{12}\mathcal{Y}^{-1/2}\mathcal{T}_z^{-1/2}\mathcal{T}_z^{1/2}\mathcal{Y}^{1/2} \\ \text{where} \quad \mathcal{Y}^{\pm 1/2} &= \mathcal{Y}(\pm\frac{\Phi_{12}}{2}) \\ \mathcal{T}_z^{\pm 1/2} &= \mathcal{T}_z(\pm\frac{O_1O_2}{2}). \end{aligned} \tag{10.28}$$

We rearrange Equation (10.28) as

$$\begin{aligned} \mathcal{M}_{12} &= \mathcal{D}_{1:12}\mathcal{M}_{12}^0\mathcal{D}_{12:2} \\ \text{where} \quad & \begin{cases} \mathcal{D}_{1:12} = \mathcal{Y}^{1/2}\mathcal{T}_z^{1/2} \\ \mathcal{D}_{12:2} = \mathcal{T}_z^{1/2}\mathcal{Y}^{1/2} \end{cases} \\ \text{and} \quad \mathcal{M}_{12}^0 &= \mathcal{D}_{1:12}^{-1}\mathcal{M}_{12}\mathcal{D}_{12:2}^{-1}. \end{aligned} \tag{10.29}$$

Looking at Figure 10.2, one notices that the (free propagator) map $\mathcal{D}_{1:12}$ takes a particle from the layout frame at O_1 to the frame at O_{12} in the center of the layout. Similarly, the map $\mathcal{D}_{12:2}$ drifts a particle from the frame at O_{12} to the layout frame at O_2. We must conclude that, *by construction*, the map $\mathcal{M}_{12}^0$, which contains **all** the dynamical information of the map $\mathcal{M}_{12}$, is a thin map which propagates the rays across an interface at O_{12}.

We have succeeded in replacing a thick map by a thin map expressed in terms of the frame of reference used to describe the graphical element. Since the map is now thin, i.e., the rays propagate at a single "time" or layout plane, the usual formula applies. For example, if the graphical element is moved as in Equation (10.8), then the thin dynamical element is transformed using

$$\begin{aligned} \mathcal{P}\left[\mathcal{M}_{12}^0\right] &= \mathcal{P}\mathcal{M}_{12}^0\mathcal{P}^{-1} \\ &= \mathcal{T}\mathcal{R}\mathcal{M}_{12}^0\mathcal{R}^{-1}\mathcal{T}^{-1}. \end{aligned} \tag{10.30}$$

The thin element frame is fixed on the actual magnet and thus the thick rotated magnet/block must be

$$\begin{aligned}\mathcal{P}\left[\mathcal{M}_{12}\right] &= \mathcal{D}_{1:12}\mathcal{P}\left[\mathcal{M}^0_{12}\right]\mathcal{D}_{12:2} \\ &= \mathcal{D}_{1:12}\mathcal{P}\mathcal{M}^0_{12}\mathcal{P}^{-1}\mathcal{D}_{12:2} \\ &= \mathcal{D}_{1:12}\mathcal{P}\mathcal{D}^{-1}_{1:12}\mathcal{M}_{12}\mathcal{D}^{-1}_{12:2}\mathcal{P}^{-1}\mathcal{D}_{12:2} \\ &= \mathcal{Y}^{1/2}\mathcal{T}_z^{1/2}\mathcal{P}\mathcal{T}_z^{-1/2}\mathcal{Y}^{-1/2} \\ &\quad\times\mathcal{M}_{12}\mathcal{Y}^{-1/2}\mathcal{T}_z^{-1/2}\mathcal{P}^{-1}\mathcal{T}_z^{1/2}\mathcal{Y}^{1/2} . \end{aligned} \quad (10.31)$$

Of course in Equation (10.31) the dynamical forms of the Euclidean transformations must be used.

Exercise 10-3
By making a block thin at the entrance face, *derive an expression for the rotated map of Figure* 10.3. *Answer:*

$$\mathcal{P}\left[\mathcal{M}_{12}\right] = \mathcal{Y}(\theta)\mathcal{M}_{12}\mathcal{T}_z^{-1}(O_1O_2)\mathcal{Y}^{-1}(\theta)\mathcal{T}_z(O_1O_2) \ . \quad (10.32)$$

Show that this expression is equal to that of Equation (10.19). *Hint: compare the Lie operators of the trailing maps.*

Our entire treatment exploits to a maximum the similarity between a graphical element and the dynamical element. This is only possible if a completely local theory is used. We will see later that a lot of complex magnets benefit from a local theory—the computation of the map $\mathcal{M}^0_{12}$ should also be completely local and dictated by questions of symmetry which are not visible at the layout level. Again none of this would be possible if we had insisted on using a Frenet-Serrat curvilinear frame in anticipation of Hamiltonian perturbation theory.

At this point, we come back to the question of "essential aspects" and in particular what can one say about the dynamical rotations $\mathcal{X}$, $\mathcal{Y}$, $\mathcal{Z}$ and the translations?

1. As we said, a bird's eye view of the layout of an actual ring would show straight beam pipes between magnets. This implies that the z-translation along the pipe must commute with the other translations. In other words, translating a drift space does not do anything even though the map through a drift cannot be an identity. In fact the displacement of a drift by translations or rotations leaves a drift invariant. This follows by construction. The actual form of the drift propagator is not important: this is an important fact since we may want to use a simpler Hamiltonian in large machines.

2. In our initial discussion of the layout, we define magnets as aggregates between straight pipes: it allowed us to derive the dynamical representation of the Euclidean group in terms of the free propagator. What if the layout was not immersed in a field-free region? This question is not totally ludicrous. For example, in the interaction region of a collider a solenoid can contain quadrupoles and bends. If one looks at the solenoid as one machine containing several elements, then it follows that these elements are immersed in a region of non-zero magnetic field. However our representation may *still*[10.6] holds, but the propagators are the finite field propagators. The map $\mathcal{T}_z(d_z)$ is the propagator along the solenoid axis for a distance d_z while the maps $\mathcal{X}(\theta)$ and $\mathcal{Y}(\theta)$ are propagators in cylindrical coordinates within the solenoid.

3. Finally, the solenoid example is the first example of a recursive definition of a layout. The fact that a ring or beam line can exist within a beam line with its own properties (layout, Euclidean group, propagators, etc.) is a somewhat novel concept which comes naturally in the local theory. It is also something that modern computer languages can handle more easily that FORTRAN.

The next issue we must address concerns the divergences in Equation (10.26). In Equation (10.26.a), the expression for the variable x^{new} has a clear divergence at $\theta = \frac{\pi}{2}$. What does it means? What does it imply? The answers to these crucial questions are in the following sections.

10.4.3.3 Making a thick block 180 degrees

The graphical representation of the Euclidean transformation acting on a magnet shape is defined irrespective of the shape of the magnet. In particular one can rotate a 180 degree bend. These bends are not academic constructions since small rings are often designed in a racetrack shape. For these bends, the representation of the Euclidean group described in Sect. 10.4.3.2, Equation (10.28), requires the use of the dynamical map $\mathcal{Y}^{\pm 1/2} = \mathcal{Y}(\pm\frac{\Phi_{12}}{2}) = \mathcal{Y}(\pm\frac{\pi}{2})$ which is ill-defined. For this reason we need to investigate the effect of an Euclidean transformation on a 180 degree block. Let us look at Figure

[10.6] Provided that the field of the solenoid does not penetrate into the quadrupole.

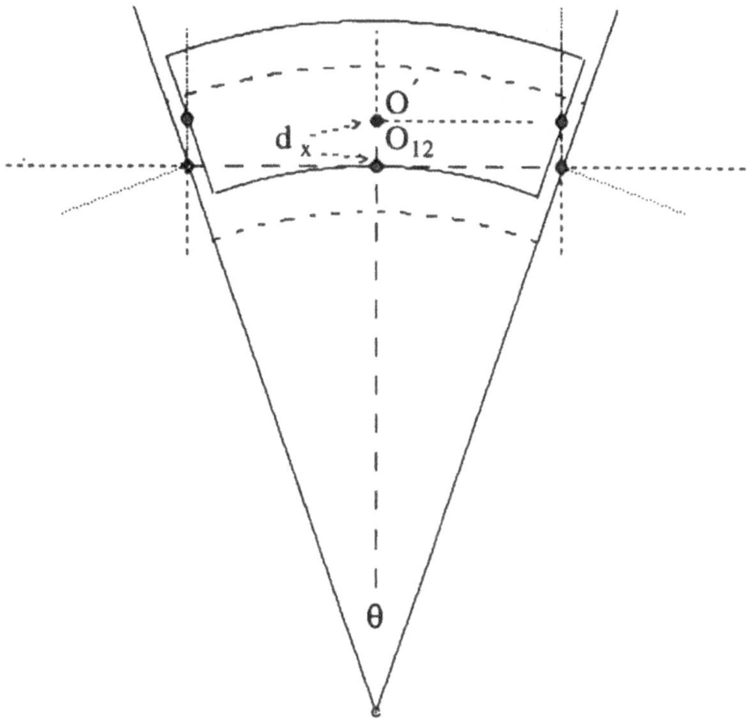

Figure 10.5: Translating a Bend by d_x.

10.5 and imagine moving the sector bend along the x-axis as defined around the point O_{12}. Following the rules of Sect. 10.4.3.2, this map is given by:

$$\begin{aligned} \mathcal{T}_x\left[\mathcal{M}_{12}\right] &= \mathcal{Y}^{1/2}\mathcal{T}_z^{1/2}\mathcal{T}_x\mathcal{T}_z^{-1/2}\mathcal{Y}^{-1/2}\mathcal{M}_{12}\mathcal{Y}^{-1/2}\mathcal{T}_z^{-1/2}\mathcal{T}_x^{-1}\mathcal{T}_z^{1/2}\mathcal{Y}^{1/2} \\ &= \mathcal{Y}^{1/2}\mathcal{T}_x\mathcal{Y}^{-1/2}\mathcal{M}_{12}\mathcal{Y}^{-1/2}\mathcal{T}_x^{-1}\mathcal{Y}^{1/2} \\ &= \mathcal{Y}^{1/2}\mathcal{T}_x\mathcal{M}_{\theta=0}^{L}\mathcal{T}_x^{-1}\mathcal{Y}^{1/2}. \end{aligned} \tag{10.33}$$

This map is, as I just said, certainly ill-defined for $\Phi_{12} = \pi$. Leaving it aside for awhile, we go back to a 180 degree bend. Suppose that we have such a map which we denote as $\mathcal{M}_{\theta=\pi}^{L}$ by analogy with the straight map. The reader may already be guessing that I will give

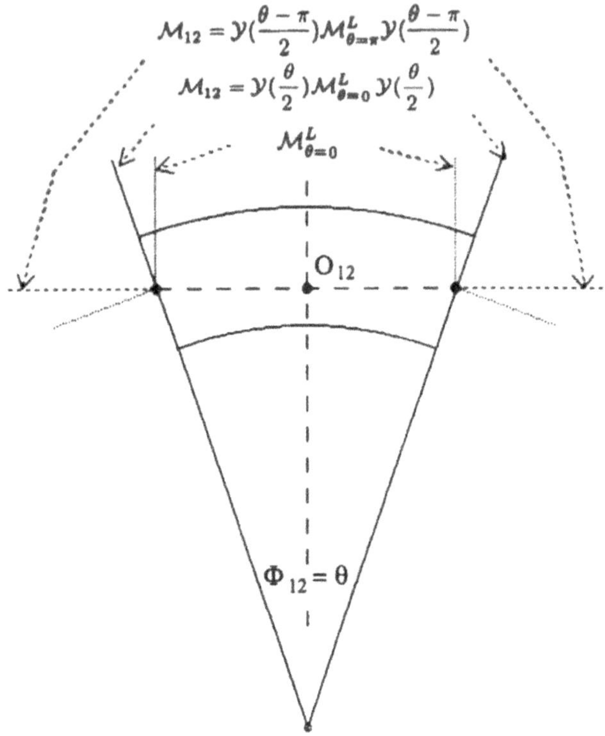

Figure 10.6: Contrasting 0 and 180 degree blocks.

a prescription for turning a layout block of arbitrary angle $\Phi_{12} = \theta$ into a 180 degree racetrack; he is certainly right. Temporarily it is better to imagine that we have an actual 180 degree bend so that no prescription is necessary.

Using purely geometrical common sense, let us look at the effect of the translation in the x-direction on a 180 degrees block, that is to say, on half of a racetrack. The reasoning that follows is well-known to anyone who plays with train sets or miniature racing cars. The entire magnet recedes from an incoming particle by an amount d_x. The particle exits the 180 degrees block a distance d_x short of the original plane. Again this is the same as setting up the rails of an electric train set: a half of a circle chunk of rails can be extended at

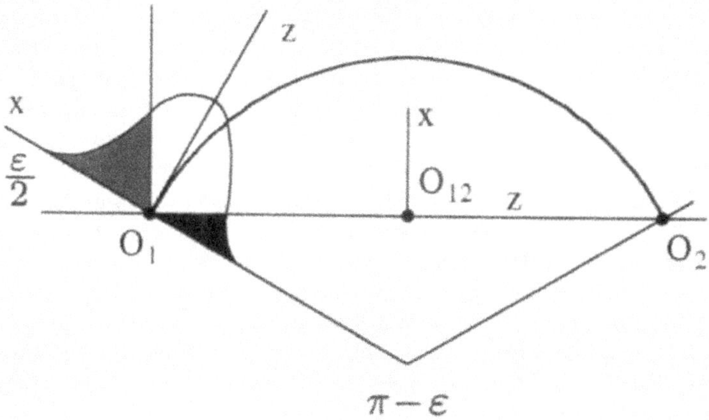

Figure 10.7: Changing Layout Plane: 0-180.

each extremity using a straight piece of rail road!

$$\mathcal{T}_x\left[\mathcal{M}_{12}\right] = \mathcal{Y}_\pi^{1/2}\mathcal{T}_z(d_x)\mathcal{M}^L_{\theta=\pi}\mathcal{T}_z(d_x)\mathcal{Y}_\pi^{1/2}$$
$$\text{where } \mathcal{Y}_\pi^{1/2} = \mathcal{Y}(\frac{\theta-\pi}{2}). \tag{10.34}$$

Having derived Equation (10.34), we can deduce the existence of the map $\mathcal{M}^L_{\theta=\pi}$ for arbitrary angles (other than zero) from Figure 10.6.

$$\mathcal{M}_{12} = \mathcal{Y}(\frac{\theta-\pi}{2})\mathcal{M}^L_{\theta=\pi}\mathcal{Y}(\frac{\theta-\pi}{2}).$$

In the general case the angle $\Phi_{12} = \theta$ of the layout block is neither zero nor π; we know that the maps involved in defining the blocks $\mathcal{M}^L_{\theta=0}$ and $\mathcal{M}^L_{\theta=\pi}$ must share a region of phase space around the origin of the layout in which all the maps involved are well-defined. Both maps must also have regions where legitimate trajectories are excluded. Let us see how this happens. In Figure 10.7 the bell curve represents a distribution of momenta in the $x-z$ plane for an incoming beam. If we rotate the entrance frame of reference by $\frac{\pi-\varepsilon}{2}$ then all the particles in red must travel backward in time! Conversely, if we choose to make the magnet a 180 degree block and rotate by $-\frac{\varepsilon}{2}$ then all the blue particles are lost for the same reason. Looking at Figure 10.7 it is clear that we must select the 180 degree bend/block for values of ε around zero. The reader will notice that, whatever the choice, some particles are lost. This should not be a worry since these particles must be near the short-term dynamic aperture, but it does imply that the choice of the layout plane cannot be completely arbitrary. Let us proceed with the full derivation of the 180 degree representation.

The simplicity of this particular example tells us that we should be able by geometrical considerations alone to figure out the equivalences between the 0 and 180 degree blocks; we should be able to move a 180 degree bend using pure geometry and a little bit of mental gymnastics in three dimensions. However it is safer and more instructive to appeal to the Lie operators themselves. We recall that the graphical element as well as the Hamiltonian described by time (i.e., *the Hamiltonian*) rotates and translates in the usual manner. The group is fully linear and thus global. In these representations the rotations $\mathcal{Y}(\frac{\pi}{2})$ and $\mathcal{Y}(\pi)$ are well-defined. This implies that one can go freely from the 0 degree block to the 180 degree block without a problem. *Even though* this is not possible in the layout representation, we will now pretend that it is and use the local isomorphism between the usual Euclidean transformations (time-based) and the layout transformations. This is done by deriving the expression for the motion of a 180 degree block using directly the thin block formula (10.31).

We use the global linear maps of Equation (10.16) to investigate the relationship between a thin block and a 180 degree block. We start with the transformation

$$x \xrightarrow{\mathcal{Y}(\frac{\pi}{2})} z \tag{10.35.a}$$

$$p_x \longrightarrow p_z \tag{10.35.b}$$

$$z \longrightarrow -x \tag{10.35.c}$$

$$p_z \longrightarrow -p_x. \tag{10.35.d}$$

Of course the map $\mathcal{Y}(-\frac{\pi}{2})$ is simply the inverse of $\mathcal{Y}(\frac{\pi}{2})$. Finally the map $\mathcal{Y}(\pi)$ is the identity in the $y - p_y$ plane and minus the identity in all other directions. To ease our notation we label these important maps as

$$\mathfrak{S} \equiv \mathcal{Y}(\frac{\pi}{2}) \tag{10.36.a}$$

$$\mathfrak{T} \equiv \mathcal{Y}(\pi). \tag{10.36.b}$$

The map $\mathfrak{T}$ which is the square of $\mathfrak{S}$ will not be used but we will discuss its significance in the conclusion of this section. To exploit the local isomorphism between the time base dynamics and the layout dynamics we write an arbitrary operator $\mathcal{P}$ acting on a layout block. It is given by Equation (10.31). Our goal is to manipulate this formula until we get it to act on the map $\mathcal{M}^L_{\theta=\pi}$. Using Figure 10.6, we can write the following relation:

$$\underbrace{\mathcal{Y}(\frac{\theta}{2})\mathcal{M}^L_{\theta=0}\mathcal{Y}(\frac{\theta}{2})}_{\mathcal{M}_{12}} = \underbrace{\mathcal{Y}(\frac{\theta-\pi}{2})\mathcal{M}^L_{\theta=\pi}\mathcal{Y}(\frac{\theta-\pi}{2})}_{\mathcal{M}_{12}}$$
$$\Downarrow$$
$$\mathcal{M}^L_{\theta=0} = \mathfrak{S}^{-1}\mathcal{M}^L_{\theta=\pi}\mathfrak{S}^{-1}. \tag{10.37}$$

The next step consists of expressing the thin map in terms of $\mathcal{M}^L_{\theta=\pi}$. This is done using Equations (10.29) and (10.37), with the result

$$\begin{aligned}\mathcal{M}^0_{12} &= \mathcal{T}_z^{-1/2}\mathcal{M}^L_{\theta=0}\mathcal{T}_z^{-1/2} \\ &= \mathcal{T}_z^{-1/2}\mathfrak{S}^{-1}\mathcal{M}^L_{\theta=\pi}\mathfrak{S}^{-1}\mathcal{T}_z^{-1/2}.\end{aligned} \tag{10.38}$$

We are ready to substitute this new expression into Equation (10.31):

$$\begin{aligned}\mathcal{P}\left[\mathcal{M}_{12}\right] &= \mathcal{D}_{1:12}\mathcal{P}\mathcal{M}^0_{12}\mathcal{P}^{-1}\mathcal{D}_{12:2} \\ &= \mathcal{D}_{1:12}\mathcal{P}\mathfrak{S}^{-1}\mathcal{M}^L_{\theta=\pi}\mathfrak{S}^{-1}\mathcal{P}^{-1}\mathcal{D}_{12:2} \\ &= \mathcal{Y}^{1/2}\mathfrak{S}^{-1}\mathfrak{S}\mathcal{T}_z^{1/2}\mathfrak{S}^{-1}\mathfrak{S}\mathcal{P}\mathfrak{S}^{-1} \\ &\quad \times\mathcal{M}^L_{\theta=\pi}\mathfrak{S}^{-1}\mathcal{P}^{-1}\mathfrak{S}\mathfrak{S}^{-1}\mathcal{T}_z^{1/2}\mathfrak{S}\mathfrak{S}^{-1}\mathcal{Y}^{1/2}.\end{aligned} \tag{10.39}$$

The above expression is not meaningful in the layout dynamics because the map $\mathfrak{S}$ is divergent. We must get rid of this map using the rules derived in the time based dynamics. This can be done using Equation (10.16) and the effect of the map $\mathfrak{S}$ described in Equation (10.35). For example, let us do this for the case of the map $\mathcal{X}$.

$$\begin{aligned}
\mathfrak{S}\mathcal{X}(\theta)\mathfrak{S}^{-1} &= \mathfrak{S}\exp\left(\theta : yp_z - zp_y :\right)\mathfrak{S}^{-1} \\
&= \exp\left(\theta\mathfrak{S} : yp_z - zp_y : \mathfrak{S}^{-1}\right) \\
&= \exp\left(\theta : \mathfrak{S}y\mathfrak{S}p_z - \mathfrak{S}z\mathfrak{S}p_y :\right) \\
&\overset{\text{Eq. (10.38)}}{=} \exp\left(\theta : -yp_x + xp_y :\right) \\
&= \mathcal{Z}(\theta).
\end{aligned} \tag{10.40}$$

The same calculation can be done on all the maps of Equation (10.16):

$$\begin{array}{ccc}
\mathcal{X}(\theta) & \xrightarrow{\mathfrak{S}} & \mathcal{Z}(\theta) \\
\mathcal{Y}(\theta) & & \mathcal{Y}(\theta) \\
\mathcal{Z}(\theta) & & \mathcal{X}(-\theta) \\
\mathcal{T}_x(d) & & \mathcal{T}_z(d) \\
\mathcal{T}_y(d) & & \mathcal{T}_y(d) \\
\mathcal{T}_z(d) & & \mathcal{T}_x(-d)
\end{array} \tag{10.41}$$

Applying these transformations to Equation (10.39) gives us

$$\begin{aligned}
\mathcal{P}_{\vec{d},\vec{\alpha}}\left[\mathcal{M}_{12}\right] = \mathcal{Y}^{1/2}_{-\pi/2}\mathcal{T}_x^{-1/2}\mathcal{P}_{\pi/2} \\
\times\mathcal{M}^{L}_{\theta=\pi}\mathcal{P}^{-1}_{-\pi/2}\mathcal{T}_x^{-1/2}\mathcal{Y}^{1/2}_{-\pi/2}
\end{aligned} \tag{10.42.a}$$

where

$$\mathcal{Y}^{1/2}_{-\pi/2} = \mathcal{Y}\left(\frac{\theta-\pi}{2}\right) \tag{10.42.b}$$

$$\mathcal{T}_x^{\pm 1/2} = \mathcal{T}_x\left(\pm\frac{O_1O_2}{2}\right) \tag{10.42.c}$$

$$\begin{aligned}
\mathcal{P}_{\pi/2} = \mathcal{T}_z(d_x)\mathcal{T}_y(d_y)\mathcal{T}_x(-d_z) \\
\times\mathcal{Z}(\theta_x)\mathcal{Y}(\theta_y)\mathcal{X}(-\theta_z)
\end{aligned} \tag{10.42.d}$$

$$\begin{aligned}
\mathcal{P}^{-1}_{-\pi/2} = \mathcal{X}(-\theta_z)\mathcal{Y}(-\theta_y)\mathcal{Z}(\theta_x) \\
\times\mathcal{T}_x(-d_z)\mathcal{T}_y(-d_y)\mathcal{T}_z(d_x).
\end{aligned} \tag{10.42.e}$$

The entire expression for $\mathcal{M}^{L}_{\theta=\pi}$ can be substituted in:

$$\begin{aligned}
\mathcal{P}_{\vec{d},\vec{\alpha}}\left[\mathcal{M}_{12}\right] = \mathcal{Y}^{1/2}_{-\pi/2}\mathcal{T}_x^{-1/2}\mathcal{P}_{\pi/2}\mathcal{T}_x^{1/2}\mathcal{Y}^{-1/2}_{\pi/2} \\
\times\mathcal{M}_{12}\mathcal{Y}^{-1/2}_{\pi/2}\mathcal{T}_x^{-1/2}\mathcal{P}^{-1}_{-\pi/2}\mathcal{T}_x^{1/2}\mathcal{Y}^{1/2}_{-\pi/2}.
\end{aligned} \tag{10.43}$$

The maps of Equation (10.43) are well-defined in the layout dynamics for a block in the neighborhood of 180 degrees.

10.4.3.4 Conclusion on the two simple representations

In our discussion we encountered two diverging maps in the layout representation $\mathfrak{S}$ and $\mathfrak{T}$. There are mathematical and, I dare to say, physical reasons for the divergence in these maps. We recall from Sect. 3.1.2, that one can use a distance along the path of the gutter or racetrack if and only if the marble does not reverse direction. If it does then the "s-dependent" Hamiltonian ceases to apply and one must go back to the time based dynamics of the laboratory or Lagrangian class common to single and collective dynamics. Another way to put it: u-turns are forbidden in the layout representation.

The map $\mathfrak{S}$ (German letter "S" as in Stop!) can be interpreted as going into a frame where time has stopped. Consider the simplest straight block/element: a drift. Normally, for a properly defined layout, the particles in the beam congregate around the momentum $\vec{p} = (0, 0, p_0)$. For these particles, the usual angle coordinate dx/dz is around zero. If we rotate by $\frac{\pi}{2}$ using $\mathfrak{S}$, the value of dx/dz becomes infinite. The choice of our layout plane, instead of being perpendicular to the direction of propagation, is now parallel! Thus the value of dx/dz (i.e., the change in x as we microscopically move the layout plane in the z-direction) is ill-defined. This divergence is reflected properly in Equations (10.23) and in (10.26).

The map $\mathfrak{T}$ corresponds to a time reversal operator with a change of z into $-z$, in fact, it is an illegal u-turn. Indeed one looks at a sector block of 180 degrees and relates its map $\mathcal{S}_{12}^{\pi}$ to a zero degree block (ignoring divergences) then one obtains the formula

$$\begin{aligned} \mathcal{S}_{12}^{\pi} &= \mathfrak{S}\mathcal{T}_z^{1/2}\mathcal{M}_{12}^{0}\mathcal{T}_z^{1/2}\mathfrak{S} \\ &= \mathfrak{S}^2 \underbrace{\mathcal{T}_x^{-1/2}\mathfrak{S}^{-1}\mathcal{M}_{12}^{0}\mathfrak{S}\mathcal{T}_x^{-1/2}}_{\text{Map formula rotated by } \pi/2} \end{aligned} \tag{10.44}$$

The maps in Equation (10.44) are purely formal. The map $\mathfrak{S}^{-1}\mathcal{M}_{12}^{0}\mathfrak{S}$ represents a fictitious thin map located in the center of the bend and acting in the original $x - y$ plane (the common layout plane of the entrance and exit). The first map $\mathcal{T}_x^{-1/2}$ brings a ray from O_1 to the center of the magnet O_{12} and the second one from the center to O_2. The map $\mathfrak{S}^2\mathcal{T}_x^{-1/2}\mathfrak{S}^{-1}\mathcal{M}_{12}^{0}\mathfrak{S}\mathcal{T}_x^{-1/2}$ looks very much like the usual representation of a thick map in terms of a thin one (modulo interchanges of x and z and minus signs) except for the factor $\mathfrak{S}^2 = \mathfrak{T}$ which represents a "time" reversal: a 180 degree block must reverse the direction of the propagation! Coming back to the trivial example of Equation (10.34) we notice that a positive displacement of the 180 degree bend required at the exit the same positive displacement instead of its inverse. This switch of sign is due to the reversal map $\mathfrak{T}$.

The occurrence of the map $\mathfrak{S}$ or $\mathfrak{T}$ in any attempt to connect zero degree and 180 degree blocks forces us to have a minimum of two representations of the Euclidean group. This cannot be avoided. The question which remains is whether or not it is enough. Right angle bends sit in between the two cases. This implies that particles with trajectories (minus) 45 degrees away from the original z-axis will be lost by the rotation into the plane of a zero (180) degree block. Can we have an infinite number of representations of the Euclidean group each tailored to the particular block angle Φ_{12} between the entrance and layout planes? Of course the answer is yes but with additional pain. The 0 and 180 degree representations are simply related as shown in Equation (10.42), by a mere relabeling scheme. This is not the case for a representation tailored to the particular layout angle Φ_{12}. It is the topic of the next section.

10.4.4 Locally Accurate Representation

Let us start[10.7] with Equation (10.31) and a new representation of the map $\mathcal{P}$. We will write $\mathcal{P}$ as

$$\mathcal{P} = \exp\left(: \vec{d} \cdot \vec{p} :\right) \exp\left(\theta : \vec{\beta} \cdot \vec{L} :\right)$$
$$\text{where } \left\|\vec{\beta}\right\| = 1. \tag{10.45}$$

Obviously a complex relation must exist between the vector $\vec{\alpha}$ of the previous sections and the new vector $\theta\vec{\beta}$. This relation depends only on the Lie algebra of the group generators.

In the following exercise we review the famous geometric properties of the vector $\theta\vec{\beta}$.

[10.7]This section finishes some incomplete work started by Healy in his thesis. See [52] chapters 11 and 12. In addition Healy's work on mispowering of ideal elements ([52] and [53]) for "matrix" codes can be tremendously simplified by the results in this chapter. Please take a look at the gruesome Lie polynomials for the mispowered ideal bend in reference [53].

Exercise 10-4

In this exercise, we display the local isomorphism between the layout and the time based dynamics. The operators involved are

$$\begin{array}{lll} & \text{Time} & \text{Layout} \\ p_z & p_z & \sqrt{(1+\delta)^2 - p_x^2 - p_y^2} \\ L_x & y p_z - z p_y & y\sqrt{(1+\delta)^2 - p_x^2 - p_y^2} \\ L_y & z p_x - x p_z & -x\sqrt{(1+\delta)^2 - p_x^2 - p_y^2} \\ L_z & x p_y - y p_x & x p_y - y p_x . \end{array} \quad (10.46)$$

1. *Show explicitly that*

$$\begin{aligned} [L_x, L_y] &= L_z \\ [L_z, L_x] &= L_y \\ [L_y, L_z] &= L_x \\ \text{or } [L_i, L_j] &= \varepsilon_{kij} L_k, \end{aligned} \quad (10.47)$$

is correct for both dynamics. The tensor ε_{kij} is the usual anti-symmetric tensor that defines the vector product.

2. *Show that* $\left[\vec{\beta}\cdot\vec{L},\ \vec{L}\right] = \vec{L}\times\vec{\beta}$ *and* $\left[\vec{\beta}\cdot\vec{L},\vec{q}\right] = \vec{q}\times\vec{\beta}$.

3. *It is well known that an arbitrary vector $\vec{q}$ can be written as* $\vec{q} = \left(\vec{q}\cdot\vec{\beta}\right)\vec{\beta} + \left(\vec{\beta}\times\vec{q}\right)\times\vec{\beta}$. *Show that*

$$\exp\left(\theta : \vec{\beta}\cdot\vec{L} :\right)\vec{q} = \left(\vec{q}\cdot\vec{\beta}\right)\vec{\beta} + \cos(\theta)\left(\vec{\beta}\times\vec{q}\right)\times\vec{\beta} + \sin(\theta)\,\vec{\beta}\times\vec{q}. \quad (10.48)$$

The same formula applies for $\vec{p}$. Hence one concludes that the operator $\exp\left(\theta : \vec{\beta}\cdot\vec{L} :\right)$ *is a rotation of angle θ around the axis described by the unit vector $\vec{\beta}$. Similarly one gets the result*

$$\exp\left(\theta : \vec{\beta}\cdot\vec{L} :\right)\vec{L} = \left(\vec{L}\cdot\vec{\beta}\right)\vec{L} + \cos(\theta)\left(\vec{\beta}\times\vec{L}\right)\times\vec{\beta} + \sin(\theta)\,\vec{\beta}\times\vec{L}. \quad (10.49)$$

In cases for which the layout angle Φ_{12} is substantial, but the displacement map $\mathcal{P}$ is near the identity (errors for example), we know intuitively that an adequate representation of the displacement should only "remove" rays with initial directions in the neighborhood of $\pm\pi/2$. The angle Φ_{12} enters artificially in the zero and π representation: these are the only angles with simple formulas and are thus used. Here we will attempt to get rid of this accidental divergence. Let us start with Equation (10.31); our goal is to rewrite $\mathcal{Y}^{\pm 1/2}\mathcal{T}^{\pm 1/2}\mathcal{P}^{\pm}\mathcal{T}^{\mp 1/2}\mathcal{Y}^{\mp 1/2}$ by moving the map $\mathcal{T}^{\mp 1/2}$ into the exponents defining the map $\mathcal{P}^{\pm}$. Of course we first analyze everything "as if" we are in the time based representation. First we commute a translation past a rotation:

$$\begin{aligned}\mathcal{R}\mathcal{T}_{\vec{d}}\vec{q} &= \mathcal{R}(\vec{q}-\vec{d})\\ &= R\vec{q}-\vec{d}\\ &= R(\vec{q}-R^{-1}\vec{d})\\ &= \mathcal{T}_{R^{-1}\vec{d}}R\vec{q}\\ &= \mathcal{T}_{R^{-1}\vec{d}}\mathcal{R}\vec{q}\\ &\Downarrow\\ \mathcal{R}\mathcal{T}_{\vec{d}} &= \mathcal{T}_{R^{-1}\vec{d}}\mathcal{R}. \end{aligned} \tag{10.50}$$

In Equation (10.50) we use the definition of a compositional map. Using this result, we rewrite $\mathcal{T}^{1/2}\mathcal{P}\mathcal{T}^{-1/2}$ as

$$\begin{aligned}\mathcal{T}^{1/2}\mathcal{P}\mathcal{T}^{-1/2} &= \mathcal{T}_{\vec{D}}\mathcal{P}\\ \text{where } \vec{D} &= (1-R_{\theta\beta}^{-1})\begin{pmatrix}0\\0\\O_{12}/2\end{pmatrix}.\end{aligned} \tag{10.51}$$

And on the exit side, we write similarly

$$\mathcal{T}^{-1/2}\mathcal{P}^{-1}\mathcal{T}^{1/2} = \mathcal{P}^{-1}\mathcal{T}_{\vec{D}}. \tag{10.52}$$

Now, we lump all the translations together:

$$\begin{aligned}\mathcal{P}\left[\mathcal{M}_{12}\right] &= \mathcal{Y}^{1/2}\mathcal{T}_{\vec{D}}\mathcal{P}\mathcal{Y}^{-1/2}\mathcal{M}_{12}\mathcal{Y}^{-1/2}\mathcal{P}^{-1}\mathcal{T}_{\vec{D}}\mathcal{Y}^{1/2}\\ &= \mathcal{Y}^{1/2}\mathcal{T}_{\vec{D}+\vec{d}}\mathcal{R}_{\theta\vec{\beta}}\mathcal{Y}^{-1/2}\mathcal{M}_{12}\\ &\quad\times\mathcal{Y}^{-1/2}\mathcal{R}_{-\theta\vec{\beta}}\mathcal{T}_{\vec{D}-\vec{d}}\mathcal{Y}^{1/2};\end{aligned} \tag{10.53}$$

we must then commute the rotation $\mathcal{Y}^{-1/2}$ across the other maps so as to cancel $\mathcal{Y}$.

Exercise 10-5

In this exercise we develop the formulas which will permit us to commute rotations past one another.

1. *Using the operators* $:L_i:$ *show that the transpose of a rotation is its inverse.*
2. *Using the similarity between Equations* (10.48) *and* (10.49) *show that if*

$$\exp\left(\gamma : \vec{\alpha}\cdot\vec{L} :\right)\vec{q} = R_{\gamma\vec{\alpha}}\vec{q}$$

then

$$\mathcal{R}_{\gamma\vec{\alpha}}\exp\left(\theta : \vec{\beta}\cdot\vec{L} :\right)\mathcal{R}^{-1}_{\gamma\vec{\alpha}} = \exp\left(\theta : \left\{R^{-1}_{\gamma\vec{\alpha}}\vec{\beta}\right\}\cdot\vec{L} :\right).$$

Show that a similar result holds for translations.

3. *Prove the above result directly using Equation* (10.49) *and the standard vector identity on the cross product, i.e.,* $\left(\vec{a}\times\vec{b}\right)\times\vec{c} = \left(\vec{a}\cdot\vec{c}\right)\vec{b} - \left(\vec{b}\cdot\vec{c}\right)\vec{a}$.

Using the results of Exercise 10-5 and Equation (10.50), one obtains

$$\begin{aligned}\mathcal{P}\left[\mathcal{M}_{12}\right] &= \mathcal{Y}^{1/2}\mathcal{T}_{\vec{D}+\vec{d}}\mathcal{Y}^{-1/2}\mathcal{R}_{\theta Y^{-1/2}\vec{\beta}}\mathcal{M}_{12} \\ &\quad\times\mathcal{R}_{-\theta Y^{1/2}\vec{\beta}}\mathcal{Y}^{-1/2}\mathcal{T}_{\vec{D}-\vec{d}}\mathcal{Y}^{1/2} \\ &= \mathcal{T}_{Y^{-1/2}\{\vec{D}+\vec{d}\}}\mathcal{R}_{\theta Y^{-1/2}\vec{\beta}}\mathcal{M}_{12} \\ &\quad\times\mathcal{R}_{-\theta Y^{1/2}\vec{\beta}}\mathcal{T}_{Y^{1/2}\{\vec{D}-\vec{d}\}} \\ &= \mathcal{T}_{\vec{\Delta}^{in}}\mathcal{R}_{\theta\vec{\beta}^{in}}\mathcal{M}_{12}\mathcal{R}_{-\theta\vec{\beta}^{out}}\mathcal{T}_{-\vec{\Delta}^{out}}.\end{aligned} \tag{10.54}$$

These new inputs are all computable using simple linear algebra:

$$\vec{\Delta}^{in} = Y^{-1/2}\left\{\vec{D}+\vec{d}\right\} \tag{10.55.a}$$

$$\vec{\beta}^{in} = Y^{-1/2}\vec{\beta} \tag{10.55.b}$$

$$\vec{\Delta}^{out} = Y^{1/2}\left\{\vec{D}-\vec{d}\right\} \tag{10.55.c}$$

$$\vec{\beta}^{out} = Y^{1/2}\vec{\beta} \tag{10.55.d}$$

$$\vec{D} = (1 - R^{-1}_{\theta\vec{\beta}})\begin{pmatrix}0\\0\\O_{12}/2\end{pmatrix}. \tag{10.55.e}$$

Finally the maps $\mathcal{R}_{\theta\vec{\beta}^{in}}$ and $\mathcal{R}_{-\theta\vec{\beta}^{out}}$ must be evaluated in their dynamical form. We can again use our local isomorphism and transform the matrix into an Euler form:

$$
\begin{aligned}
R_{\theta\vec{\beta}^{in}} &\overset{\substack{\text{Linear}\\\text{Algebra}}}{=} Z(\theta_z^{in})Y(\theta_y^{in})X(\theta_x^{in}) \\
&\Downarrow \\
\mathcal{R}_{\theta\vec{\beta}^{in}} &= \mathcal{X}(\theta_x^{in})\mathcal{Y}(\theta_y^{in})\mathcal{Z}(\theta_z^{in}).
\end{aligned}
\tag{10.56}
$$

Notice that to leading order $\theta_x^{in} = \theta\beta_x^{in}$, $\theta_y^{in} = -\theta\beta_y^{in}$, and $\theta_z^{in} = \theta\beta_z^{in}$. A similar formula applies for the exit angles.

The passage from the single exponent representation to the Euler representation can be done analytically or numerically. In any event, it is a good exercise to write down a code which compares the zero, 180 and local dynamical representations to satisfy oneself that the theory works (the isomorphism) and the results are correct. The author has done it and the serious users of these formulas should also do it.

I should be pointed out that, while the above gymnastic will convince us of the validity of our local isomorphisms, it is probably better in an actual code to use the geometrical approach of Exercise 10-2 to implement the dynamical rotation of Equation (10.54).

10.5 CONCLUSION: GRAPHICAL=DYNAMICAL?

The title of this section reflects the content of the section before this book went into print. At that time it was my intention to show on a geometrical model how the thin block approach to the derivation of the dynamical representation has geometrical equivalents. One can construct maps in the regular geometrical affine space which manipulates figures and pictures using the exact sequence of maps given by Equation (10.31).

However, as I showed in Sect. 10.4.4, the locally accurate representation gives the dynamical map from the entrance layout plane to entrance magnet plane ($\mathcal{T}_{\vec{\Delta}^{in}}\mathcal{R}_{\theta\vec{\beta}^{in}}$) as well as the exit plane version of this map ($\mathcal{R}_{-\theta\vec{\beta}^{out}}\mathcal{T}_{-\vec{\Delta}^{out}}$). It also turns out that one can compute these maps directly from geometry and, using the results of Exercise 10-2, immediately write down the locally accurate dynamical representation.

So what is the purpose of all of this? In part the answer lies in the huge emphasis place on large machine with small bending radii. In these machines, as we will see, tricks are used to compute the maps of bends and to displace them. These tricks are not derived from a solid theoretical foundation, but from "back-of-the-envelop" arguments. Nothing wrong one might say... However when the need to treat "small machine" arises, one must go back to first principles. In this chapter I showed how the techniques which are rigorous and good for large machine (thin block representation) can be modified to give rise to small machine expressions.

In Chapter 14, I will show how the present results lead to correct derivations of the "back-of-the-envelop" formulas used in large machines. So, in conclusion, my goal here was to establish a solid foundation which takes into account our predilection for "large machine" but does not preclude the correct treatment necessary in machines with small bending radii.

CHAPTER 11
SYMPLECTIC INTEGRATION

The study of beam dynamics presupposes the existence of a model Hamiltonian within a block of the layout (i.e., the Hamiltonian of the magnet) and a representation of the map generated by this model from entrance plane i to exit plane $i+1$. This statement is acceptable to all fields of charged particle dynamics whenever single particle dynamics is useful: rings, linacs, spectrometers, electron microscopes and even light optics. Hence the discussion in Chapter 10 is a common trunk to all these fields of charged particle optics.

Unfortunately this is approximately where the various fields sharply divide into branches having little in common. In fact they have dogmas and superstitions which are contradictory. We have already addressed the cause of this phenomenon in Sect. 8.2. Mainly, the ring physicists study the iterates of a map while physicists from the other fields send externally defined rays through a system and measure the output.[11.1] Single pass physicists submit their study to a strict error analysis—neglecting effects which are not measurable in the output plane. In rings, by contrast, we have argued that microscopic effects, unmeasurable at the single magnet level, such as radiation, can be critical in determining the long term stability of an electron ring.

Not surprisingly, the experimental differences and goals lead to a totally different sense of what is important and what is negligible. Personally I have heard a lot of questions directed towards us by the single pass community that reflect this state of affairs:

1. Why do you compute certain effects to high order (such as Taylor series maps), when the individual magnets are represented with models of utter triviality (a single sextupole kick for example)?

[11.1] Linac physicists are in a special class because they tend to study a quasi-periodic structure that is often connected to a ring. Hence they share the same language as the ring physicists; they are often practitioners of both trades!

2. How can you use simplistic formulas like the first order quadrupole fringe effect of Lee-Whitting which is at least "off by 10%?" And worse, how can you neglect fringe fields altogether in most of your simulations?

3. Why do you care about Lie polynomials since, in the end, you want the coordinates of the ray?

These questions are only matched by opposite statements from our community such as: "Quadrupoles are linear!" It is interesting to point out that in most electron rings ignoring "nonlinearity in quadrupoles," i.e., the small angle approximation, is a far bigger effect than neglecting radiation when analyzed as single pass systems.

A proper understanding of these issues is important, not because it improves our diplomatic relations with the single pass community, but because it helps us understand what are the dogmas of our own field. As the experienced reader will see very soon, I am not rejecting the standard dogmas of particle accelerator dynamics, but I am insisting that they should be discussed, so that neglected effects are properly reinserted when necessary.

11.1 "STANDARD DOGMA"

I will start with a description of my interpretation of the dogmatic sets of rules to which I mostly agree. As you will see, accelerator physicists take, quite appropriately, a point of view similar to that of engineers in constructing a model for the machine to be built or to be probed. One sets up an incomplete and grossly simplistic model which mimics, to the best of our knowledge, the relevant features of the operation of the actual system.

I will here quote Talman and Peggs[54]:

> Instead of using approximate formulae to perform tracking through "exact" (i.e., thick) elements, it is possible to perform exact tracking through (i.e., thin) elements.[11.2] The approximation can be improved by breaking thick elements into several thin elements.[11.3] Long-term precision in such a program is only compromised by round-off error

[11.2] Note of the Author: That is to say "inexact" elements.

[11.3] N.A. Integration is performed.

> in the computer. Being exact, such a calculation is globally symplectic, though not locally symplectic (unless extended to six dimensions).[11.4] A program called TEAPOT uses this method. [11.5]

Another statement from Ron Ruth, which I quote from memory, stipulates that

> "One can tolerate longitudinal approximations,[11.6] but **uncontrolled** transverse approximations[11.7] are very dangerous. "Symplectification" of an approximate one-turn Taylor series is not the same as symplectic integration and the two procedures should not be confused.

The most important concept here is that of "uncontrolled transverse approximations" because the purpose of ring studies is to determine the effect of errors and new magnets on the transverse stability of the ring. We are trying to answer whether or not certain physical effects will eject the particle out of the machine. Hence it is highly disturbing if we must, at every change of the parameters, perform complex checks on the transverse validity of our map. This does not imply that the usage of integrators guarantees reliable results. It is by now well-known that the blind use of the small angle approximation, leading to the usual "kick codes" with "linear elements" is a constant source of confusion in small machines because many accelerator designers and theorists have forgotten the very existence of the small angle approximation. It thus follows that one must also ask serious questions before using a symplectic integrator:

- Can we use the small angle approximation, i.e., is it a controlled transverse approximation?
- Do we need fringe field effects?
- How many integrations steps are needed in each magnet? (How many thin lenses in Talman's language.)

[11.4] N.A. This statement is the result of a misunderstanding by Talman and Peggs. Please ignore it except for the fact that their code is indeed symplectic and in fact "locally" symplectic despite their doubts. All of this will become clear to the reader in Chapter 12.

[11.5] N.A. Reference [6].

[11.6] N.A. Finite number of integration steps.

[11.7] N.A. Taylor series expansion of the map in Lie form or any other form.

Fortunately it is a result of my own experience and that of other physicists in the field, that these questions need to be asked at the onset of a study and rarely afterwards.

When using a transverse approximation of the one-turn map, this question must be addressed virtually each time parameters are changed considerably. Hence the central dogma of ring dynamics is that the most reliable model consists of

1. A model with controlled transverse approximations such as
   (a) Small angle approximation or so-called large machine Hamiltonian leading to the "linear quadrupoles and linear bends."
   (b) Exclusion of the fringe field with the notable exception of vertical focusing in bends.
   (c) Approximate misalignments of magnets based also on the small angle approximation (small bending radii).
2. Usage of (explicit) symplectic integration whenever possible accompanied by
   (a) A reinterpretation of the model so as to include in its definition the number of integration steps as well as the order of the method. In other words, the "thin lens" machine becomes the "exact" machine as in Talman's TEAPOT[6] philosophy. This permits a strong minimization of the number of steps.
   (b) Small nonsymplectic effects of physical origin (i.e., radiation) are then introduced on top of the symplectic integrator and thus their relative effect is clearly visible.
3. High order integrators nearly reaching machine precision must be used if everything else fails.

I adhere to this philosophy[11.8] *completely in the sense that I presuppose always the existence of an integrator, high order predictor-*

[11.8]This dogma cannot be justified except on linear systems where it is rigorous. Indeed, I became a follower of this approach after a question posed by my thesis advisor Alex Dragt to his previous student David Douglas. Dragt was interested to know how the small angle standard kick code (used by Douglas) could produce the correct chromatic behavior while truncating the Hamiltonian in the leading order in angle ($\sin\alpha = \tan\alpha = \alpha = x' = p_x$). The answer was that such a code contained all the correct δ dependence except for terms proportional to the bending angle which, for the SSC, was *always* within the small angle approximation. The third order Tay-

corrector if need be, against which other approximations can be tested so that they themselves become controlled approximations.

Any attempt to popularize an understanding of the theory based on finite maps and the necessary introduction of Lie methods going along with it (i.e., the Courant-Snyder invariant!) is viewed and misunderstood by many as an insidious way to introduce uncontrolled approximate maps. Quite characteristically, in their review, which I just quoted, Peggs and Talman spend most of their time telling the readers what "Lie methods are useless for" while correctly concluding that "the main virtue is its systematic handling of distortion functions." In fact the entire first part of their paper based on the "iterative map" is what map/Lie methods can handle with most power and elegance. Their criticism of the Lie methods would be tantamount to my criticism of their approximate analytical results[11.9] on the basis that they should not be used for tracking.

So I state again: *Once the integrator exists, other approximations can be tested so that they themselves become controlled approximations.*

In conclusion, I consider the study of approximate one-turn maps with transverse approximations to be important, but careful tests must be performed against the integrator. This is true if the map is used to perform perturbative calculations, and even more so, if used for tracking purposes. Hence our discussion in rings must start with the integrator. Other types of approximation will co-exist with the integrator without ever supplanting it completely.[11.10]

11.2 EXPLICIT SYMPLECTIC INTEGRATORS

Explicit symplectic integrators are based on four simple statements:

---

lor series approximation in canonical variables, which was used in Dragt's code MARYLIE[5], was bound to fail beyond the second order chromaticity because it involved an uncontrolled δ expansion.

[11.9] The authors give formulas in one degree of freedom for the tune shift due to sextupoles (second order) and the discrete one-resonance Hamiltonian near a third integer in what I call the super naïve approximation. See Equation (5.53).

[11.10] If, let us say, a tenth degree, properly symplectified Taylor series map extracted from the integrator by TPSA, and thus with exact coefficients, produces tracking results in gross disagreement with the code that generated it, how can one believe any tenth degree map for that ring. That is the fundamental barrier faced by those who wish to use simulation tools based on transverse Taylor series maps only.

1. If one has the exact solution for a Hamiltonian, then it is symplectic.

2. If we do not have the exact solution for the exact Hamiltonian, we may rewrite it as a sum of solvable parts, and thus producing, according to 1, symplectic maps.

3. A symplectic (longitudinally) approximate map for the total Hamiltonian can be found using all the maps of item 2.

4. The order of this integrator can be increased using results of the theory of free Lie algebra.

Let us examine Talman's quotation of Sect. 11.1 in the light of symplectic integration. Let us start with the thick Hamiltonian for a combined function sector bend filling the layout block between entrance plane 1 and exit plane 2:

$$H_1 = -\ (1 + \frac{x}{\rho_1})\sqrt{(1+\delta)^2 - p_x^2 - p_y^2} \\ + b_1(x + \frac{x^2}{2\rho_1}) + V(x, y; \rho_1). \tag{11.1}$$

The parameter ρ_1 is related to the distance between the entrance and exit frame using the formula $O_1O_2 = 2\rho_1 \sin\left(\frac{\Phi_{12}}{2}\right)$. The map through this magnet is gotten by integrating for an arc length $L_{12} = \Phi_{12}\rho_1$.

In general there is no exact solution for the map generated by Equation (11.1). However, the individual Hamiltonians

$$D_1 = -\ (1 + \frac{x}{\rho_1})\sqrt{(1+\delta)^2 - p_x^2 - p_y^2} \tag{11.2}$$

and

$$K_1 = b_1(x + \frac{x^2}{2\rho_1}) + V(x, y; \rho_1) \tag{11.3}$$

are exactly solvable. This is immediately obvious for the case of K_1 which is just a multipole kick. As for D_1, it is none other than our favorite drift in polar coordinates, the dynamical rotation $\mathcal{Y}$ in the $x - z$ plane shifted around the design arc of circle at ρ_1.

Let us now assume that the Hamiltonian of the magnet is replaced by a drift in polar coordinates (D_1) and a kick (K_1) at the center of the layout. The new Hamiltonian T_1 is

$$T_1 = D_1 + \delta(s - \frac{L_{12}}{2})L_{12}K_1. \tag{11.4}$$

The compositional map for this Hamiltonian is given by

$$\begin{aligned}\mathcal{T}_{12}^2 &= \exp\left(- : \frac{L_{12}}{2} D_1 :\right) \exp\ (: -L_{12}K_1 :) \exp\left(- : \frac{L_{12}}{2} D_1 :\right) \\ &= \mathcal{Y}(\frac{\Phi_{12}}{2}) \exp\ (: -L_{12}K_1 :) \mathcal{Y}(\frac{\Phi_{12}}{2}). \end{aligned} \tag{11.5}$$

It is quite clear that the compositional map $\mathcal{T}_{12}^2$ is a symplectic approximation of the map $\mathcal{M}_{12}$:

$$\mathcal{T}_{12}^2(L_{12}) = \mathcal{M}_{12}(L_{12}) + \text{Order}(L_{12}^3). \tag{11.6}$$

At this stage there are two ways to look at the approximate solution of $\mathcal{T}_{12}^2$:

1. One step of a second order symplectic integrator for H_1.
2. A new model for the map of the magnet differing from the original model in powers of the integration (longitudinal) variable only.

I say emphatically and without reservation that accelerator physicists always view the map of the integrator as a new model for the magnet. This they do consciously as Talman did in defending his TEAPOT code[6] or unconsciously when they run standard kick codes. Historically kick codes that integrate linear elements using the drift-kick integration split were criticized for their inaccuracies because they did not produce the accurate lattice functions of the thick representation (matrices). This happens because of an *inconsistent* refusal to re-fit the linear parameters of the systems (quadrupole strengths, etc...) so as to produce, with the drift-kick integrator, a machine whose tune and lattice functions match that of the matrix-kick integrator. I say that this refusal is inconsistent because sextupoles where virtually always represented by a single thin lens kick by the detractors of the drift-kick integrators. The decision to use a single kick for most sextupoles is a modeling decision based on a particular understanding of their effects on the complex dynamical properties of the lattice. It is clear that if the parameters of the lattice are changed—putting the center of the sextupoles π apart—the modeling of sextupoles by a single thin lens becomes highly questionable. These dogmatic and unfair criticisms towards a drift-kick integrator lead to the rather defensive exposition of Talman and collaborators in references [6, 54].

Talman's approach to modeling restricts even further the type of maps one may use in the integrator. Talman views the "new ring" as made of thin magnets. In other words the new magnets are restricted to be limiting cases of actual thick magnets. If many steps are needed (i.e., many thin lenses), then the new magnet can be viewed as a collection of thin lenses (laminations?) separated by drift space. Pushing this "physicalization" of the integrator further, Talman even computes the new closed orbit created by each thin lens. The design orbit and the layout arcs become part of a polygon. For example, a ring with 36 bends, if each one is represented by one thin lens, possesses a "tridecahexagonal" shaped closed orbit.

We have argued that the layout planes and thus the reference arcs of circle joining them should be totally disconnected from the model of the map. In reference [6], Talman and Schachinger fail to recognize the virtue of this separation, not by ignoring the layout as is done traditionally, but by insisting that the design values of the bends determine the closed reference polygon of the ring. But, more important, in giving a physical meaning to all the drifts and the kicks of the integrator, they reject implicitly the usage of higher order integrators. Higher order combinations of drifts and kicks require maps which "drift backwards" as high order Runge-Kutta methods require evaluations of forces at specific times which do not grow monotonically with each evaluation. For example the 4th order version of Equation (11.5), first derived by Ruth, is given by the combination

$$
\begin{aligned}
\mathcal{T}_{12}^{4} = & \exp\left(-:d_1 L_{12} D_1:\right) \exp\left(:-k_1 L_{12} K_1:\right) \exp\left(-:d_2 L_{12} D_1:\right) \\
& \times \exp\left(:-k_2 L_{12} K_1:\right) \qquad (11.7) \\
& \times \exp\left(-:d_2 L_{12} D_1:\right) \exp\left(:-k_1 L_{12} K_1:\right) \exp\left(-:d_1 L_{12} D_1:\right).
\end{aligned}
$$

The coefficients d_i and k_i are

$$
\begin{aligned}
d_1 &= \frac{1}{2(1+\alpha)}\,, \quad d_2 = \alpha d_1 \\
k_1 &= \frac{1}{(1+\alpha)}\,, \quad k_2 = (\alpha - 1) k_1 \\
\alpha &= 1 - 2^{1/3}. \qquad (11.8)
\end{aligned}
$$

The reader notices that d_2 and k_2 are negative as advertised.

The motivation for using higher order integrators is speed. Let us assume that in order to reach a similar modeling accuracy, the second order integrator $\mathcal{T}_{12}^{2}$ requires N_2 steps while the fourth order needs N_4 steps. The speed of computation will be proportional to the total number of maps involved and thus approximately, the model replacement should obey the rules:

$$\text{if } N_2 < 3N_4 \Rightarrow \mathcal{M}_{12} \rightarrow \left(T_{12}^2(\frac{L_{12}}{N_2})\right)^{N_2}$$
$$\text{if } N_2 > 3N_4 \Rightarrow \mathcal{M}_{12} \rightarrow \left(T_{12}^4(\frac{L_{12}}{N_4})\right)^{N_4} . \tag{11.9}$$

It is our experience based on tracking with integrators that the strict Talman interpretation of the modeling is not needed. Going to higher order integrators simply improves the accuracy of the model without introducing any nasty effects.[11.11] However, the essential aspect of Talman's statement remains: the integrator map is re-interpreted as the "exact" solution. The reader is invited to take a look at Sect. 3.3 of Chapter 3 where we illustrated these ideas with a simple model.

11.3 GENERALIZED YOSHIDA THEORY

In this section I will discuss symplectic integration following the recent work of Yoshida[55], not because I am trying to ignore the historical importance of Ruth's work but because Yoshida's Lie algebraic formulation is the most powerful, surpassing his predecessors, the author included!

11.3.1 Symmetrized Integrators

The starting point of Yoshida's theory is the existence of symmetrized integrators. Consider the following product of Lie transformations

$$\mathcal{T}(\varepsilon) = e^{\varepsilon \widehat{A}_1} e^{\varepsilon \widehat{A}_2} \ldots e^{\varepsilon \widehat{A}_n} \ldots e^{\varepsilon \widehat{A}_2} e^{\varepsilon \widehat{A}_1} \tag{11.10}$$

which we decide to re-write to some order in ε as

$$\mathcal{T}(\varepsilon) = e^{\varepsilon \widehat{C}(\varepsilon)}. \tag{11.11}$$

[11.11] It must be said that models which allow self-consistent time tracking as well as "s" tracking must be made of integrators for which $s(t)$ is monotonic in the interior of the magnets. Thus higher order integrators are forbidden. Talman's code TEAPOT falls in the category of acceptable models: Machida (see [48]) successfully turned it into a time tracking code for the purpose of space charge calculation.

Theorem 4 *The Taylor series expansion of the operator $\widehat{C}(\varepsilon)$ contains no odd powers of ε.*

Hence if theorem 4 is true a symmetrized integrator of order $2k+1$ is automatically of order $2k+2$. Let us prove the theorem by computing the inverse of the map $\mathcal{T}(\varepsilon)$ in two different ways:

$$\begin{aligned}\mathcal{T}^{-1}(\varepsilon) &= \left\{e^{\varepsilon\widehat{A}_1}e^{\varepsilon\widehat{A}_2}\cdots e^{\varepsilon\widehat{A}_n}\cdots e^{\varepsilon\widehat{A}_2}e^{\varepsilon\widehat{A}_1}\right\}^{-1}\\ &= e^{-\varepsilon\widehat{A}_1}e^{-\varepsilon\widehat{A}_2}\cdots e^{-\varepsilon\widehat{A}_n}\cdots e^{-\varepsilon\widehat{A}_2}e^{-\varepsilon\widehat{A}_1}\\ &= \mathcal{T}(-\varepsilon),\end{aligned} \tag{11.12}$$

and now using the single exponent Lie representation,

$$\begin{aligned}\mathcal{T}^{-1}(\varepsilon) &= \left\{e^{\varepsilon\widehat{C}(\varepsilon)}\right\}^{-1}\\ &= e^{-\varepsilon\widehat{C}(\varepsilon)}.\end{aligned} \tag{11.13}$$

It follows that the operator $\widehat{C}$ obeys the rule

$$\begin{gathered}\widehat{C}(-\varepsilon) = \widehat{C}(\varepsilon)\\ \Downarrow\\ \widehat{C}(\varepsilon) = \sum_{n=0}^{\infty}\varepsilon^{2n}\widehat{C}_{2n}.\end{gathered} \tag{11.14}$$

Let us now look now at the second order integrator of Equation (11.5) in the light of theorem 4.

$$\begin{aligned}\mathcal{T}_{12}^{2}(L_{12}) &= e^{-:\frac{L_{12}}{2}D_1:}e^{:-L_{12}K_1:}e^{-:\frac{L_{12}}{2}D_1:}\\ &= \exp\left(:-L_{12}H_1:+L_{12}\left\{L_{12}^2\widehat{C}_2+L_{12}^4\widehat{C}_4+\cdots\right\}\right).\end{aligned} \tag{11.15}$$

The exact form of the operators $\widehat{C}_{2n}$ is not relevant as we will now see. Yoshida's original idea was to try to create a higher of order symmetric integrator using the integrator $\mathcal{T}_{12}^2$. Since we know from theorem 4 that one should symmetrize the ansatz, the next order integrator might be

$$\mathcal{T}_{12}^{4}(L_{12}) = \mathcal{T}_{12}^{2}(x_0L_{12})\mathcal{T}_{12}^{2}(x_1L_{12})\mathcal{T}_{12}^{2}(x_0L_{12}), \tag{11.16}$$

which according to Equation (11.15) must have a Lie exponent given by

$$\begin{aligned}\mathcal{T}_{12}^{4}(L_{12}) &= \exp\left(:-x_0 L_{12} H_1 : + x_0^3 L_{12}^3 \widehat{C}_2 + \cdots\right)\\ &\quad \exp\left(:-x_1 L_{12} H_1 : + x_1^3 L_{12}^3 \widehat{C}_2 + \cdots\right)\\ &\quad \exp\left(:-x_0 L_{12} H_1 : + x_0^3 L_{12}^3 \widehat{C}_2 + \cdots\right)\\ &= \exp\left(:-(2x_0 + x_1) L_{12} H_1 : + (2x_0^3 + x_1^3) L_{12}^3 \widehat{C}_2 + \cdots\right).\end{aligned} \tag{11.17}$$

Equating (11.17) with the original map gives us a trivial set of equations for x_0 and x_1:

$$2x_0 + x_1 = 1 \text{ and } 2x_0^3 + x_1^3 = 0. \tag{11.18}$$

This set of equations has one real root and two complex ones. Let us concentrate on the real solution

$$x_0 = \frac{1}{2 - 2^{1/3}} \text{ and } x_1 = \frac{-2^{1/3}}{2 - 2^{1/3}}, \tag{11.19}$$

for which it is easy to verify that the integrator we just obtained is in fact the Ruth integrator $\mathcal{T}_{12}^{4}$ of Equation (11.7).

The derivation of Yoshida, being based on Lie operators, applies to Hamiltonians far more complex than the original $H(\vec{q}, \vec{p}) = T(\vec{p}) + V(\vec{q})$ of Ruth. Moreover, Ruth's original derivation is horrible, generating gigantic equations for the various coefficients of each map (200 terms!). Yoshida's derivation is simple and is clearly the beginning of an iterative scheme.

Indeed, let us assume that we have a symmetric integrator $\mathcal{T}_{12}^{2k}$ of order $2k$. We can construct an ansatz for a $2k+2$ integrator

$$\mathcal{T}_{12}^{2k+2}(L_{12}) = \mathcal{T}_{12}^{2k}(x_0 L_{12}) \mathcal{T}_{12}^{2k}(x_1 L_{12}) \mathcal{T}_{12}^{2k}(x_0 L_{12}), \tag{11.20}$$

from which follows immediately

$$2x_0 + x_1 = 1 \text{ and } 2x_0^{2k+1} + x_1^{2k+1} = 0. \tag{11.21}$$

A real solution can be found

$$x_0 = \frac{1}{2 - \sqrt[2k+1]{2}} \text{ and } x_1 = -\frac{\sqrt[2k+1]{2}}{2 - \sqrt[2k+1]{2}}. \tag{11.22}$$

The sixth order integrator has nine second order maps. Yoshida found more efficient solutions using less second order maps:

$$\begin{aligned} \mathcal{T}^6_{12}(L_{12}) = \mathcal{T}^2_{12}(w_3 L_{12})\mathcal{T}^2_{12}(w_2 L_{12})\mathcal{T}^2_{12}(w_1 L_{12}) & \\ \mathcal{T}^2_{12}(w_0 L_{12}) & \qquad (11.23) \\ \mathcal{T}^2_{12}(w_1 L_{12})\mathcal{T}^2_{12}(w_2 L_{12})\mathcal{T}^2_{12}(w_3 L_{12}) & \end{aligned}$$

where the coefficients w_i are given in table V.

TABLE V. Yoshida's Sixth Order Integrators.

	Solution A	Solution B	Solution C
w_1	$-0.117767998417887\ 10^{1}$	$-0.213228522200144\ 10^{1}$	$0.152886228424922\ 10^{-2}$
w_2	0.235573213359357	$0.426068187079180\ 10^{-2}$	$-0.214403531630539\ 10^{1}$
w_3	0.784513610477560	$0.143984816797678\ 10^{1}$	$0.144778256239930\ 10^{1}$
$w_0 = 1 - 2\,(w_1 + w_2 + w_3)$			

Remarkably Koseleff[29] proved that these are the only real solutions. A naïve application of the BCH formula, as I used in reference [56], indicates that there are eight commutators to remove and thus a sixth order integrator requires two more maps than Yoshida's solution. This analysis was naïve in that it failed to recognize the special nature of the problem and the potential nonlinear relationship between the coefficients. Koseleff's work, based on the theory of algebraic variety, attacked this problem successfully. It also follows from Koseleff that special integrators for quadratic kinetic energies[11.12] will not, at the sixth order, reduce the number of independent equations. Working in astronomy, where high order integrators are commonly used, Yoshida also found five eight order integrators.

11.3.2 Implications of Yoshida's Derivation

For most accelerator applications one could live quite happily with Ruth's fourth order and the connection to Lie operators found initially by Neri and the author. Thus Yoshida's derivation, although of supreme simplicity and elegance, may at first glance be of no practical use. But, Yoshida's method generalizes the applicability of the fourth order Ruth integrator further.

[11.12] If the Hamiltonian has the form $H(\vec{q},\vec{p}) = T_{ij}p_ip_j + V(\vec{q})$ then the commutators $[V,[V,[V,[V,T_{ij}p_ip_j]]]]$ and $[T_{ij}p_ip_j,[V,[V,[V,T_{ij}p_ip_j]]]]$ vanish "for free." See reference [56] for a basis of independent commutators.

Yoshida's derivation seems to depend, at first glance, on the particular form of $\mathcal{T}_{12}^2$—a symmetrized product of two Lie operators. However, the only crucial point in deriving Yoshida's formulas was the absence in Equation (11.15) of operators proportional to even powers of L_{12}. It follows from our proof of theorem 4 that the only property necessary to insure the absence of even powers, is obtained by equating the first and last terms of (11.12):

$$\mathcal{T}^{-1}(t) = \mathcal{T}(-t). \tag{11.24}$$

This property is called time reversability or, in rings, s-reversability.

Theorem 5 *The formulas of Yoshida apply to any quadratic approximation obeying Equation* (11.24).

This is a just corollary of theorem 4. However the consequences are profound. And we now list a few of them.

11.3.2.1 Multi-map integrators

Consider an s-independent Hamiltonian which can be rewritten as the sum of Q solvable parts:

$$H = \sum_{n=1}^{Q} H_Q. \tag{11.25}$$

Then consider the symmetrized map

$$\mathcal{T}^2(L) = \exp\left(-\frac{L}{2} : H_Q :\right) \exp\left(-\frac{L}{2} : H_{Q-1} :\right) \cdots \exp\left(-\frac{L}{2} : H_{Q-1} :\right) \exp\left(-\frac{L}{2} : H_Q :\right). \tag{11.26}$$

It then follows that the map $\mathcal{T}^4(L)$ given by

$$\mathcal{T}^4(L) = \mathcal{T}^2(x_0 L)\mathcal{T}^2(x_1 L)\mathcal{T}^2(x_0 L), \tag{11.27}$$

is the fourth order multi-map version of Ruth's integrator. Higher order maps are created out of the higher integrators of Yoshida recursively from Equation (11.26).

11.3.2.2 Integrators for s-dependent Hamiltonians

Consider the Hamiltonian in (11.25) and let us assume furthermore that all the terms are s-dependent and exactly solvable under the incorrect assumption that s is a parameter different from the integration parameter (in other words, the map $\exp(-L : H_i :)$ is exactly solvable even though the exact solution is the time-ordered exponential $\mathrm{Texp}(-\int_{s_0}^{s+L} ds : H_i :)$).

Although it is easy to write a first order approximation of the map from s to $s + ds$ as

$$\mathcal{T}^1_{s\to s+ds} = \prod_k^Q \exp\left(ds : H_k(s) :\right) \tag{11.28}$$

the generalization of Yoshida's method gives us an immediate prescription to include the s dependence beyond first order. Recall the trick used in Sect. 6.2.2, Equation (6.39) which consists in extending the phase space so as to make the Hamiltonian formally time-independent. Repeating this procedure, we replace the s-Hamiltonian H by a σ independent Hamiltonian K where s and p_s are new canonical variables.

$$H(\vec{q},\vec{p};s) \longrightarrow K(\vec{q},\vec{p},s,p_s) = H(\vec{q},\vec{p};s) + p_s \tag{11.29}$$

The new Hamiltonian is integrated from $\sigma = 0$ to $\sigma = L$ with the initial value $s(0) = s_0$. The reader can check that the compositional map $\exp\left(-L : K :_\sigma\right)$ is the exact solution in extended phase space where the subscript σ refers to extended Lie transforms.

We are in a position to create a simple symmetrized quadratic approximation for $\exp\left(-L : K :_\sigma\right)$ namely

$$\begin{aligned}\mathcal{T}^2(L) = \exp\left(-\frac{L}{2} : p_s :_\sigma\right)\exp\left(-\frac{L}{2} : H_Q :_\sigma\right)\exp\left(-\frac{L}{2} : H_{Q-1} :_\sigma\right) \\ \times \qquad\qquad\qquad\qquad \\ \cdots\exp\left(-\frac{L}{2} : H_{Q-1} :_\sigma\right)\exp\left(-\frac{L}{2} : H_Q :_\sigma\right)\exp\left(-\frac{L}{2} : p_s :_\sigma\right)\end{aligned} \tag{11.30}$$

In Ruth's original paper[57] and in the Lie operator extension [58] published with the author, the phase extension trick fails if more than two maps are required. In fact Ruth's original work implies that an extension of phase space works only for Hamiltonians of the form $T(\vec{p}) + V(\vec{x};s)$ by extending the kinetic energy term into $T(\vec{p}) + p_s$

thus preserving the standard "drift-kick" split. We know now thanks to Yoshida's theory that this extension is more general.

What if everything else fails? Of course, we can use implicit integrators which have apparently no connection with Yoshida's work...Or do they?

11.3.2.3 Implicit integrators and Yoshida's method

Implicit integrators are usually avoided by accelerator physicists but are often studied in other fields. In particular the Spanish school of Sanz-Serna[59] and collaborators have worked extensively on explicit and implicit formulation of symplectic Runge-Kutta algorithms. When explicit methods fail and symplecticity is still desired, it may be better to try a low order implicit symplectic integrator instead of a high order nonsymplectic scheme.

Again it is trivial to generate a first order scheme, one simply uses the following first order mixed variable generating function

$$S(\vec{q},\vec{p}^f\,;s) = \vec{q}\cdot\vec{p}^f + ds\; H(\vec{q},\vec{p}^f\,;s) \tag{11.31}$$

which leads to the implicit algorithm

$$\vec{q}^f = \vec{q} + ds\,\left(\frac{\partial H}{\partial \vec{p}^f}\right)(\vec{q},\vec{p}^f\,;s) \tag{11.32.a}$$

$$\vec{p}^f = \vec{p} - ds\,\left(\frac{\partial H}{\partial \vec{q}}\right)(\vec{q},\vec{p}^f\,;s). \tag{11.32.b}$$

The computer program requires a Newton search in order to solve this equation to machine precision and thus create a symplectic map to machine precision.

The Yoshida extension[60] to implicit integrators requires that the map for one step obeys Equation (11.24). Such a symplectic implicit map exists and was apparently first used by none other than Poincaré himself:

$$\vec{q}^f = \vec{q} + ds\,\left(\frac{\partial H}{\partial \vec{z}_2}\right)\left(\frac{\vec{q}+\vec{q}^f}{2},\frac{\vec{p}+\vec{p}^f}{2};s+\frac{ds}{2}\right) \tag{11.33.a}$$

$$\vec{p}^f = \vec{p} - ds\,\left(\frac{\partial H}{\partial \vec{z}_1}\right)\left(\frac{\vec{q}+\vec{q}^f}{2},\frac{\vec{p}+\vec{p}^f}{2};s+\frac{ds}{2}\right) \tag{11.33.b}$$

where H is a function of $\vec{z} = (\vec{z}_1, \vec{z}_2)$.

The map generated by Equation (11.33) is symplectic and s-reversible.

Exercise 11-1
Prove the two previous assertions. This map is often called in the literature the mid-point rule.

The production of higher order integrators is immediate.

11.3.2.4 Scovel's general extension

Now let us take any operator equation whatsoever (time-independent will be shown to be general enough) acting on some abstract space:

$$\frac{d\mathcal{M}}{dt} = \widehat{H}\mathcal{M}, \tag{11.34}$$

and let us assume that we have a first order solution to Equation (11.34) called $\mathcal{F}$. Then it follows that the map

$$\mathcal{T}^2(t) = \mathcal{F}(\frac{t}{2})\mathcal{F}^{-1}(-\frac{t}{2}) \tag{11.35}$$

is not only a second order approximation but also a time-reversible one. Thus higher order approximations can be constructed using $\mathcal{T}^2(t)$.

Exercise 11-2
Using the results of this section and Equation (11.32), *construct another implicit integrator which can used Yoshida's formulas. You may want to check your results on trivial Hamiltonians.*

11.3.2.5 General time extension

In Sect. 11.3.2.4 we neglected the possibility of time dependence. A priori it is not clear that the extension of phase space used in Sect. 11.3.2.2 will work because there is no concept of phase space for arbitrary operators. Hence the extension of Equation (11.29) to arbitrary operators is meaningless. However, if we look at the Lie operators instead of the Hamiltonian function itself we get the "generalizable" procedure immediately:

$$\begin{aligned} :-K:_\sigma &= :-H: + :-p_s:_\sigma \\ &\Downarrow \\ &= :-H: + \frac{\partial}{\partial s}. \end{aligned} \tag{11.36}$$

Hence Equation (11.34) is extended trivially into

$$\frac{d\mathcal{M}}{d\sigma} = \left(\widehat{H} + \partial_t\right)\mathcal{M}, \tag{11.37}$$

where the new map $\mathcal{M}_\sigma$ acts on the extended space. In other words, if the original map $\mathcal{M}$ was acting on some vector $|\psi\rangle \in \mathfrak{V}$ then the map $\mathcal{M}_\sigma$ acts on $(|\psi\rangle, t) \in \mathfrak{V} \times \mathbb{R}$. Again the entire result of this extension is to provide a procedure which tells the integrator at what time should each step of a Yoshida integrator be evaluated.

Yevick and Glaser[61], who have independently derived Yoshida's formulas for applications dealing with unitary operators, published a complex paper which dealt with the time-dependent problem. While their results may apply to maps which have interfaces (i.e., not differentiable in the time-variable), since the number of interfaces is finite usually (i.e., number of layout planes in our case), one may ask why they did not attempt a phase-space like extension of their operators. The reason is sociological: phase space extension is a common trick used in classical mechanics and in particular by users of Lie methods (astronomers). It is **not** commonly found in quantum mechanical text books!

Exercise 11-3

Consider the trivial differential equation:

$$\frac{dx}{ds} = a(s)x \tag{11.38}$$

1. *Show that the solution is*

$$x(s) = \exp\left(\int_{s_0}^{s} a(s')ds'\right) x(s_0). \tag{11.39}$$

2. *Under what condition is this solution valid if $a(s)$ is an abstract operator. Hint: Look at BCH formula.*

3. *Apply Yoshida's fourth order integrator to Equation (11.38) by extending the operator with $\frac{\partial}{\partial s}$. Answer: Yoshida's method reduces to an approximation for the integral $\int_{s_0}^{s} a(s')ds'$:*

$$\int_{s_0}^{s} a(s')ds' = \Delta \sum_{n=0}^{N-1} \left\{ x_0 \left\{ a\left(s_0+\Delta\left[\frac{x_0}{2}+n\right]\right)+a\left(s_0+\Delta\left[x_1+\frac{3}{2}x_0+n\right]\right) \right\} + x_1 a\left(s_0+\Delta\left[\frac{1}{2}x_1+x_0+n\right]\right) \right\}$$

$$\text{where } \Delta = \frac{s-s_0}{N} \tag{11.40}$$

and x_0 and x_1 are given by Equation (11.19).

4. *Show explicitly (on a special choice of $a(s)$) that the approximate solution* (11.40) *is quartic in the time step.*

11.4 BIASED SYMPLECTIC INTEGRATION

All explicit symplectic schemes involve breaking the Hamiltonian into exactly solvable parts and treating each part on an equal footing. This may be a waste of computer resources. To see how this might be, let us go to a different problem with apparently no relationship with symplectic integration. It is that of the Neuffer[62] correction scheme or its mirror image the Ritson[63] tracking code. Let us consider a problem defined abstractly in terms of the compositional map:

$$\begin{aligned} \frac{d\mathcal{M}}{ds} &= \mathcal{M} : -H : \\ &= \mathcal{M} : -\varepsilon_0 H_0 - \varepsilon_1 v(s) V : . \end{aligned} \tag{11.41}$$

Here we assume that the unperturbed motion (H_0) is known exactly. Our goal is to find a solution for the map $\mathcal{M}$ from $s = 0$ to $s = L$ where L is a short distance. Originally, in the Neuffer correction scheme, the map was integrated over half a FODO cell. The function $v(s)$ would then represent a multipole distribution with potential V; this distribution would include measured error fields and the correctors. The goal of the Neuffer scheme was to choose the correctors so as to minimize the effect of V.

To proceed further we need to solve for the perturbed map. Using the results of perturbation theory (see Sect. 8.4.1.1), the map $\mathcal{M}$ is known to second order in ε through the formula

$$\begin{aligned}\mathcal{M}_0(0 \to L; \varepsilon_1) &= \mathcal{M}_0(0 \to L/2) \\ &\times \exp\Big\{ :-\varepsilon_1 \textstyle\int_0^L ds\, \mathcal{M}_0^{-1}(s \to L)\, v(s)\, V \\ &+ \tfrac{\varepsilon_1^2}{2} \textstyle\int_0^L ds \int_s^L ds' \left[\mathcal{M}_0^{-1}(s \to L) v(s) V, \mathcal{M}_0^{-1}(s' \to L) v(s') V\right] : \Big\}\end{aligned} \tag{11.42}$$

The map of Equation (11.42), if it could be evaluated exactly, would be better than one step of a second order integrator over the same distance because it is to all orders exact in the smallness parameter ε_0. In other words, it is more accurate than the map

$$\exp\left(: -\frac{L}{2}\varepsilon_0 H_0 : \right) \exp\left(: -L\varepsilon_1 v(\frac{L}{2}) V : \right) \exp\left(: -\frac{L}{2}\varepsilon_0 H_0 : \right)$$

which is accurate to the combined order $\varepsilon_0^{3-m}\varepsilon_1^m$ where $0 \leq m \leq 3$.

The idea of a biased integrator is to derive an integration scheme which is first order in ε_1 but is correct to all orders ε_0. Hence, we look at the map

$$\exp\{: -\varepsilon_1 \int_0^L ds\, \mathcal{M}_0^{-1}(s \to L)\, v(s)\, V\}$$

whose Lie transform we rewrite, assuming no s-dependence in H_0, as

$$\begin{aligned} f_1 &= -\varepsilon_1 \int_0^L ds \exp\left(: \{L - s\}\, \varepsilon_0 H_0 : \right) v(s)\, V \\ &= -\varepsilon_1 e^{:\varepsilon_0 L H_0:} \sum_{n=0}^{\infty} \frac{: -\varepsilon_0 H_0 :^n}{n!} V \underbrace{\int_0^L ds\, s^n\, v(s)}_{v_n} \\ &= -\varepsilon_1 e^{:\varepsilon_0 L H_0:} \sum_{n=0}^{\infty} v_n \frac{: -\varepsilon_0 H_0 :^n}{n!} V\,. \end{aligned} \tag{11.43}$$

The coefficients v_n are form factors which depend on the potential distribution $v(s)$ only. The idea behind the Neuffer correction scheme or biased explicit integration, is to recognize that an expansion in the form factors v_n is essentially an expansion in the phase advance. Thus, if we intend to correct the lattice, we simply use the correctors to cancel as many form factors as possible. The so-called Simpson rule scheme of Neuffer consists in correcting a constant field by using three localized (Dirac delta) multipole correctors placed symmetrically:

$$v(s) = V_0 + \delta(s)c_1 + \delta(s - L/2)c_2 + \delta(s - L)c_1 \tag{11.44}$$

which, if substituted in the definition for the v_n, gives the result:

$$\begin{aligned}
v_0 &= V_0 L + 2c_1 + c_2 = 0 \\
v_1 &= V_0 \frac{L^2}{2} + c_1 L + c_2 \frac{L}{2} = 0 \\
v_2 &= V_0 \frac{L^3}{3} + c_1 L^2 + c_2 \frac{L^2}{4} = 0 \\
v_3 &= V_0 \frac{L^4}{4} + c_1 L^3 + c_2 \frac{L^3}{8} = 0.
\end{aligned}$$

This admits the solution

$$c_1 = -\frac{1}{6} V_0 L \text{ and } c_2 = -\frac{4}{6} V_0 L, \tag{11.45}$$

from which it derives its name: the 1-4-1 or Simpson's rule.

Conversely, when performing integration, we may want to replace a map whose form factors are known by a series of kicks with identical or nearly identical form factors. Thus a magnet with time-independent Hamiltonian $H_0 + V$ can be integrated using the map

$$\exp\left(:-\tfrac{L}{6}V:\right)\exp\left(:-\tfrac{L}{2}H_0:\right)\exp\left(:-\tfrac{4L}{6}V:\right)\exp\left(:-\tfrac{L}{2}H_0:\right)\exp\left(:-\tfrac{L}{6}V:\right).$$

The combination and variation on this theme are infinite: form factors, leading order tune shifts, etc...can all be maintained at their original values by the appropriate selection of the coefficient c_i's. It is a powerful tool which complements symplectic integration by taking advantage of the localized nature of our magnets.

11.5 CONCLUSION

The introduction of Lie methods to explicit symplectic integration has done two things:

1. In a characteristic way it has simplified the formulation and thus provided us with simpler derivations. Ruth's initial derivation of the fourth order method covered pages of output from an algebraic manipulator. The solution had to be found numerically using a Newton search method coupled with a random selection of initial guesses. Only later did Ruth find the analytical expression for each coefficient in Equation (11.7). But, on a practical level, in the theory of rings where the Talman reinterpretation

of the integrator is used, the fourth order Ruth integrator once obtained by any method, seemed and still seems today, quite sufficient.

2. Therefore in our field the greatest contribution of the Lie methods to symplectic integration resides in the vastly expanded domain of applicability of Ruth's formula. First of all, even without Yoshida, it became obvious that Ruth's formulas could be used in "small machines" where the exact Hamiltonian (as in Equation (11.1)) does not have the trivial $T(\vec{p}) + V(\vec{x})$ "drift-kick" split of Ruth's original paper. This initial Lie connection revealed that Ruth's integrators where connected to nonsymplectic integrators and groups other than the symplectic group. Then, thanks to Yoshida's formulation, Ruth's formula not only becomes part of a recursive series, but was shown to be related to multi-map formulations and implicit integrators. This is truly a triumph of inter-disciplinary physics!

We are ready to use this new knowledge to attack head-on the issue of "small" machine integrators. This is the topic of the next chapter.

INTERLUDE XII
TIME REVERSIBLE SYSTEMS

It is well-known for a symmetric interaction region that the Twiss parameter α vanishes at the interaction point. A more general result can be stated as follows.

Theorem 6 (Time-reversible) *The Lie operator of a symmetric system composed of time reversible maps does not have terms of odd powers in the momenta at the symmetry point.*

First we consider the time reversible components of such a system. By time reversability it is assumed that if the following map $\mathcal{T}_{\mathcal{R}}$ is applied to a system

$$\mathcal{T}_{\mathcal{R}} \equiv \begin{cases} t \rightarrow -t \\ (p_x, p_y) \rightarrow -(p_x, p_y) \end{cases} \qquad (11.46)$$

then outgoing rays will retrace their paths backwards in time. It is a non-symplectic transformation which nevertheless preserves the Hamiltonian nature of the flow. Let us apply the transformation $\mathcal{T}_{\mathcal{R}}$ of (11.46) on an arbitrary Lie transform $: f(\vec{q},\vec{p}) :$. It is given by

$$\mathcal{T}_{\mathcal{R}} : f(\vec{q},\vec{p}) : \mathcal{T}_{\mathcal{R}}^{-1} = : -f(\vec{q},-\vec{p}) : . \tag{11.47}$$

This formula can be derived by considering $-f$ to be a Hamiltonian or by simply applying the general formula of Equation (4.29) for a similarity transformation on a general vector field.

In an ideal machine made out of ideal quadrupoles and dipoles, each element is time-reversible because the dependence on the momentum is quadratic. Thus if the map for an element of the beam line is $\mathcal{M}_{i\to i+1}$, we assume that the inverse is given by

$$\mathcal{M}_{i\to i+1}^{-1} = \mathcal{T}_{\mathcal{R}} \mathcal{M}_{i\to i+1} \mathcal{T}_{\mathcal{R}}^{-1}. \tag{11.48}$$

The rest of the proof follows the steps in Sect. 11.3.1 on symmetrized integrators.

$$\begin{aligned}
&\text{If } \mathcal{M} = \mathcal{M}_1\mathcal{M}_2\cdots\mathcal{M}_2\mathcal{M}_1 \\
&\qquad\Downarrow \\
\mathcal{T}_{\mathcal{R}}\mathcal{M}\mathcal{T}_{\mathcal{R}}^{-1} &= \mathcal{T}_{\mathcal{R}}\mathcal{M}_1\mathcal{M}_2\cdots\mathcal{M}_2\mathcal{M}_1\mathcal{T}_{\mathcal{R}}^{-1} \\
&= \mathcal{T}_{\mathcal{R}}\mathcal{M}_1\mathcal{T}_{\mathcal{R}}^{-1}\mathcal{T}_{\mathcal{R}}\mathcal{M}_2\mathcal{T}_{\mathcal{R}}^{-1}\cdots\mathcal{T}_{\mathcal{R}}\mathcal{M}_2\mathcal{T}_{\mathcal{R}}^{-1}\mathcal{T}_{\mathcal{R}}\mathcal{M}_1\mathcal{T}_{\mathcal{R}}^{-1} \\
&= \mathcal{M}_1^{-1}\mathcal{M}_2^{-1}\cdots\mathcal{M}_2^{-1}\mathcal{M}_1^{-1} \\
&= \{\mathcal{M}_1\mathcal{M}_2\cdots\mathcal{M}_2\mathcal{M}_1\}^{-1} \\
&= \mathcal{M}^{-1}.
\end{aligned} \tag{11.49}$$

It follows that the Lie operator of the map for the full beam line cannot contain odd powers of the transverse momentum. Given $\mathcal{M} = \exp(: h(\vec{q},\vec{p}) :)$, then

$$\begin{aligned}
\mathcal{M}^{-1} &= \mathcal{T}_{\mathcal{R}}\mathcal{M}\mathcal{T}_{\mathcal{R}}^{-1} \\
\exp(: -h(\vec{q},\vec{p}) :) &= \exp\left(\mathcal{T}_{\mathcal{R}} : h(\vec{q},\vec{p}) : \mathcal{T}_{\mathcal{R}}^{-1}\right) \\
\exp(: -h(\vec{q},\vec{p}) :) &= \exp(: -h(\vec{q},-\vec{p}) :) \\
&\Downarrow \\
h(\vec{q},\vec{p}) &= h(\vec{q},-\vec{p}).
\end{aligned} \tag{11.50}$$

This proves the generalized theorem on symmetrized beam lines. In general, the time reversal symmetry will not erase completely the driving terms for a given resonance, but it will on average reduce it by a factor of two. It is easy to check that the imaginary part of an eigenmonomial will contain odd powers of the momentum and is thus suppressed. In the single Lie exponent representation, the sine terms of the resonances are gone.

END OF INTERLUDE

CHAPTER 12
"SMALL" RINGS: USING THE CORRECT HAMILTONIAN

Tracking codes using symplectic integration can be divided into two categories: those with the small angle approximations and those without. The first set contains most tracking algorithms used through the years. They are easily identifiable by one "medical" diagnostic: ideal bends, quadrupoles and thus drifts are linear elements in the transverse variable, and, if self-consistently implemented, the longitudinal variable depends quadratically on the transverse variables. They are also connected to the approximate quadrupole kick formula

$$\begin{aligned}\Delta \tan\left(\alpha_x\right) = \Delta x' &= -\frac{k}{1+\delta}x \\ = \Delta y' &= \frac{k}{1+\delta}y.\end{aligned} \tag{12.1}$$

This formula, equivalent to linearity in quadrupoles, is another symptom of the small angle approximation. These codes are called "large machine" codes because it is traditionally for these machines that the small angle approximation has been working best.

It is in smaller (and rarer!) rings that this approximation has traditionally failed. The success of the small angle approximation has been responsible for the commonly spread belief (in rings and linacs only to be fair) that quadrupoles and bends are truly linear. As a result, integrators for "small" machine are rare. The code TEAPOT[6] does not use the small angle approximation, but, unfortunately concocted for the defunct SSC (paradox of paradoxes), it misses several items needed in small machines (quadrupole fringe field for example). Another code called SAD, written in Japan, correctly includes large angle and fringe effects. These became necessary as the not so small "B-factories" started to display small machine diseases. Of course the author himself has played around with small machine codes starting with off-shoots of TEAPOT.

In any event, with the modern tools of Chapter 11, it is relatively simple to discuss the integration through the "correct" Hamiltonian (or "exact" to paraphrase Talman) describing the body of each type magnet. In Chapter 14 we will correctly and self-consistently take the small angle limit.

A final caveat: what about fringe fields? In turns out that Maxwellian effects at the end of magnets is an important effect mostly in bends where it provides vertical focusing. In straight layout elements it is an effect which has opposite signs at the entrance and exit layout plane to leading order. In fact the lack of cancellation is proportional to angles which are small whenever the small angle approximation is valid. Anyway we will devote a complete chapter to the issue of first order fringe fields.

12.1 STRAIGHT OR CARTESIAN ELEMENTS

Just before we start I must remind the reader of a fundamental tenet of this book.

12.1.1 Reminder: Straight Magnet$\not\Rightarrow$Straight Layout

Since we separate layout and magnet at this stage we must define the word "straight" completely locally. Imagine that the magnet is sitting on a bench and being measured; how will you decide if the magnet is straight or curved? The properties of the magnetic field inside this magnet would teach you whether the word straight or curved is appropriate, *irrespective of the final layout position of this magnet*. For example, in Sect. 1.3, where we introduced the fundamental global/local dichotomy, we talked about the ALS bend: a shifted quadrupole whose purpose is to provide 10 degrees of bending (the layout angle Φ_{12} must be 10 degrees) and some focusing. Despite its primordial bending "raison d'être," keeping the beam in the tunnel/layout, this magnet is best described in Cartesian coordinates since the magnetic field as well as the effective boundary of the magnet is well-defined using a longitudinal Cartesian variable z. All the various splits of the Hamiltonian presented here can be used for a Cartesian bend as well as for the straight layout element. The reader will discover that, in practical terms however, not all the splits are equivalent—some are preferable in a curved layout and some are preferable in a straight layout (see Sect. 12.1.5).

12.1.2 The Hamiltonian H

The Hamiltonian H for a straight magnetic element of length L is

$$H = -\sqrt{(1+\delta)^2 - p_x^2 - p_y^2} + V(x,y). \tag{12.2}$$

where the function $V(x, y)$ is related to the z-component of the magnetic vector potential. Of course this description is valid only for the body of such a magnet as fringe effects are unavoidable near the front and back end of the magnet. To be precise the function V is given, in the harmonic expansion, by the formula

$$V(x,y) = \Re \sum_{n=1}^{\infty} \frac{(ia_n + b_n)}{n} (x+iy)^n \tag{12.3}$$

where the coefficients a_n and b_n are related to a measurement of the transverse field in the mid-plane $y = 0$. The formulas are simply

$$B_x(x,0) = -\left.\frac{\partial V}{\partial y}\right|_{y=0} = Brho \sum_{n=1}^{\infty} a_n x^{n-1} \tag{12.4.a}$$

$$B_y(x,0) = \left.\frac{\partial V}{\partial x}\right|_{y=0} = Brho \sum_{n=1}^{\infty} b_n x^{n-1}, \tag{12.4.b}$$

$$Brho = \frac{p_0}{q}.$$

The Hamiltonian H is scaled by the "design momentum" p_0, the coefficients a_n and b_n are scaled by the so-called magnetic rigidity *Brho* which is the ratio of the design momentum over the electric charge. I purposely wrote "*Brho*" in the Roman alphabet to emphasize that neither B nor the *rho* of *Brho* have anything to do with the local curvature of the layout or the local magnetic field.

For an arbitrary V, the reader can convince himself that H is not solvable in terms of simple functions. Hence the crucial question of explicit symplectic integration is posed: what are the possible splits and which one is most appropriate.

12.1.3 The Drift-Kick Split

The first obvious split is given by

$$\begin{aligned} H &= H_1 + H_2 \\ H_1 &= -\sqrt{(1+\delta)^2 - p_x^2 - p_y^2} \\ H_2 &= V(x,y). \end{aligned} \tag{12.5}$$

and the maps for each term are—first the map $\mathcal{T}_z(L)$ of Equation (10.23)—and then a simple multipole kick given by

$$\begin{aligned}
\mathcal{K}(L) &= \exp\ (-L : H_2 :) \\
p_x^{new} &= p_x - L\partial_x V = p_x - \frac{LB_y}{Brho} \\
p_y^{new} &= p_y - L\partial_y V = p_y + \frac{LB_x}{Brho}.
\end{aligned} \tag{12.6}$$

All other variables are left constant by the kick map $\mathcal{K}(L)$.

The two maps are simple, but their combination is necessarily nonlinear. In this split of H the linear properties of the quadrupoles change as we go from H to the integrator. Thus the Talman reinterpretation of the integrator as the new model is essential.

12.1.4 The Matrix-Kick Split

Another potential split of H separates "the quadratic part" from H itself.

$$\begin{aligned}
H &= H_1 + H_2 + H_3 \\
H_1 &= \frac{p_x^2 + p_y^2}{2(1+\delta)} + V_2(x, y) \\
H_2 &= \sqrt{(1+\delta)^2 - p_x^2 - p_y^2} - \frac{p_x^2 + p_y^2}{2(1+\delta)} \\
H_3 &= V(x, y) - V_2(x, y).
\end{aligned} \tag{12.7}$$

In the above split, the quadratic part of H is exactly solvable in terms of a matrix. Thus, if the straight element is placed in a layout straight block, the matrix part will describe the paraxial optics of the exact Hamiltonian H.

The code SAD[64] of KEK is, to my knowledge, the only serious code using this particular split. Personally I have programmed "patched up jobs" to large machine codes using this split. Indeed the Hamiltonian H_1 in Equation (12.7) is precisely the Hamiltonian for a quadrupole in the small angle approximation. The term H_2 is a correction term removing the small angle approximation. Thus it is easy to "patch up" a standard matrix-kick code and test the validity of its approximate model by adding the momentum dependent map produced by H_2.

Variations on this theme are virtually infinite. Personally, I consider the drift-kick split of Sect. 12.1.3 to be the most efficient and simplest. The reader will also notice that the solution for H_1 is not a nice *real* function of the quadrupole parameters a_2 and b_2. The typical implementation requires logical branching depending on the sign of these parameters or the use of complex numbers. This is annoying if we include TPSA calculations through polymorphism. In particular, derivatives around the normal quadrupole value $b_2 = 0$ are extremely common and prevented by branching. This is one of the reasons I caution against the use of precomputed matrices in large machine codes.

12.1.5 The Cartesian Bend-Kick Split

The reason for deriving a Bend-Kick type integrator is grounded in the possibility that a straight magnet will be used to provide zeroth order bending to the beam—in other words, this magnet is located in a curved layout block, i.e., $\Phi_{12} \neq 0$. This is not academic at all: it is the case of rectangular magnets and the ALS shifted quadrupole bend. The reader will notice that there is an intrinsic matching problem with this kind of bend as displayed in Figure 12.1. The Hamiltonian H describes the map from the plane $1'$ to the plane $2'$. If we view the magnet as a straight machine, it is composed of only parallel layout planes. In other words, if N integration steps are needed, then one can view (à la Talman) the magnet as a straight beam line with N thin bends. Whatever the view taken, one cannot escape the fact that this straight magnet demands a local frame of reference for integration which does not match the layout in which it must be ultimately lowered. But this is no problem since the map is immersed in a free field region, the matching is done by the same elements of the dynamical Euclidean group which we studied in Chapter 10. Thus the full map, from entrance to exit layout plane, will be

$$\mathcal{M}_{12} = \mathcal{Y}(\frac{\Phi_{12}}{2})\mathcal{F}_{1'}\mathcal{M}_{1'2'}\mathcal{F}_{2'}\mathcal{Y}(\frac{\Phi_{12}}{2}). \tag{12.8}$$

The maps $\mathcal{F}$'s are fringe effects which we will discuss later. They usually provide vertical focusing, and in small rings, a fair amount of vertical chromaticity.

The body of such a magnet, the map $\mathcal{M}_{1'2'}$, is also represented by Equation (12.2) and can be integrated using any of the above integrators. However we can attempt the following split as well:

$$H = -\sqrt{(1+\delta)^2 - p_x^2 - p_y^2} + \frac{x}{\rho_c} + V(x,y)$$

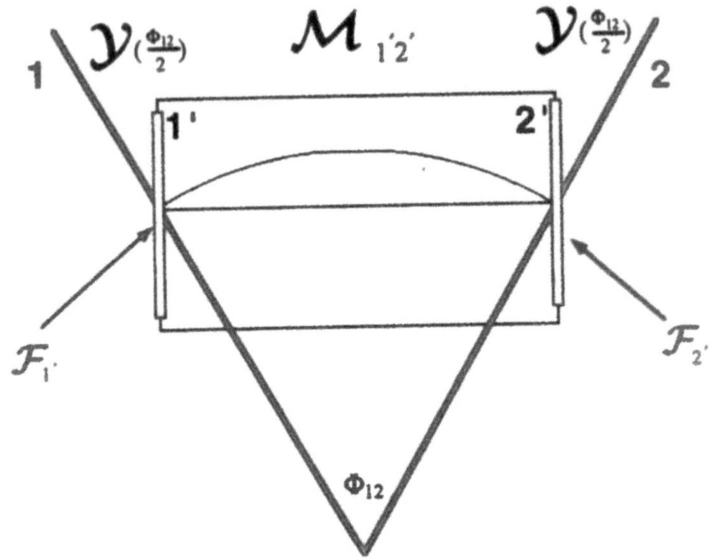

Figure 12.1 : Straight Magnet in a Curved Block.

$$
\begin{aligned}
&= H_1 + H_2 \\
H_1 &= -\sqrt{(1+\delta)^2 - p_x^2 - p_y^2} + \frac{x}{\rho_c} \\
H_2 &= V(x,y).
\end{aligned}
\tag{12.9}
$$

The Hamiltonian H_1 models an ideal rectangular bend in Cartesian coordinates. If the value of the field is set to values related to the layout plane

$$\rho_c = \frac{O_1O_2}{2\sin(\Phi_{12})}, \tag{12.10}$$

then the map $\mathcal{M}_{1'2'} = \exp(: -O_1O_2\ H_1 :)$ once substituted in Equation (12.8), will ensure that $\mathcal{M}_{12}$ sends the origin of the layout into itself for the ideal constant field bend. In the $x - p_x$ plane, the origin $(0,0)$ is first transformed into $(0, \sin(\Phi_{12}/2))$ by the rotation $\mathcal{Y}(\Phi_{12}/2)$. Then the ideal map $\mathcal{M}_{1'2'}$ will change the coordinates at the entrance into the exit coordinates $(0, -\sin(\Phi_{12}/2))$. The last rotation $\mathcal{Y}(\Phi_{12}/2)$ will give us the origin $(0,0)$ once more.

For the non-ideal case, the step of integration involves the map

$$\mathcal{T}_2 = \exp\left(-:\tfrac{\mathcal{O}_1\mathcal{O}_2}{2N}V:\right)\exp\left(-\tfrac{\mathcal{O}_1\mathcal{O}_2}{N}:H_1:\right)\exp\left(-:\tfrac{\mathcal{O}_1\mathcal{O}_2}{2N}V:\right), \qquad (12.11)$$

or equivalently,

$$\mathcal{T}_2 = \exp\left(-\tfrac{\mathcal{O}_1\mathcal{O}_2}{2N}:H_1:\right)\exp\left(-:\tfrac{\mathcal{O}_1\mathcal{O}_2}{N}V:\right)\exp\left(-\tfrac{\mathcal{O}_1\mathcal{O}_2}{2N}:H_1:\right). \qquad (12.12)$$

The above discussion hinges on our ability to compute the map $\exp\left(-\frac{\mathcal{O}_1\mathcal{O}_2}{N} : H_1 :\right)$ of the ideal rectangular bend. We delay this discussion to Sect. 12.3.1 because this map can be obtained using the formula for a sector bend (in cylindrical coordinates) by taking the appropriate limit. Therefore, let us discuss the more familiar cylindrical geometry now.

12.2 BENDS WITH CYLINDRICAL SYMMETRY

One can imagine a perfect cyclotron field: a bend of angle 2π with total azimuthal symmetry and the fields are invariant (in cylindrical coordinates) as we move on a circle centered at the origin of the cyclotron. Just as we might do with a big wheel of parmesan, we slice a sector of angle θ from the wheel. It is clear that at a large radii the fields will still be invariant along the arc of circle. Just as in the case of a straight magnet, we expect some fringe effects at the beginning and at the end of the slice, but we again neglect these effects in the following discussion.

12.2.1 The Hamiltonian

This Hamiltonian was seen before in our introduction to symplectic integration and, in particular, in the discussion of Talman's TEAPOT philosophy.

$$\begin{aligned} H = &-(1+\frac{x}{\rho_c})\sqrt{(1+\delta)^2 - p_x^2 - p_y^2} \\ &+b_1(x+\frac{x^2}{2\rho_c}) + V(x,y;\rho_c). \end{aligned} \qquad (12.13)$$

I have labeled the radius of the frame of reference ρ_c and it may not necessarily coincide with the layout's radius of curvature. The symmetry of the magnet and Maxwell's equations impose a more complex relationship between a measurement of the field in the mid-plane, ($B_x(x,0)$ and $B_y(x,0)$), and the potential $V(x,y;\rho_c)$. This relation is explicit in the dipole term $b_1(x+\frac{x^2}{2\rho_c})$ which reverts to the straight magnet result as ρ_c becomes infinite. A very useful program can be written using a TPSA library to relate mid-plane fields and the potential. This is the topic of Interlude XIV.

12.2.2 The Drift-Kick Split

This split was already discussed in relation with Talman's quotation. In effect, it is a drift in polar coordinates followed by a kick. The Hamiltonian is split into

$$
\begin{aligned}
H &= H_1 + H_2 \\
H_1 &= -(1+\frac{x}{\rho_c})\sqrt{(1+\delta)^2 - p_x^2 - p_y^2} \\
H_2 &= b_1(x+\frac{x^2}{2\rho_c}) + V(x,y;\rho_c).
\end{aligned}
\tag{12.14}
$$

The drift in polar coordinates which is given by Equation (10.26) can be interpreted geometrically as shown on Figure 12.2. The particle drifts in Cartesian coordinates; the initial frame of reference defining the momenta is unchanged. This is indicated by the green frame following the particle trajectory. The ray intercepts the new plane which makes an angle θ with the original plane. One can then rotate the green frame into the final red frame. This drift and this final rotation are not symplectic maps, but their combination is. This is the origin of "globally symplectic, though not locally symplectic (unless extended to six dimensions)" of Talman and Schachinger[6, 54]. In fact, the map is "locally" symplectic if restricted to four or even two dimensions! Figure 12.3 shows the correct geometrical Hamiltonian representation: the green frames continuously rotate so as to make an angle θ for all values of θ. The variable θ is the time-like variable of the Hamiltonian generating the map $\mathcal{Y}(\theta)$ of the dynamical Euclidean group (see Exercise 10-2 for yet another geometrical representation which is very useful for the dynamical representation of the Euclidean group). The integration step is simply

$$
\begin{aligned}
\mathcal{T}_2(\frac{L}{N}) &= \exp\left(:-\frac{L}{2N}H_1:\right)\exp\left(:-\frac{L}{N}H_2:\right)\exp\left(:-\frac{L}{2N}H_1:\right) \\
&= \mathcal{Y}(\frac{L}{2N\rho_c})\exp\left(:-\frac{L}{N}H_2:\right)\mathcal{Y}(\frac{L}{2N\rho_c}).
\end{aligned}
\tag{12.15}
$$

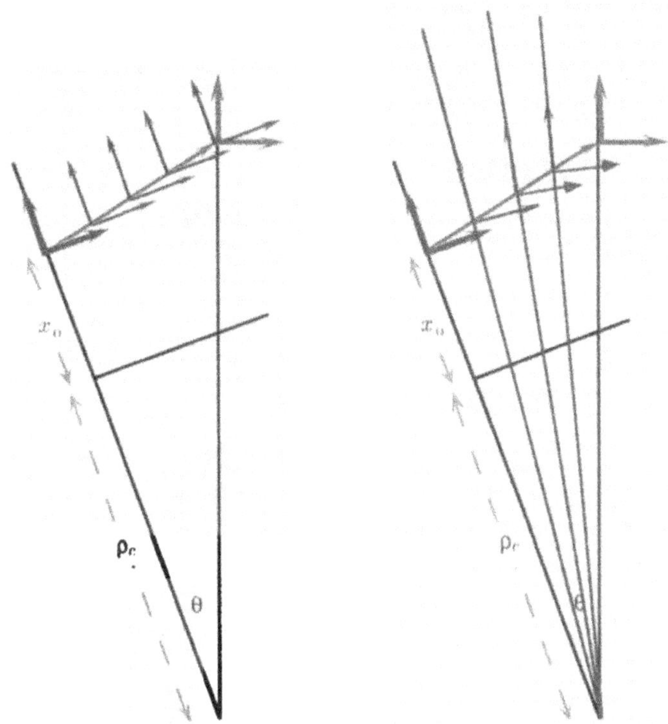

Figures 12.2 and 12.3: The Geometrical ("Propagate First")
and
the Hamiltonian Interpretation of $\mathcal{Y}(\theta)$.

This map has one bad feature: if applied to an ideal bend whose bending angle and radius match the layout angle, the origin of phase space at the entrance plane is not mapped into the origin of the exit plane. There are many ways to cure this problem: one adopted by Talman consists in considering the ring as a polygon and modifying the design bending field appropriately. I discourage this practice, but it is certainly a possibility. Clearly if the order is increased, the polygon becomes a fractal figure of no physical reality; in any event one loses the arc of the circle completely.

Another possibility: one subtracts the residual design orbit by pre-calculating it for all bending magnets which are to be placed in a bend/layout position. However, maybe the cleanest method is to investigate a split of the integrator for which this problem does not occur. This is now explained.

12.2.3 The Bend-Kick Split

Ideal bends should be exactly solvable since particles in constant magnetic fields have circular cork-screw trajectories. This split of H

$$\begin{aligned} H &= H_1 + H_2 \\ H_1 &= -(1+\frac{x}{\rho_c})\sqrt{(1+\delta)^2 - p_x^2 - p_y^2} + b_1(x + \frac{x^2}{2\rho_c}) \\ H_2 &= V(x, y; \rho_c), \end{aligned} \tag{12.16}$$

gives us an elegant solution for sector bends lowered into a curved layout of angle Φ_{12} and layout planes with origins O_1 and O_2. One can select the field b_1 of the Hamiltonian H_1 to be just $\frac{1}{\rho_c}$ and

$$\rho_c = \frac{O_1O_2}{2\sin(\Phi_{12})}. \tag{12.17}$$

Hence the bend part of the integrator map will send the origin of the entrance layout into the origin of the exit layout. Any magnetic field error is handled by the kick Hamiltonian H_2. All of this hinges on our ability to derive an expression for the map of an ideal sector, so here is the result:

$$x(s) = \frac{\rho_c}{b_1}\left(\frac{1}{\rho_c}\sqrt{(1+\delta)^2 - p_x(s)^2 - p_y^2} - \frac{dp_x(s)}{ds} - b_1\right) \tag{12.18.a}$$

$$p_x(s) = p_x\cos\left(\frac{s}{\rho_c}\right) + \left(\sqrt{(1+\delta)^2 - p_x^2 - p_y^2} - b_1(\rho_c + x)\right)\sin\left(\frac{s}{\rho_c}\right) \tag{12.18.b}$$

$$\begin{aligned} y(s) &= y + \frac{p_y s}{b_1\rho_c} \\ &\quad + \frac{p_y}{b_1}\left(\sin^{-1}\left(\frac{p_x}{\sqrt{(1+\delta)^2 - p_y^2}}\right) - \sin^{-1}\left(\frac{p_x(s)}{\sqrt{(1+\delta)^2 - p_y^2}}\right)\right) \end{aligned} \tag{12.18.c}$$

$$p_y(s) = p_y \tag{12.18.d}$$

$$\delta(s) = \delta \tag{12.18.e}$$

$$\begin{aligned} \ell(s) &= \ell + \frac{(1+\delta)s}{b_1\rho_c} \\ &\quad + \frac{(1+\delta)}{b_1}\left(\sin^{-1}\left(\frac{p_x}{\sqrt{(1+\delta)^2 - p_y^2}}\right) - \sin^{-1}\left(\frac{p_x(s)}{\sqrt{(1+\delta)^2 - p_y^2}}\right)\right). \end{aligned} \tag{12.18.f}$$

Because this sector map is central, we will give a name to its associated compositional map: $\mathcal{W}(L, b_1; \rho_c)$. The integrator step which sends the layout origin into itself can be written as

$$\mathcal{T}_2\left(\frac{L}{N}\right) = \mathcal{W}\left(\frac{L}{2N}, \frac{1}{\rho_c}; \rho_c\right) \exp\left(-\frac{L}{N} :H_2:\right) \mathcal{W}\left(\frac{L}{2N}, \frac{1}{\rho_c}; \rho_c\right). \tag{12.19}$$

Some readers, facing the complex expressions in (12.18), may appropriately suggest that the kick map be put at the edge of each integration step. In the case when only one step is used it might increase the speed of computation by a factor of two. This map,

$$\mathcal{T}_2\left(\frac{L}{N}\right) = \exp\left(-:\frac{L}{2N} V:\right) \mathcal{W}\left(\frac{L}{N}, \frac{1}{\rho_c}; \rho_c\right) \exp\left(-:\frac{L}{2N} V:\right), \tag{12.20}$$

is also a quadratic, time-reversible, operator. In an abstract sense, the drift-kick and bend-kick splits are similar in spirit. The first one involves a dynamical rotation in a free medium and the second split uses a dynamical representation of the $x - z$ rotation in a constant B_y medium.

INTERLUDE XIII
EUCLIDEAN TRANSFORMATIONS IN SOLENOIDS

This interlude is to be written by the reader and published! Here is the question: our ability to compute $\mathcal{W}$ opens a new area: is it possible to derive the full rotation group in a constant field medium, thus allowing rotations and translations in the body of a solenoid?

The ideal solenoid, in Cartesian coordinates, is exactly solvable. The Hamiltonian is given by

$$H = -\sqrt{(1+\delta)^2 - \left(p_x + \frac{b_z}{2} y\right)^2 - \left(p_y - \frac{b_z}{2} x\right)^2}. \tag{12.21}$$

The field in the longitudinal direction is given by $B_z = b_z\ Brho$. Therefore, in the light of Exercise 10-2, we conclude that the $x - z$ and $y - z$ rotations of the Euclidean solenoidal group will at most require the solutions of transcendental equations of trigonometric functions. See Sect. 12.5 for more on the solenoid and the exact solution.

Good luck!
END OF INTERLUDE

The potential term in a cylindrical bend depends on the bending radius of the cylindrical coordinate frame used to describe the field. This requires us to solve Maxwell's equations and to relate it to a measurement of the magnetic field in the midplane. This can be done elegantly with a TPSA routine attached to a tracking code. We now examine this in the following interlude.

INTERLUDE XIV
THE CURVATURE DEPENDENCE OF $V(X, Y; \rho_C)$ USING TPSA

This interlude describes a calculation often done by algebraic manipulators while it should in fact be done in a TPSA environment. We now describe this calculation.[12.1] The reader will notice that the procedure is also implementable in a manipulator, but the advantage (rational coefficients instead of floating points) is of no use. It is also true that one can regain the rational results anyway with a TPSA calculation done with a sufficient number of digits.

Inside its body, a bend has usually either a Cartesian symmetry or a cylindrical symmetry. As we have been pointing out, this has nothing to do with the actual "design" orbit which passes through this bend. It has to do with the internal symmetries of the field and eventually with the fabrication techniques used in building the device. In the case of the Cartesian bend, we expect the field to be translationally invariant along the longitudinal z-axis. For example, the reduced vector potential of a constant field bend is simply:

$$a_z = -b_0 x.$$

However we know already that in a sector bend with cylindrical symmetry, the correct value for a vector potential producing a constant vertical field is given by

$$a_s = -\frac{V}{1 + \frac{x}{\rho_c}} = -\frac{b_0}{1 + \frac{x}{\rho_c}} \left(x + \frac{x^2}{2\rho_c} \right).$$

It turns out that this term could be explained by pure geometry. In a parallel face bend of constant field, there cannot be any horizontal focusing because of the transverse translational invariance in the x-direction. In a sector bend, the wedge shape will produce focusing because a particle

[12.1] Our treatment follows somewhat that of reference [65]. It has been pointed out to me by C. Iselin that the new version of MAD in C++ contains an elegant TPSA algorithm for getting the curvature dependence directly from Equation (12.33). This confirms the spirit behind this interlude: it is just normal to write such an algorithm if a code has access to a TPSA library.

which travels at large values of x must traverse a thicker magnet. This is reflected by the presence of $\frac{b_0 x^2}{2\rho_c}$ in the cylindrical vector potential. Of course, one can derive this result using Maxwell's equations in cylindrical coordinates. The use of Maxwell's equations is the only way to proceed with the computation of the correct value of $V(x, y; \rho_c)$ when dealing with multipolar terms beyond the dipole.

We start from the relation $\vec{b} = \vec{\nabla} \times \vec{a}$. In cylindrical coordinates, the reduced magnetic field is given by

$$b_x = \frac{1}{\rho}\frac{\partial}{\partial y}(\rho a_s) \tag{12.22.a}$$

$$b_y = -\frac{1}{\rho}\frac{\partial}{\partial x}(\rho a_s) \tag{12.22.b}$$

$$b_s = 0 \tag{12.22.c}$$

$$\text{and } \rho \equiv \rho_c + x. \tag{12.22.d}$$

Using the condition $\vec{\nabla} \times \vec{b} = 0$ (no current), the equation for the vector potential reduces to:

$$\left\{\nabla_\perp^2 - \frac{1}{\rho}\frac{\partial}{\partial x}\right\}(\rho a_s) = 0. \tag{12.23}$$

The operator $\nabla_\perp^2$ is the transverse Laplacian $\frac{\partial^2}{\partial x^2} + \frac{\partial^2}{\partial y^2}$. Denoting the product $(1 + x/\rho_c)a_s$ by the letter $F (F = -V(x, y; \rho_c)\ )$, we obtain the following equation for F:

$$\left\{(\rho_c + x)\left(\nabla_\perp^2 - \frac{\partial}{\partial x}\right) - \frac{\partial}{\partial x}\right\} F = 0. \tag{12.24}$$

As was done in the case of the Cartesian multipole, we expand the function F in homogeneous polynomials:

$$F = \sum_{k=1}^{\infty} F_k. \tag{12.25}$$

A recursive system of equations can be written for the F_k's by substituting into Equation (12.24). One gets

$$\nabla_\perp^2 F_{k+1} = \rho_c^{-1}\left(\frac{\partial}{\partial x} - x\nabla_\perp^2\right) F_k. \tag{12.26}$$

The first thing to notice is that our constant bend vector potential is an exact solution of this recursion relation if we set $F_1 = b_0 x$ and $F_2 =$

$b_0 \frac{x^2}{2\rho_c}$. Second, because this is a recursive relation, we can solve the recursion by assuming that the lowest order in the recursion is a straight multipole solution:

$$F_n^n = \underbrace{a_n s_n}_{A_n^n} + \underbrace{b_n n_n}_{B_n^n} = -\Re \frac{1}{n} \left(i a_n + b_n \right) \left(x + i y \right)^n . \quad (12.27)$$

Hence our general solution for F becomes a huge double summation:

$$\begin{aligned} F &= \sum_{n=1}^{\infty} \sum_{k=n}^{\infty} B_n^k + A_n^k \\ &= \sum_{n=1}^{\infty} A_n + B_n. \end{aligned} \quad (12.28)$$

Next, let us scale the x and y variables by ρ_c^{-1}. A new universal operator emerges:

$$\nabla_{\perp}^2 F_{k+1} = \left(\frac{\partial}{\partial x} - x \nabla_{\perp}^2 \right) F_k. \quad (12.29)$$

We are now ready to think "TPSA/LIELIB" on this problem. First, let us define the inverse of the Laplacian operator $\nabla_{\perp}^2$. If we define the following "pseudo-resonance" variables

$$r_{\pm} = x \pm i y,$$

then the Laplacian is $\nabla_{\perp}^2 = 4 \partial_1 \partial_2$. Thus we can define a possible inverse for the Laplacian as follows:

$$\nabla_{\perp}^{-2} r_+^{k_+} r_-^{k_-} = \frac{1}{4 \left(k_+ + 1 \right) \left(k_- + 1 \right)} r_+^{k_+ + 1} r_-^{k_- + 1}. \quad (12.30)$$

It should be said that one can always add to Equation (12.30) an element of the kernel (i.e., an harmonic multipole of the straight system). These multipoles are made out of monomials in one variable only. We can use a normalizing library such as LIELIB[35] and a TPSA package to construct a routine for the inverse Laplacian. The reader is probably disturbed by the gauge freedom. For example, if we can introduce an arbitrary sextupole when computing the cubic term generated by the curvature and the quadrupole, what is then the meaning of "sextupole?" As in the normal form, we need to relate these arbitrary objects to a physical measurement. In our case, we will choose a measurement of the field in the mid-plane. So, in the mean time, we can view our choice of gauge as one convenient choice for mathematical reasons.[12.2]

Hence, in the resonance basis, the skew and normal components of the harmonic ansatz are given by the formulas

$$s_n = -i\frac{r_+^n - r_-^n}{2n}$$
$$n_n = -\frac{r_+^n + r_-^n}{2n}.$$

Defining a new operator

$$\widehat{S} = \mathcal{C}^{-1} \underbrace{\nabla_\perp^{-2}}_{\text{As in (12.30)}} \mathcal{C}\left(\frac{\partial}{\partial x} - x\nabla_\perp^2\right), \tag{12.31}$$

where $\mathcal{C}$ is the transformation into the resonance basis, we conclude that the functions A^n and B^n are given compactly by

$$\begin{aligned} F &= (1-\widehat{S})^{-1}\left\{\sum_{n=1}^{\infty} a_n s_n + b_n n_n\right\} \\ &= \sum_{n=1}^{\infty} \underbrace{a_n (1-\widehat{S})^{-1} s_n}_{A_n} + \sum_{n=1}^{\infty} \underbrace{b_n (1-\widehat{S})^{-1} n_n}_{B_n}. \end{aligned} \tag{12.32}$$

This operation can be carried out in a TPSA sense to an arbitrary jet order N_o. At most it will require the evaluation of N_o terms in the expansion of $(1-\widehat{S})^{-1}$.

The solution given in Equation (12.32) contains free gauge terms of unknown physical meaning. As an example of the use of TPSA to help us fix these terms, we will assume that we are given from the "experimentalist" an expansion of the horizontal and vertical fields in the mid-plane $y = 0$. Thus the functions

$$b_x^{y=0} = \sum_{n=1}^{No} a_n x^{n-1}$$
$$b_y^{y=0} = \sum_{n=1}^{No} b_n x^{n-1} \tag{12.33}$$

[12.2]The imposition of a gauge can be nontrivial if we only know the boundary conditions at infinity. This is the case of space charge problems. See reference [66] section 8.4.

are input for the code. We can relate these to our solutions using Equation (12.22):

$$b_x^{y=0} = \left.\frac{1}{1+x}\frac{\partial F}{\partial y}\right|_{y=0} = \sum_{\substack{n=1,N_0\\ m=1,n}} a_m \mathfrak{A}_{nm} x^{n-1} \quad (12.34.a)$$

$$b_y^{y=0} = -\left.\frac{1}{1+x}\frac{\partial F}{\partial x}\right|_{y=0} = \sum_{\substack{n=1,N_0\\ m=1,n}} b_m \mathfrak{B}_{nm} x^{n-1}. \quad (12.34.b)$$

The matrices $\mathfrak{A}$ and $\mathfrak{B}$ are lower triangular. Using Equation (12.34) we can reexpress the potential F in terms of the physical measurements given by (12.33):

$$F(x,y;\rho_c=1) = \sum_{\substack{n=1,N_0\\ m=1,n}} \mathfrak{A}^{-1}_{nm} a_m \left(1-\widehat{S}\right)^{-1} s_n + \sum_{\substack{n=1,N_0\\ m=1,n}} \mathfrak{B}^{-1}_{nm} b_m \left(1-\widehat{S}\right)^{-1} n_n. \quad (12.35)$$

We can regain the dependence on ρ_c at the end of the calculation because the operator $\widehat{S}$ raises the degree of a homogeneous polynomial by one unit exactly and the matrices $\mathfrak{A}^{-1}$and $\mathfrak{B}^{-1}$ are lower triangular. Thus the exact expression given by

$$\begin{aligned} F(x,y;\rho_c) &= -V(x,y;\rho_c) \\ &= \sum_{\substack{n=1,N_0\\ m=1,n\\ k=0,N_0-n}} \frac{1}{\rho_c^{n-m+k}} \mathfrak{A}^{-1}_{nm} a_m \widehat{S}^k s_n \\ &\quad + \sum_{\substack{n=1,N_0\\ m=1,n\\ k=0,N_0-n}} \frac{1}{\rho_c^{n-m+k}} \mathfrak{B}^{-1}_{nm} b_m \widehat{S}^k n_n. \quad (12.36) \end{aligned}$$

can be guessed by looking at each monomial and the field coefficient that multiplies it. For example, if the "quadrupolar" coefficient b_2 multiplies x^4, we know that it should be divided by ρ_c^2.

The advantages of a TPSA based program are enormous. First of all, it can be incorporated easily into a tracking code. Second, because it is in the tracking code, the user can decide on the "physical relation" equivalent to Equation (12.33) and generate the appropriate expressions. For example we could have chosen some bizarre units for the magnetic fields such as

$$b_x^{y=0} = \sum_{n=1}^{No} a_n \left\{ \frac{x}{r_0} \right\}^{n-1} \tag{12.37}$$

where the field is measured for each multipole at some convenient radius r_0. Needless to say that more complex relationships involving Fourier transforms could be put in "on the spot" by the user and the expression correctly inverted by a TPSA routine.

Finally, it is clear that the magnetic field itself is available. This is useful in radiation calculations.
END OF INTERLUDE

12.3 LIMITING CASES OF THE SECTOR MAP $\mathcal{W}$

There are several useful maps which can be constructed using Equation (12.18). They are important for the proper treatment of the arbitrary bend which is discussed later.

12.3.1 The Ideal Parallel Face Straight Bend

The body of a parallel face bend was first mentioned in Sect. 12.1.5. It is gotten through an obvious geometrical limit: the layout parameter ρ_c goes to infinity and the arc length s becomes the Cartesian length L of the straight magnet. Thus the body map of an ideal parallel face bend is given by:

$$\mathcal{V}(L, b_1) = \lim_{\substack{\rho_c \to \infty \\ s = L}} \mathcal{W}(s, b_1; \rho_c), \tag{12.38}$$

whose transfer map is thus

$$x(L) = x + \frac{1}{b_1}\left(\sqrt{(1+\delta)^2 - p_x(L)^2 - p_y^2} - \sqrt{(1+\delta)^2 - p_x{}^2 - p_y^2}\right) \tag{12.39.a}$$

$$p_x(L) = p_x - b_1 L \tag{12.39.b}$$

$$y(L) = y + \frac{p_y}{b_1}\left(\sin^{-1}\left(\frac{p_x}{\sqrt{(1+\delta)^2 - p_y^2}}\right) - \sin^{-1}\left(\frac{p_x(L)}{\sqrt{(1+\delta)^2 - p_y^2}}\right)\right) \tag{12.39.c}$$

$$p_y(L) = p_y \tag{12.39.d}$$

$$\delta(L) = \delta \tag{12.39.e}$$

$$\ell(L) = \ell + \frac{(1+\delta)}{b_1}\left(\sin^{-1}\left(\frac{p_x}{\sqrt{(1+\delta)^2 - p_y^2}}\right) - \sin^{-1}\left(\frac{p_x(L)}{\sqrt{(1+\delta)^2 - p_y^2}}\right)\right). \tag{12.39.f}$$

12.3.2 The Ideal Wedge

The ideal parallel face bend and the sector bend are related to each other by a missing wedge as indicated in Figure 12.4. This wedge has positive bending in the upper part and negative bending in the lower part. It is obtained using the limit

$$\mathcal{U}(\theta, b_1) = \lim_{\substack{\rho_c \to 0;\, s \to 0 \\ \frac{s}{\rho_c} = \theta}} \mathcal{W}(s, b_1; \rho_c). \tag{12.40}$$

Technically the map for negative x is meaningless since x is a radius, but it does make sense by analytic continuation. The expression for the transfer map is given by

$$\begin{aligned}
x(\theta) &= x\cos(\theta) \\
&\quad + \frac{x p_x \sin(2\theta) + \sin^2(\theta)\left(2x\sqrt{(1+\delta)^2 - p_x^2 - p_y^2} - b_1 x^2\right)}{\sqrt{(1+\delta)^2 - p_x(\theta)^2 - p_y^2} + \sqrt{(1+\delta)^2 - p_x^2 - p_y^2}\cos(\theta) - p_x\sin(\theta)} && (12.41.a) \\
p_x(\theta) &= p_x\cos(\theta) + \left(\sqrt{(1+\delta)^2 - p_x^2 - p_y^2} - b_1 x\right)\sin(\theta) && (12.41.b) \\
y(\theta) &= y + \frac{p_y\theta}{b_1} \\
&\quad + \frac{p_y}{b_1}\left(\sin^{-1}\left(\frac{p_x}{\sqrt{(1+\delta)^2 - p_y^2}}\right) - \sin^{-1}\left(\frac{p_x(\theta)}{\sqrt{(1+\delta)^2 - p_y^2}}\right)\right) && (12.41.c) \\
p_y(\theta) &= p_y && (12.41.d) \\
\delta(\theta) &= \delta && (12.41.e) \\
\ell(\theta) &= \ell + \frac{(1+\delta)\theta}{b_1} \\
&\quad + \frac{(1+\delta)}{b_1}\left(\sin^{-1}\left(\frac{p_x}{\sqrt{(1+\delta)^2 - p_y^2}}\right) - \sin^{-1}\left(\frac{p_x(\theta)}{\sqrt{(1+\delta)^2 - p_y^2}}\right)\right). && (12.41.f)
\end{aligned}$$

The expression in Equation (12.41.a) has been transformed so as to remove all artificial divergences at the origin.

12.3.3 The Dynamical Rotation $\mathcal{Y}(\theta)$

The reader might have noticed that the wedge map of Sect. 12.3.2 is a rotation in a constant B_y field. It is a dynamical rotation in the $x - z$ plane in this medium. Hence the transfer map associated with $\mathcal{Y}(\theta)$, given by Equation (10.26), can also be evaluated using Equation (12.41). From a computer standpoint it is not advisable to use Equation (12.41) for $\mathcal{Y}(\theta)$ since it has vanishing denominators at $b_1 = 0$. The problem resides in the arcsine functions for which it is extremely messy to rewrite Equation (12.41) in terms of expressions which are all regular at the origin and at $b_1 = 0$.

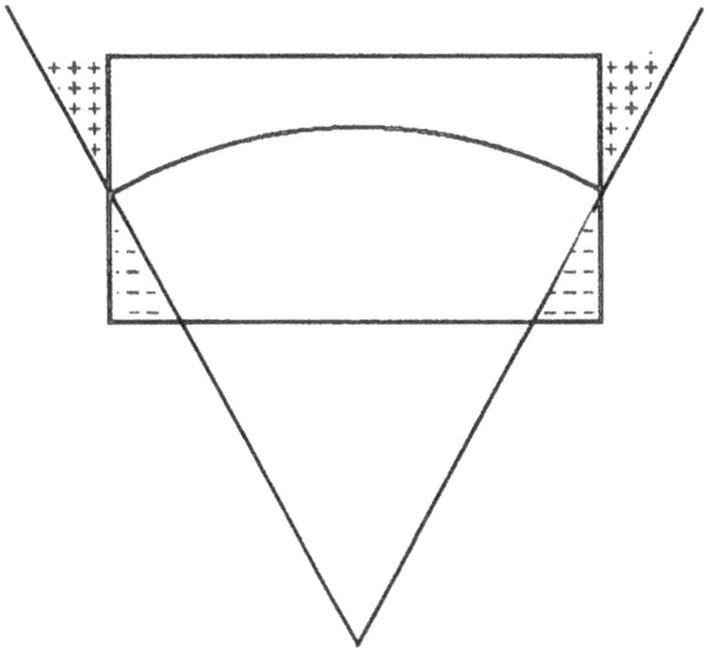

Figure 12.4 : Using a Wedge to go from Sector to Parallel Face Bend.

Exercise 12-1

The reader is encouraged to derive Equation (10.26) *from* (12.41). *Hint: If* $b_1 = 0$ *then the expression*

$$\sin^{-1}\left(\frac{p_x}{\sqrt{(1+\delta)^2 - p_y^2}}\right) - \sin^{-1}\left(\frac{p_x(\theta)}{\sqrt{(1+\delta)^2 - p_y^2}}\right)$$

is of the form $\sin^{-1}\sin(\gamma) - \sin^{-1}\sin(\gamma + \theta)$ *where* θ *is the bending angle.*

This concludes our collection of limiting cases of the ideal sector bend map. Now we proceed to apply these maps to the general bending magnet.

12.4 BENDS WITH ARBITRARY GEOMETRY

The case of the ideal bend is puzzling since it is obvious that the body of a bend can be represented in two different ways. Using the map of Equation (12.41) and Figure 12.4, we can write the ideal sector map as

$$\mathcal{W}(\rho_c\Phi_{12},\tfrac{1}{\rho_c};\rho_c) = \mathcal{U}(\tfrac{\Phi_{12}}{2};\tfrac{1}{\rho_c})\mathcal{V}(O_1O_2;\tfrac{1}{\rho_c})\mathcal{U}(\tfrac{\Phi_{12}}{2};\tfrac{1}{\rho_c}), \tag{12.42}$$

and conversely, the body of the parallel face ideal bend as

$$\mathcal{V}(O_1O_2;\tfrac{1}{\rho_c}) = \mathcal{U}(-\tfrac{\Phi_{12}}{2};\tfrac{1}{\rho_c})\mathcal{W}(\rho_c\Phi_{12},\tfrac{1}{\rho_c};\rho_c)\mathcal{U}(-\tfrac{\Phi_{12}}{2};\tfrac{1}{\rho_c}). \tag{12.43}$$

In fact, consider the body map of an arbitrary ideal bend. Using the wedges, we can construct it using a sector bend as displayed in Figure 12.5 or a parallel face bend as in Figure 12.6.

How is one supposed to decide how to represent the ideal arbitrary bend? Does it really matter? The answer to this question is to be found outside single particle dynamics: in the symmetry of the magnet and its resulting Maxwellian fields. To illustrate this we consider the field due to two plates of opposite magnetic polarity. If the plates are rectangles placed on top of each other and perfectly aligned, we expect the field to be constant over a rectangle aligned and centered between the plates. We now tilt the plates in the $x-y$ plane; the field will develop a gradient. In the body of the magnet the field will be constant along the z-axis. This is the case of the shifted quadrupole mentioned repeatedly in this book. Both configurations are displayed in Figure 12.7. It is quite clear that an arbitrary magnet whose body has rectangular symmetries should be integrated in the body in rectangular coordinates. The quadratic integrator step for the body map $\mathcal{M}_{1'2'}$ will be, according to Equation (12.11) or (12.12):

$$\mathcal{T}_2 = \exp\left(-:\tfrac{O_1O_2}{2N}(V-\tfrac{x}{\rho_c}):\right)\mathcal{V}(\tfrac{O_1O_2}{N};\tfrac{1}{\rho_c})\exp\left(-:\tfrac{O_1O_2}{2N}(V-\tfrac{x}{\rho_c}):\right). \tag{12.44}$$

It important to realize that the use of rectangular coordinates for the body is done *irrespective* of the entrance or exit angles. In a very precise simulation, accurate fringe maps will be needed to handle the corrections at both ends of the magnet. In the absence of any specific model one can put wedges to accommodate the end-shape. For example, for entrance and exit angles ε_1 and ε_2 respectively, as in Figure 12.6, the map can be written as

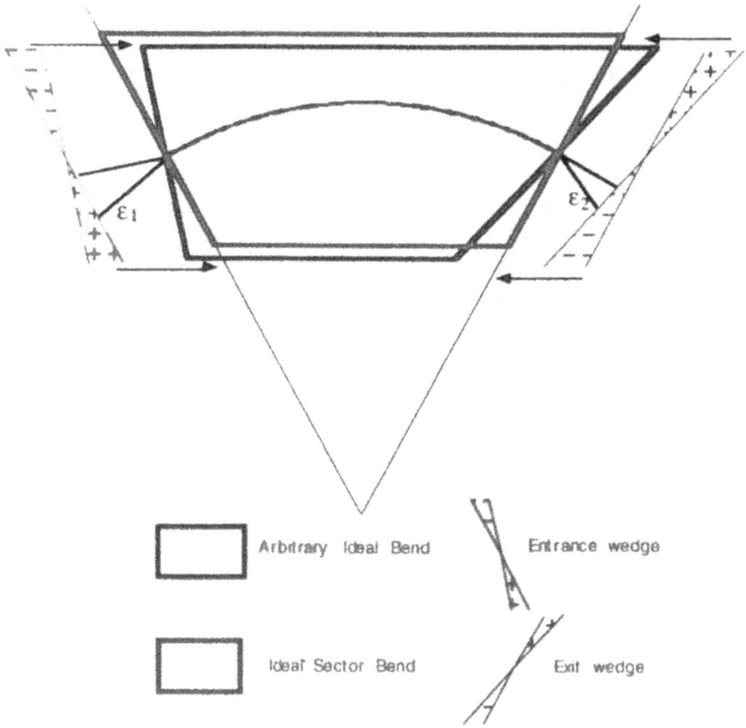

Figure 12.5 : Arbitrary Ideal Bend Using Sector Bend.

$$\begin{aligned}\mathcal{M}_{12} = &\mathcal{Y}(\varepsilon_1)\mathcal{F}_1\mathcal{U}(\frac{\Phi_{12}}{2} - \varepsilon_1)\\ &\times\mathcal{M}_{1'2'}\\ &\times\mathcal{U}(\frac{\Phi_{12}}{2} - \varepsilon_2)\mathcal{F}_2\mathcal{Y}(\varepsilon_2).\end{aligned} \tag{12.45}$$

Of course depending on the exact form of the fringe maps $\mathcal{F}_1$ and $\mathcal{F}_2$ this expression may even be exact. Usually only approximate fringes are used and thus this representation is not completely correct.

It is possible to have bends of arbitrary entrance and exit angles with cylindrical symmetry. In Figure 12.8 we display such an oddity. The field resulting from this magnet will be invariant along an arc of a circle centered at the "origin" of the donut. Integration for the

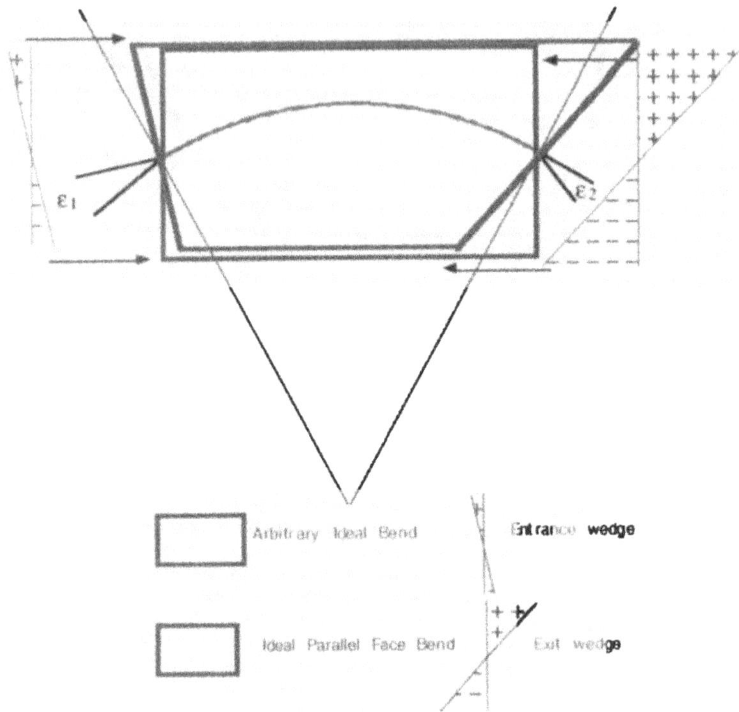

Figure 12.6 : Arbitrary Ideal Bend Using Parallel Bend

body will require cylindrical coordinates as well as a vector potential in cylindrical coordinates—one must use Equation (12.13). For bends with cylindrical symmetry, the equivalent of Equation (12.45) is

$$\begin{aligned} \mathcal{M}_{12} = & \mathcal{Y}(\varepsilon_1)\mathcal{F}_1\mathcal{U}(-\varepsilon_1) \\ & \times \mathcal{M}_{1'2'} \\ & \times \mathcal{U}(-\varepsilon_2)\mathcal{F}_2\mathcal{Y}(\varepsilon_2), \end{aligned} \tag{12.46}$$

where each step of integration for the body map $\mathcal{M}_{1'2'}$ is now given by Equation (12.19) or (12.20).

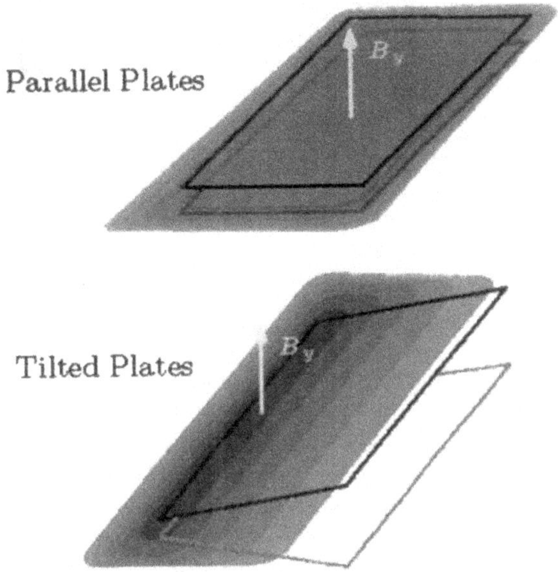

Figure 12.7 : The Effect of Rectangular Geometry in Bends.

12.5 SOLENOIDS AND WIGGLERS: STRAIGHT ELEMENTS

I do not want to embark on a complex description of wigglers, undulators and solenoids. However the decoupling of the layout and the magnet are most important in clarifying the integration methods which should be used in dealing with these elements.

12.5.1 Dump the Torsion once and for all PLEASE!

It is my experience that confusion reigns when dealing with these elements precisely because their internal symmetries (dictating the z-axis as the integration variable), their layout positioning and their "design orbit" are asking for incompatible frames of reference. In the standard theory, one always looks for a "design orbit" through the element which is then used for the theoretical description and tracking.

Figure 12.8 : The Effect of Sector Geometry in Bends.

This orbit and the matrix around it become crucial if one needs to match the solenoid or wiggler. For example, in the case of an helical wiggler, the orbit may "cork-screw" around leaving us with serious descriptive headaches. Detection solenoids with crossing angles at the interaction point (IP) will also produce complex non-planar "design orbits." What a mess, but fortunately for the global/local theory, what a *totally irrelevant* mess.

For both wigglers and solenoids the Hamiltonian can be written conveniently in Cartesian or cylindrical coordinates. These cylindrical coordinates are very different from the sector bend because the plane using polar coordinates is the transverse $x - y$ plane. For example it is well known that the solenoid can be represented by the Hamiltonian

$$H = -\sqrt{(1+\delta)^2 - (p_x - a_x)^2 - (p_y - a_y)^2} - a_z, \quad (12.47)$$

where the scaled vector potential component (defined in Equation (12.4)) can be written in terms of the on-axis z-component of the magnetic field—$B_z(0, 0, z)$:

$$a_x = -\frac{yU}{Brho}$$
$$a_y = \frac{xU}{Brho}$$
$$U = \sum_{n=0}^{\infty} \frac{(-1)^n r^{2n} B_z^{[2n]}}{2^{2n+1} n!(n+1)!}$$

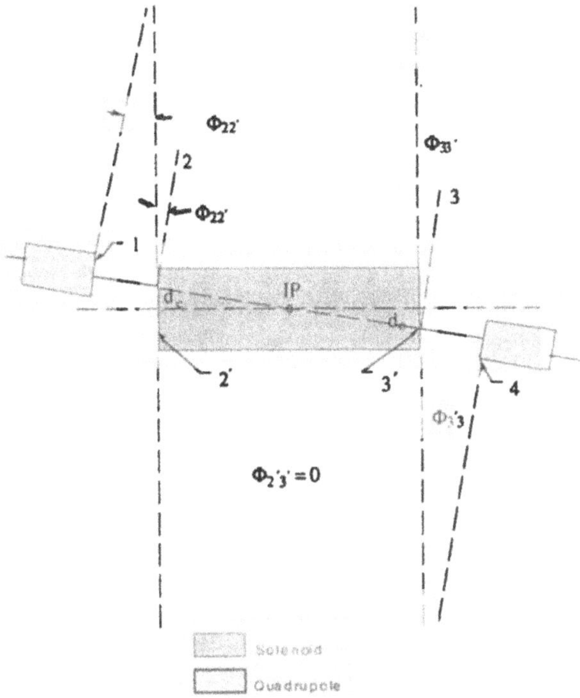

Figure 12.9 : Placing a Solenoid in the Layout.

$$B_z^{[2n]} = \frac{d^{2n}}{dz^{2n}} B_z(0, 0, z).$$

The important point of this section is to realize that the z-variable should be the integration variable as one would do in electron microscopy with round lenses. The Cartesian map from the entrance plane to exit plane (denoted by $2'$ and $3'$ on Figure 12.9) must be matched to the exit plane of the first quadrupole (denoted by 1) and the entrance plane of the second quadrupole (denoted by 4). These planes are parallel to the layout planes 2 and 3. As usual this is done with dynamical Euclidean transformations:

$$\mathcal{M}_{23} = \mathcal{Y}(\Phi_{22'}) \mathcal{T}_x(-d_e) \mathcal{M}_{2'3'} \mathcal{T}_x(-d_o) \mathcal{Y}(\Phi_{3'3}). \qquad (12.48)$$

Here the crossing angle was put in the horizontal $x - z$ plane. The position of the unprimed plane is the result of a "civil engineering" layout program which positions the tunnel and the element within it. The choice of the primed frame of reference for integration through the solenoid originates in our understanding of "Maxwell's Equations" in a solenoid: a purely local decision which must be matched to the layout.

At no point in this calculation does our knowledge of the final closed orbit (i.e., some sort of cork-screw through the solenoid) interferes with our sensible choice of integration coordinates. All quantities computed by this method (Taylor series map as well as moments maps in the presence of stochastic radiation) are carried from plane 1 to plane 4 automatically thanks to automatic differentiation. Matching this solenoid (in the sense of lattice functions) can be done with complete ease in a bug-free context thanks again to automatic differentiation and associated Lie algebraic libraries (LIELIB)[35].

The need for continuous torsion, that is to say, for a reference curve possessing torsion, is totally unnecessary in accelerator physics. It would become necessary if the beam pipe itself or the tunnel would have the cork-screw geometry—locally if Maxwell's equations were easily solved in these complex curvilinear variables. Thank God, these types of magnets do not exist. In any case, the decision to go "cork-screw" would still be local. Global torsion is sometimes necessary in a nonplanar ring due to the noncommutative properties of the rotation group in three dimensions. One can show that rings made out of horizontal and vertical 90 degree bends, may still need a 90 degree transverse rotation at the point of closure so as to insure that the frames of reference smoothly match. This topological torsion can be put anywhere in the lattice.

Finally, the need for torsion in a helical wiggler is even more contemptible since helical wigglers are straight elements often placed in *planar* rings.

12.5.2 Exact Solution of the Ideal Solenoid

We start with the Hamiltonian of the ideal solenoid given by Equation (12.21) (in Interlude XIII). We first introduce a canonical transformation of the transverse variables:

$$x^{new} = \frac{1}{b_z}\left(p_y - \frac{b_z}{2}x\right) \tag{12.49.a}$$

$$p_x^{new} = p_x + \frac{b_z}{2}y \tag{12.49.b}$$

$$y^{new} = \frac{1}{b_z}\left(-p_y - \frac{b_z}{2}x\right) \tag{12.49.c}$$

$$p_y^{new} = p_x - \frac{b_z}{2}y. \tag{12.49.d}$$

The new Hamiltonian, which we may call K, is just

$$K = -\sqrt{(1+\delta)^2 - p_x^2 - b_z^2 x^2}. \tag{12.50}$$

Automatically, we have three constants of the motion: K itself, y, and p_y. Secondly it is easy to get a second order differential equation for x and p_x. We can also "cheat" by noticing that this Hamiltonian can be put easily into action-angle variables:

$$\begin{aligned} x &= \sqrt{2J/\left|b_z\right|}\cos(\phi) \\ p_x &= -\sqrt{2\left|b_z\right|J}\sin(\phi) \\ \Rightarrow K &= \sqrt{(1+\delta)^2 - 2\left|b_z\right|J}. \end{aligned} \tag{12.51}$$

So, we conclude that the solution is

$$\phi(z) = \phi(0) + z\frac{\partial K}{\partial J} = \phi(0) - \frac{z\left|b_z\right|}{K}. \tag{12.52}$$

In terms of our new set of Cartesian variables, the solution is

$$\begin{aligned} x^{new}(z) &= x_0^{new}\cos(\omega z) + \frac{p_{x;0}^{new}}{b_z}\sin(\omega z) \\ p_x^{new}(z) &= p_{x;0}^{new}\cos(\omega z) - b_z x_0^{new}\sin(\omega z) \\ \text{where } \omega &= -\frac{b_z}{K}. \end{aligned}$$

The solution for the motion of the original transverse variables can be expressed in terms of a "matrix" with amplitude dependent coefficients:

$$S(z) = \begin{pmatrix} \frac{1+c}{2} & \frac{s}{b_z} & \frac{s}{2} & \frac{1-c}{b_z} \\ \frac{sb_z}{4} & \frac{1+c}{2} & \frac{b_z(c-1)}{4} & \frac{s}{2} \\ -\frac{s}{2} & \frac{c-1}{b_z} & \frac{1+c}{2} & \frac{s}{b_z} \\ \frac{b_z(1-c)}{4} & -\frac{s}{2} & \frac{sb_z}{4} & \frac{1+c}{2} \end{pmatrix}. \tag{12.53}$$

The functions c and s are

$$c(z) = \cos\left(-\frac{zb_z}{H}\right) \text{ and } s(z) = \sin\left(-\frac{zb_z}{H}\right). \tag{12.54}$$

The change in path length is given by

$$\ell(z) = \ell + z\frac{1+\delta}{H}, \tag{12.55}$$

a formula identical to that of a drift!

CHAPTER 13
FRINGE EFFECTS IN RING DYNAMICS

Fringe effects, with the notable exception of vertical focusing, are usually neglected in rings. We will see that they are connected to the small angle approximation—where this approximation is very good, we often find these fringe effects to be negligible.

But before engaging in this topic, we distinguish various "fringe fields." Let us take the Hamiltonian for a normal quadrupole:

$$H = -\sqrt{(1+\delta)^2 - p_x^2 - p_y^2} + \frac{k_0 + \Delta k(s)}{2}(x^2 - y^2), \quad (13.1)$$

where $\Delta k(s)$ is a function which vanishes in the body of the quadrupole and is $-k_0$ far away from the quadrupole. This Hamiltonian is based on the vector potential

$$a_z = -\frac{k_0 + \Delta k(s)}{2}(x^2 - y^2) \quad (13.2)$$

which cannot be correct in a current-free region. The usual ideal magnet approximation consists of two potentially dangerous approximations:

1. $k_0 + \Delta k(s)$ by $\overline{k}_0$ over an effective length L. Without a loss of generality we can redefine $\Delta k(s)$ so that $\overline{k}_0 = k_0$.
2. In that process, we ignore completely Maxwell's equations at the boundary. Maxwell's equations are thus satisfied piecewise.

The compositional map for this approximation is simply

$$\mathcal{M} = \exp\left(L : \left\{\sqrt{(1+\delta)^2 - p_x^2 - p_y^2} - \frac{k_0}{2}(x^2 - y^2)\right\} :\right).$$

We can try to improve on this by integrating (13.1) thus ignoring Maxwell's equations. Using the results of Sect. 11.3.2.2, we can even apply explicit symplectic integration to this Hamiltonian. The crucial question is this: under the Talman's philosophy of tracking we are replacing a single ideal quadrupole by a collection of tiny quadrupoles closely bunched together. Are we truly changing the global properties of the ring once lattice functions and tunes are refitted? Probably not under usual circumstances! In a sense, no new "ring physics" is introduced by the correct integration of Equation (13.1).

Thus comes the next question: how about solving Maxwell's equations and the compositional map equation to leading order at the boundary? Are we getting new physics? We will see that the answer is yes.

Thus the fringe field strategy in rings is as follows:

- First ignore all fringe fields and see what happens!
- Introduce first order Maxwellian maps (they are thin) at the end of the magnets; the tracking code should have these maps routinely available because, as we will see, they do not depend on the profile $\Delta k(s)$ in leading order.
- If you think that a correct modeling is necessary, realistic fringe field maps can be introduced as exact thin maps thanks once more to our knowledge of the dynamical Euclidean group. Therefore the layout planes can be preserved.

We now study these issues.

13.1 THE BEND: FRINGE MAP IN THE HARD EDGE LIMIT

The expression derived in this section is to leading order in the field and all orders in the other variables. The formula applies exactly to a bend with an infinitely long face in the transverse x-direction.

Consider an ideal bending magnet whose magnetic field in the midplane is given by

$$\forall x \; B_y(x,0,z) = Brho \; b(z). \tag{13.3}$$

Ideally this can only be true in an infinitely long parallel face bend. If the plates are sufficiently close to each other, this is a reasonable approximation near the center of the pipe at the entrance and exit face. The scaled vector potential for this field is given by

$$a_z = -x\, b(z). \tag{13.4}$$

It is easy to check that this vector potential does not obey Maxwell's equations—$\nabla \times \nabla \times (0, 0, -x\, b(z)) \neq 0$. However we can introduce an x-component to the vector potential of the form

$$a_x = \sum_{n=1}^{\infty} a_x^n(z)\, y^{2n}, \tag{13.5}$$

and require that $(a_x, 0, a_z)$ be curl-free. The solution is found immediately to be

$$a_x = \sum_{n=1}^{\infty} \frac{(-1)^n b_z^{[2n-1]}}{(2n)!} y^{2n}. \tag{13.6}$$

13.1.1 Integration of the Map

This result is substituted into the straight Hamiltonian which is then expanded to leading order in these new Maxwellian terms:

$$\begin{aligned} H &= -\sqrt{(1+\delta)^2 - (p_x - a_x)^2 - p_y^2} + x\, b_z \\ &= -\sqrt{(1+\delta)^2 - p_x{}^2 - p_y^2} + x\, b_z \\ &\quad - \frac{p_x a_x}{\sqrt{(1+\delta)^2 - p_x{}^2 - p_y^2}} + \cdots. \end{aligned}$$

The map through the fringe $\mathcal{F}$ is obtained to first order in the function b_z using the techniques of Sect. 8.4.1.1 and Equation (8.25):

$$\begin{aligned} \mathcal{F} &= \exp\,(:f:) \\ f &= \lim_{\varepsilon \to 0} \int_{-\varepsilon}^{\varepsilon} -x' \sum_{n=1}^{\infty} \frac{(-1)^n b_z^{[2n-1]}}{(2n)!} (y + (s+\varepsilon) y')^{2n} ds \\ (x', y') &= \frac{1}{\sqrt{(1+\delta)^2 - p_x^2 - p_y^2}} (p_x, p_y). \end{aligned}$$

This integral can be carried out exactly in the hard-edge limit ($b_z' = b_0 \delta(z)$) using integration by parts. The answer is

$$f_\pm = \mp b_0 \frac{x'}{1 + y'^2} \frac{y^2}{2}. \tag{13.7}$$

The $\mp$ sign refers to the entrance and exit fringes respectively.

13.1.2 Physical Interpretation

Let us apply this result to the paraxial matrix of a particle on the ideal trajectory. If the entrance angle of the bend is ε_1 as in Figures 12.5 and 12.6, the Lie exponent f should be expressed in coordinates perpendicular to the face of the magnet. This is precisely the frame where the fringe maps were inserted in Equations (12.45) and (12.46). The map for the entrance face becomes

$$\begin{aligned} \mathcal{F}_+ &= \exp\left(: -b_0 \frac{\tan(\varepsilon_1) + x'}{1 + y'^2} \frac{y^2}{2} : \right) \\ \text{since} \quad & x' \to \tan(\varepsilon_1) + x', \end{aligned} \tag{13.8}$$

which has the linear (paraxial) part

$$\begin{aligned} \mathcal{F}_+^1 &= \exp\left(: -b_0 \tan(\varepsilon_1) \frac{y^2}{2} : \right) \\ \mathcal{F}_+^1 p_y &= p_y - b_0 \tan(\varepsilon_1)\, y. \end{aligned} \tag{13.9}$$

This is the famous vertical focusing term. The exit term is given by

$$\begin{aligned} \mathcal{F}_- &= \exp\left(: b_0 \frac{-\tan(\varepsilon_2) + x'}{1 + y'^2} \frac{y^2}{2} : \right) \\ \text{since} \quad & x' \to -\tan(\varepsilon_2) + x', \end{aligned} \tag{13.10}$$

resulting in a similar expression for the paraxial map

$$\begin{aligned} \mathcal{F}_-^1 &= \exp\left(: -b_0 \tan(\varepsilon_2) \frac{y^2}{2} : \right) \\ \mathcal{F}_-^1 p_y &= p_y - b_0 \tan(\varepsilon_2)\, y. \end{aligned} \tag{13.11}$$

It should be noted that the sector bend still has a fringe field in the hard edge approximation but it is an entirely nonlinear effect producing vertical chromaticity. Furthermore, the entrance and exit field in this particular case tend to cancel each other. If we assume that a bend is mostly a drift from a dynamic point of view, we can factor the total map as follows:

$$\begin{aligned} \mathcal{M} &= \mathcal{F}_+ \mathcal{W} \mathcal{F}_- \\ &= \mathcal{F}_+ \mathcal{W} \mathcal{F}_+^{-1} \end{aligned}$$

$$\approx \exp\left(: f - \mathcal{W}f :\right)\mathcal{W}$$
$$\approx \exp\left(: -\frac{b_0}{2}\frac{x'}{1+y'^2}\left\{y^2 - (y+Ly')^2\right\} :\right)\mathcal{W}$$
$$\approx \exp\left(: \frac{b_0 L}{2}\frac{x'}{1+y'^2}\left(2y'\frac{y}{L} + y'^2\right) :\right)\mathcal{W},$$

where $\mathcal{W}$ is the sector bend compositional map given by Equation (12.18). The final expression depends on powers of the momentum in all planes and is small if the small angle approximation is valid. This explains why the fringe field of sector bends as well as the fringe field of straight elements has been neglected in simulations where large machine (small angle) codes were used. In small machines the vertical chromaticity due to Maxwellian effects in bends can be tremendous and indeed of the order of the ideal machine chromaticity.

13.1.3 Symplectic Evaluation

The transfer map $\exp\left(: f_\pm :\right)\vec{I}$ is hard to evaluate exactly. However because our treatment is first order in b_0, we can replace the map produced by the Lie transform by an equivalent map generated using a mixed variable characteristic function. We choose it to be

$$S_\pm(\vec{q}^f, \vec{p}) = \vec{q}^f \cdot \vec{p} - f_\pm(\vec{q}^f, \vec{p}), \tag{13.12}$$

which leads us to a simple set of equations to solve for the map through the fringe:

$$x = x^f \mp b_0 \frac{y^{f2}}{2}\frac{\partial}{\partial p_x}\left(\frac{x'}{1+y'^2}\right) \tag{13.13.a}$$
$$p_x^f = p_x \tag{13.13.b}$$
$$y = y^f \mp b_0 \frac{y^{f2}}{2}\frac{\partial}{\partial p_y}\left(\frac{x'}{1+y'^2}\right) \tag{13.13.c}$$
$$p_y^f = p_y \mp b_0 y^f \left(\frac{x'}{1+y'^2}\right) \tag{13.13.d}$$
$$\delta^f = \delta \tag{13.13.e}$$
$$\ell = \ell^f \mp b_0 \frac{y^{f2}}{2}\frac{\partial}{\partial \delta}\left(\frac{x'}{1+y'^2}\right). \tag{13.13.f}$$

To solve the system of equations (13.13), one first solves the quadratic equation (13.13.c) for y^f. The result is substituted in the remaining equations which are then trivially solved. This provides a complete symplectic representation of the hard edge fringe effects in a dipole.

13.1.4 Quadrupole Trick for a Bend: Void of Physical Interpretation

This exercise points out another confusion which reigns supreme. One can show that the total *linear* map for an arbitrary bend of entrance and exit angles ε_1 and ε_2 can be written as the product of a thin quadrupole kick, a sector bend matrix and another thin quadrupole kick. Thus the linear part $\mathcal{L}_{12}$ of Equation (12.46) can be factorized as

$$\mathcal{L}_{12} = \mathcal{Q}_1 \mathcal{W}_L \mathcal{Q}_2, \tag{13.14}$$

where

$$\mathcal{Q}_i = \exp\left(: \frac{b_0 \tan(\varepsilon_i)}{2}(x^2 - y^2) : \right)$$

and

$$\mathcal{W}_L = \text{Linear part of } \mathcal{W}.$$

The map in (13.14) must be equal to the paraxial expansion of the map in (12.46). Moreover the map $\mathcal{Q}_1$ is the linear expansion of $\mathcal{Y}(\varepsilon_1)\mathcal{F}_1\mathcal{U}(-\varepsilon_1)$ while $\mathcal{Q}_2$ is the linear part of $\mathcal{U}(-\varepsilon_2)\mathcal{F}_2\mathcal{Y}(\varepsilon_2)$.

It should be clear to the reader that the quadrupole trick has *no physical interpretation* whatsoever. It is completely absurd to think that the particle feels a kind of quadrupole field as it enters an arbitrary bend. Moreover it is even more absurd and dangerous to deduce some physical or phenomenological extension of the quadrupole trick. It is a mathematical formula void of physical content. In contrast, the product of the three maps $\mathcal{Y}(\varepsilon_1)\mathcal{F}_1\mathcal{U}(-\varepsilon_1)$ reflects the correct physical picture. The coordinate frame is first rotated from the direction into the local magnet face frame. Then the particle feels a mostly vertical force due to the sudden change in field. Finally it goes through the magnet which we elected to represent in terms of a wedge and a sector bend. The wedge focusing, which is mostly horizontal, happens *after* the vertical focusing.

13.2 MULTIPOLE FRINGE IN THE HARD EDGE LIMIT

In this section we will only examine the leading order effect in both the field gradient and the transverse x and y phase space. We will see that an ideal $2n$-pole whose vector potential in the z-direction is of degree n produces a fringe effect of degree $n+2$. Thus a quadrupole whose zeroth order purpose is to produce linear focusing will produce a cubic effect in the hard edge limit.

Due to the general complexity of the problem we will not attempt to go beyond degree $n+2$ for a $2n$-pole. Also the case of the solenoid is not studied since in the hard edge limit it involves the square of a Dirac delta function which is not well-defined under the integral sign.

The calculation starts with the assumption of invariance under a $\frac{2\pi}{n}$ rotation around the z-axis. For the infinitely long multipole, a solution of Maxwell's equations is given by Equation (12.3) which we rewrite in polar coordinates

$$\begin{aligned} a_z(x,y) &= \frac{A_z}{Brho} = -\Re\sum_{n=1}^{\infty}\frac{c_n}{n}r^n e^{in\phi} \\ ib_x + b_y &= \sum_{n=1}^{\infty} c_n r^{n-1} e^{i(n-1)\phi} \\ c_n &= ia_n + b_n. \end{aligned} \tag{13.15}$$

The scaled field ($\vec{b} = \frac{\vec{B}}{Brho}$) can be derived from a potential in a current-free region. In turn, this potential must satisfy Laplace's equation. Thus we have

$$\begin{aligned} \vec{\nabla}\Phi &= \vec{b} \\ \nabla^2\Phi &= 0. \end{aligned} \tag{13.16}$$

Because the Laplacian commutes with ∂_ϕ we know that it is possible to find solutions whose ϕ dependence will be that of a harmonic multipole. Thus we try solutions of the form:

$$\begin{aligned} \Phi_n &= \Re e^{in\phi}\Psi_n(r,z) \\ \Psi_n(r,z) &= \sum_{m=0}^{\infty}\psi_{n;m}(z)r^m. \end{aligned} \tag{13.17}$$

Exercise 13-1

What follows can be found in most texts of electron microscopy. The reader is invited to work it out. Here are a few useful formulas in cylindrical coordinates which might save time for the reader:

$$\vec{\nabla}\Phi = \partial_r\Phi\ \vec{e}_r + \frac{1}{r}\partial_\phi\Phi\ \vec{e}_\phi + \partial_z\Phi\ \vec{e}_z$$

$$\nabla^2\Phi = \frac{1}{r}\partial_r\ (r\partial_r\Phi) + \frac{1}{r^2}\partial_r^2\Phi + \partial_z^2\Phi$$

$$\vec{\nabla}\times\vec{A} = \left(\frac{1}{r}\partial_\phi A_z - \partial_z A_\phi\right)\ \vec{e}_r + (\partial_z A_r - \partial_r A_z)\ \vec{e}_\phi + \frac{1}{r}\ (\partial_r(rA_\phi) - \partial_\phi A_r)\ \vec{e}_z.$$

1. *Assuming that one is supposed to match the general z-dependent solution to an infinite length multipole, show that*

$$\Psi_n = \sum_{k=0}^{\infty}\frac{\psi_{n;n}^{[2k]}(-1)^k n!\ r^{2k+n}}{2^{2k}k!(k+n)!} \tag{13.18}$$

where $\psi_{n;n}^{[2k]} = \frac{d^{2k}}{dz^{2k}}\psi_{n;n}$.

2. *Matching this to the infinite length solution, show that*

$$\psi_{n;n}^{[0]} = -\frac{ic_n}{n}, \tag{13.19}$$

and thus

$$\Psi_n = -\frac{i}{n}\sum_{k=0}^{\infty}\frac{c_n^{[2k]}(-1)^k n!\ r^{2k+n}}{2^{2k}k!(k+n)!}\ ;\ n\geq 2. \tag{13.20}$$

3. *Check that*

$$a_r^n = \frac{A_r^n}{Brho} = \Re\frac{ir}{n}\partial_z\Psi_n e^{in\phi}$$

$$a_z^n = \frac{A_z^n}{Brho} = -\Re\frac{ir}{n}\partial_r\Psi_n e^{in\phi}.$$

13.2.1 Integration of the Map

To compute the map we proceed as in the case of the bend. The Hamiltonian is expanded to first order in the field:

$$H = -\sqrt{(1+\delta)^2 - (p_x - a_x)^2 - (p_y - a_y)^2} - a_z$$
$$H \approx -\sqrt{(1+\delta)^2 - p_x^2 - p_y^2} - \frac{p_r a_r}{\sqrt{(1+\delta)^2 - p_x^2 - p_y^2}} - a_z. \tag{13.21}$$

More compactly we rewrite this Hamiltonian[13.1] as

$$H \approx -\sqrt{(1+\delta)^2 - p_x^2 - p_y^2} - r' a_r - a_z. \tag{13.22}$$

Although we could integrate by parts all the relevant terms as we did in the case of the bend, here we will only keep terms of leading order in the transverse variables. It should be noticed that the integration by parts changes position variables into angle variables, which in most accelerators are negligible and therefore it provides an additional justification for neglecting terms of higher transverse degree.

We must substitute our expressions for the vector potential in Equation (13.22). But first, let us write the leading terms

$$a_r^n = \Re \frac{c_n'}{n^2} r^{n+1} e^{in\phi}$$
$$a_z^n = \Re \left\{ -\frac{c_n}{n} + \frac{c_n''(n+2)}{4n^2(n+1)} r^2 \right\} r^n e^{in\phi} \tag{13.23}$$

which are then substituted into H (dropping the sum over n):

$$H \approx \underbrace{-\sqrt{(1+\delta)^2 - p_x^2 - p_y^2}}_{D} \underbrace{-r' \Re \frac{c_n'}{n^2} r^{n+1} e^{in\phi}}_{V_1}$$
$$\underbrace{-\Re \left\{ -\frac{c_n}{n} + \frac{c_n''(n+2)}{4n^2(n+1)} r^2 \right\} r^n e^{in\phi}}_{V_0+V_2}. \tag{13.24}$$

[13.1] Notice that $\sqrt{(1+\delta)^2 - p_x^2 - p_y^2} = \sqrt{(1+\delta)^2 - p_r^2 - p_\phi^2/r^2}$, $r p_r = x p_x + y p_y$, and $p_\phi = L_z = x p_y - y p_x$.

The map for the fringe is again given by

$$\mathcal{F} = \exp(: f :)$$
$$f = -\lim_{\varepsilon\to 0}\int_{-\varepsilon}^{\varepsilon} T_z(s+\varepsilon)V\, ds \tag{13.25}$$
$$\text{where } T_z(s+\varepsilon) = \exp\left(: (s+\varepsilon)\sqrt{(1+\delta)^2 - p_x^2 - p_y^2} :\right).$$

In the hard edge limit, f_0, the integral of V_0, which is a step function, vanishes in the limit of vanishing ε. The integral of V_1 contains a Dirac delta function and is just

$$f_1 = \pm r'\Re\frac{c_n}{n^2}r^{n+1}e^{in\phi}. \tag{13.26}$$

The integral of V_2 is performed by parts:

$$\begin{aligned}
f_2(\varepsilon) &= \Re\int_{-\varepsilon}^{\varepsilon} T_z(s+\varepsilon)\left\{\frac{c_n''(n+2)}{4n^2(n+1)}r^2\right\}r^n e^{in\phi}ds \\
&= \Re\, T_z(s+\varepsilon)\left\{\frac{c_n'(n+2)}{4n^2(n+1)}r^2\right\}r^n e^{in\phi}\Big|_{-\varepsilon}^{\varepsilon} \\
&\quad -\Re\int_{-\varepsilon}^{\varepsilon}\frac{dT_z(s+\varepsilon)}{ds}\left\{\frac{c_n'(n+2)}{4n^2(n+1)}r^2\right\}r^n e^{in\phi}ds \\
&\overset{\varepsilon\to 0}{=} \pm\Re\frac{c_n(n+2)}{4n^2(n+1)}\left[D, r^{n+2}e^{in\phi}\right].
\end{aligned} \tag{13.27}$$

In deriving this formula we made use of the map equation for the drift $dT_z/dz = T_z : -D :$. The final expression for f is

$$\begin{aligned}
f &= f_0 + f_1 + f_2 \\
f_\pm &= \pm\Re\frac{c_n(n+2)}{4n^2(n+1)}\left[D, r^{n+2}e^{in\phi}\right] \pm \Re r'\frac{c_n}{n^2}r^{n+1}e^{in\phi} \\
&= \pm\Re\frac{c_n r^n e^{in\phi}}{4(n+1)D}\left\{p_r r + i\frac{n+2}{n}p_\phi\right\} \\
&= \pm\Re\frac{c_n r^n e^{in\phi}}{4(n+1)D}\left\{xp_x + yp_y + i\frac{n+2}{n}(xp_y - yp_x)\right\}.
\end{aligned} \tag{13.28}$$

As usual the $\pm$ sign refers to the entrance and exit fringe.

13.2.2 Symplectic Evaluation

The reader may have realized that the map $\exp(: f :)$ is not solvable exactly. This is a situation similar to the case of the bend fringe map in Sect. 13.1.3 and again we will use a characteristic function to get a solution that is accurate to first order in c_n. However the presence of the Hamiltonian $D = -\sqrt{(1+\delta)^2 - p_x^2 - p_y^2}$ makes it impossible (or at least very hard) to obtain an exactly symplectic first order map. But, we must remember that this solution was also leading order in the transverse variables and it is therefore permissible to replace D by $-(1+\delta)$.

$$
\begin{aligned}
f_\pm &\approx \mp\Re\frac{c_n(x+iy)^n}{4(n+1)(1+\delta)}\left\{xp_x + yp_y + i\frac{n+2}{n}(xp_y - yp_x)\right\} \\
&\overset{\text{of the form}}{=} \frac{p_x f^x + p_y f^y}{1+\delta}.
\end{aligned} \tag{13.29}
$$

The functions f^x and f^y depend on position only. The characteristic function is chosen so as to depend on the final momenta

$$
S(\vec{x}, \vec{p}^f) = \vec{x}\cdot\vec{p}^f - \frac{p_x^f f^x + p_y^f f^y}{1+\delta}. \tag{13.30}
$$

The resulting system is

$$
x^f = x - \frac{f^x}{1+\delta} \tag{13.31.a}
$$

$$
p_x = p_x^f - \frac{p_x^f(\partial_x f^x) + p_y^f(\partial_x f^y)}{1+\delta} \tag{13.31.b}
$$

$$
y^f = y - \frac{f^y}{1+\delta} \tag{13.31.c}
$$

$$
p_y = p_y^f - \frac{p_x^f(\partial_y f^x) + p_y^f(\partial_y f^y)}{1+\delta} \tag{13.31.d}
$$

$$
\delta^f = \delta \tag{13.31.e}
$$

$$
\ell = \ell^f + \frac{p_x^f f^x + p_y^f f^y}{(1+\delta)^2}. \tag{13.31.f}
$$

It is solved by inverting Equations (13.31.b) and (13.31.d) simultaneously for the final transverse momenta. The rest follows trivially.

13.2.3 A Famous Case: Lee-Whiting's Quadrupole Map

The results presented here for the straight elements were first derived by Milutinovic[67] and the author. However the most important application, the quadrupole fringe field map, was first correctly derived by Lee-Whiting in a paper on quadrupole fringe effects. Lee-Whiting took the hard-edge limit of his more complex formulas just out of curiosity. He never really expected this limit to be of any use. The reader should remember our discussion about the contrast between single pass systems and rings: their demands on absolute accuracy are usually much higher and certainly very different.

To obtain the quadrupole fringe field one substitutes $c_n = b_2$ in Equation (13.29). The Lie transform becomes

$$f_{\pm} = \pm \frac{b_2}{12(1+\delta)} \left\{ y^3 p_y - x^3 p_x + 3x^2 y p_y - 3y^2 x p_x \right\}. \quad (13.32)$$

The resulting transfer map is given by $\exp(:f_{\pm}:)\vec{I} = \vec{I} + \left[f_{\pm}, \vec{I}\right] + \cdots$

$$x^f = x \pm \frac{b_2}{12(1+\delta)} \left\{ x^3 + 3y^2 x \right\} \quad (13.33.a)$$

$$p_x^f = p_x \pm \frac{b_2}{4(1+\delta)} \left\{ 2xyp_y - x^2 p_x - y^2 p_x \right\} \quad (13.33.b)$$

$$y^f = y \mp \frac{b_2}{12(1+\delta)} \left\{ y^3 + 3x^2 y \right\} \quad (13.33.c)$$

$$p_y^f = p_y \mp \frac{b_2}{4(1+\delta)} \left\{ 2xyp_x - y^2 p_y - x^2 p_y \right\} \quad (13.33.d)$$

$$\delta^f = \delta \quad (13.33.e)$$

$$\ell^f = \ell - \frac{f_{\pm}}{(1+\delta)}. \quad (13.33.f)$$

The first four lines of Equation (13.33) are Lee-Whiting's original 1970 result. The change of the path length is a consequence of the δ dependence. Quite amusingly several scientists in our field managed over the years to rederive incorrectly Lee-Whiting's result. For example see references [68, 69]. The first of these references is particularly revealing of our "multipole" mentality.

Let us discuss a few features of these hard edge fields using the quadrupole as an example by responding to two questions often asked by skeptics. I should add that the questions and comments which follow are applicable to higher-order multipole fringe fields as well as to the bend fringe.

1. Why are these fields usually negligible at least in "large" machines?

2. These formulas involve a "jump" of the position variables; how can that be physical? Should they be neglected?

I will not answer fully question 1, partly out of pure ignorance I must admit! Certainly many authors have given approximate scaling laws for the tunes, beta functions, and dispersion for large machines made of FODO cells. In such machines the tunes, the chromaticities, the quadrupoles and sextupoles strengths are tied to each other. Thus the unavoidable nonlinearities due to the quadrupole fringes, as well as those present in the ideal elements, are linked to the chromaticity sextupoles. It is easy to imagine, as the ring gets bigger, that the relative effect of the fringes and sextupoles must tend towards a limit. Of course, the cultural ignorance of both the quadrupole fringes and nonlinearity of the ideal elements, suggests that the effect of the sextupoles dominates in the limit of large FODO rings. In particular, let us see how the validity of the small angle approximation might be connected to neglecting fringe fields. We start with the full map for the quadrupole in the hard edge limit

$$\begin{aligned} \mathcal{Q}_{tot} &= \exp\left(: f_+ :\right) \mathcal{Q} \exp\left(: -f_+ :\right) \\ &= \exp\left(: f_+ - \mathcal{Q} f_+ :\right) \mathcal{Q}, \end{aligned} \tag{13.34}$$

and we approximate $\mathcal{Q}$ as

$$\begin{aligned} \mathcal{Q} &= \exp\left(: -L\sqrt{(1+\delta)^2 - p_x^2 - p_y^2} - L\frac{b_2}{2}(x^2 - y^2) :\right) \\ &\approx \exp\left(: -L\frac{p_x^2 + p_y^2}{2} :\right) \exp\left(: -L\frac{b_2}{2}(x^2 - y^2) :\right). \end{aligned} \tag{13.35}$$

We substitute this approximate linear quadrupole in Equation (13.34) and examine the total Lie transform of the fringes:

$$\begin{aligned} f_+ - \mathcal{Q} f_+ \approx & \frac{Lb_2}{4}(x^2 + y^2)(p_x^2 - p_y^2) \\ & - \frac{Lb_2^2}{12}(x^4 + y^4 + 6x^2y^2) + O(L^2). \end{aligned} \tag{13.36}$$

The first term in this expression is obviously small whenever the angles are small. The second term requires the kind of scaling analysis which we will not provide in this book. Amusingly this second term looks like an octupole vector potential except for the presence of $6x^2y^2$ instead of $-6x^2y^2$ in the case of the octupole. So there is some ground to the statement "quadrupole fringe fields look like an octupole" even though they do not!

This brings us to the next topic, the kick of the fringe does indeed change both positions and momenta, an unphysical feature at first glance since it implies a jump in the variable x and y.[13.2] Concerning question 2, Lee-Whiting in his original paper remarked:

> Eq. (8b) was derived[13.3] by Gluckstern[70] by assuming that $k'(z)$ had the form of a Dirac Delta-function. There does not appear to be any good reason for including the terms in (8b) while ignoring the terms in (8a).

Indeed, there is not any good reason, *especially* in ring dynamics and in quasi-periodic systems. In rings the relative sizes of the positions and momenta are determined internally; in the case of uncoupled linear dynamics the lattice functions β, α and γ are the relevant functions. If a nonlinear map is added, such as the quadrupole hard edge map, one should measure its effect in Floquet variables—for example by applying a Courant-Snyder transformation to it. We know that in scaled variables, the average sizes of the normalized position and momentum are identical because the ray moves on a circle in phase space. For simplicity, if we look at a place where the Twiss α functions vanish, it is easy to check that the normalized x and p_x (for $y = p_y = 0$) are perturbed by terms of similar average strengths. Moreover, if one neglects the effect on the positions, the resulting map is **grossly** nonsymplectic. This is not permissible in rings.

Concerning the "jump" in the position variables, one should remember that we took a limit. Strictly speaking the field decreases from its full value over a distance λ which is hopefully small, i.e., the phase advance and the lattice functions do not vary considerably over this distance. Thus the map which has "physical" reality is the full map through the interface of width λ. At the entrance face, this map is

$$\mathcal{T}_z(\frac{\lambda}{2}) \exp\,(: f_+ :)\, \mathcal{Q}(\frac{\lambda}{2}; b_2). \tag{13.37}$$

Suppose we knew the exact map $\mathcal{F}(\lambda)$ through an entrance fringe region of thickness λ, then we could rewrite it as follows:

$$\mathcal{T}_z(\frac{\lambda}{2}) \underbrace{\mathcal{T}_z(-\frac{\lambda}{2})\mathcal{F}_+(\lambda)\mathcal{Q}^{-1}(\frac{\lambda}{2}; b_2)}_{\approx \exp(:f_+:)} \mathcal{Q}(\frac{\lambda}{2}; b_2). \tag{13.38}$$

[13.2] Of course if you are a fan of science fiction this is not a problem: Scottie beam me up!

[13.3] Equation (8b) of Lee-Whiting is equivalent to our Equations (13.33.b) and (13.33.d).

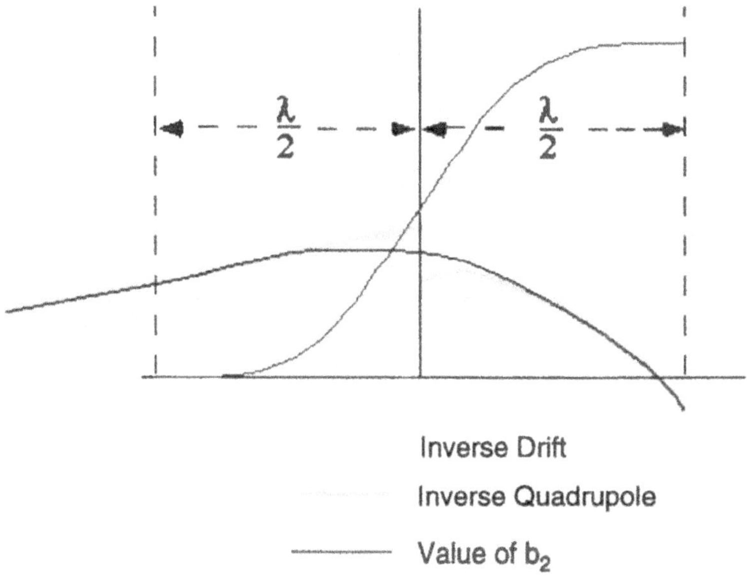

Figure 13.1 : Thin Fringe Field Map.

Thus the issue is whether or not the hard edge map agrees pretty well with the real thick map $\mathcal{F}_+(\lambda)$ after it has been "sandwiched" between an inverse drift and an inverse ideal quadrupole. On Figure 13.1, we display an "exaggerated realistic" fringe map. The thin black curve is supposed to represent the exact trajectory through the fringe. The jump from the yellow line to the green line is the exact thin map $\mathcal{T}_z(-\frac{\lambda}{2})\mathcal{F}_+(\lambda)\mathcal{Q}^{-1}(\frac{\lambda}{2}; b_2)$ which, like our approximate thin map, must jump at the interface.

13.3 "IMPORTATION" OF REALISTIC FRINGE MAPS

So far we have ignored the inclusion of more realistic fringe calculations in our theoretical framework. In single particle systems, this issue is not ignorable. Papers on this topic abound in electron microscopy, spectrometer design and cyclotron design. In the field of rings, the issue of fringe fields has been swept under the rug until recently. Small rings and strongly focusing interaction regions have required the inclusion of Lee-Whiting's quadrupole map in simulation codes. For these reasons we shall examine the proper methods to include a general fringe within our ideal layout. This section is not on the computation of the map. I will assume that the map for the fringe region is given to us as a "black box," the issue is how to incorporate this corrective field exactly and correctly on the top of the ideal magnet. Again we shall assume that overlapping fields between magnets are not present, otherwise the original layout *must be redefined*.

I will first discuss the importation into one's tracking code of a model for the full magnet that includes fringe fields. The particular "importation problem" is entirely solvable in terms of the dynamical Euclidean transformation described in Chapter 10 as we will now show.

13.3.1 Importing Full Magnet Maps

In this section we will give a few examples of importation. The reader, if confronted with an actual importation, will hopefully remember the dynamical Euclidean group!

Let us start with a magnet with internal straight symmetry. This could be a regular straight magnet which belongs to a straight layout slot or an ALS-type bending magnet as shown on Figure 12.1. So we assume here that the map is provided in a Cartesian frame because the fields, including the complex fringe fields, are easily expressed in Cartesian coordinates. The correct map $\mathcal{Q}_{tot}$, provided by some "sophisticated" code, must be inserted into the code as if ideal. This defines for us the importation problem: how to insert a complex map with bizarre fringe regions into a code which expects it to fit into the layout? This is done by "sandwiching" the real map with negative drift. Hence the map to be put in the layout is simply

$$\mathcal{Q}_{12} = \mathcal{T}_z(-\lambda)\mathcal{Q}_{tot}\mathcal{T}_z(-\lambda). \tag{13.39}$$

The next example is that of a Cartesian Bend with fringe fields. It is the same element as above from a local point of view, but it is shifted and placed in a curved layout slot. Hence the map to be lowered in the ideal layout position is given by

$$\mathcal{M}_{12} = \mathcal{Y}(\frac{\Phi_{12}}{2})\mathcal{T}_x(-d_x)\mathcal{Q}_{12}\mathcal{T}_x(d_x)\mathcal{Y}(\frac{\Phi_{12}}{2}), \tag{13.40}$$

where the map $\mathcal{Q}_{12}$ is given by Equation (13.39).

It should be clear from this discussion that the entire "importation" problem, that is the replacement of ideal maps which fit snugly

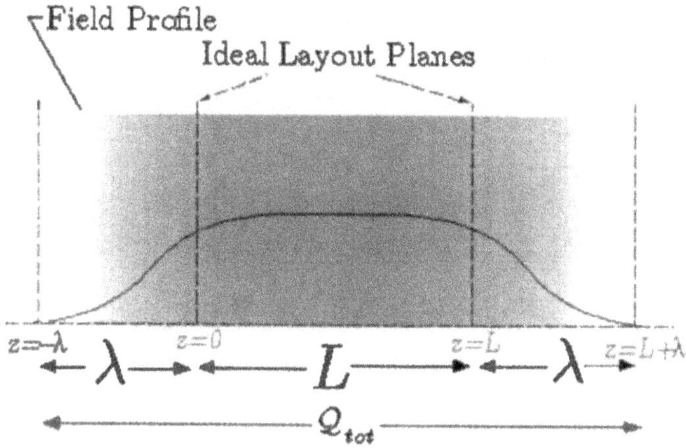

Figure 13.2 : Thick Fringe Field Map in a Straight Layout.

and naturally into their layout position by nonideal maps, is a geometrical exercise with the dynamical Euclidean group. The variations on this theme are infinite and will depend on the bizarre local nature of the magnet that is imported into the code.

We have already mentioned the importation of fringes alone. The thin hard edge fringe is a special case. We discuss this problem briefly in the next section for the case of a realistic extended fringe map.

13.3.2 Importing Fringe Maps Only

We are motivated by the existence in most simulation codes of models for the ideal elements. Thus it is natural to "add" fringe maps to a code for two reasons: first the fringe is usually a small spatial perturbation and second, in part because it is localized, we would like to be able to add to the ideal magnet the fringe effect as we did easily in the hard edge limit.

The solution to this is again the dynamical Euclidean group, at least "*in theory*." In practice since one side of the fringe will be embedded in the ideal magnet, unless we have the dynamical Euclidean group represented in this medium, we have no practical way to import this fringe into our code.

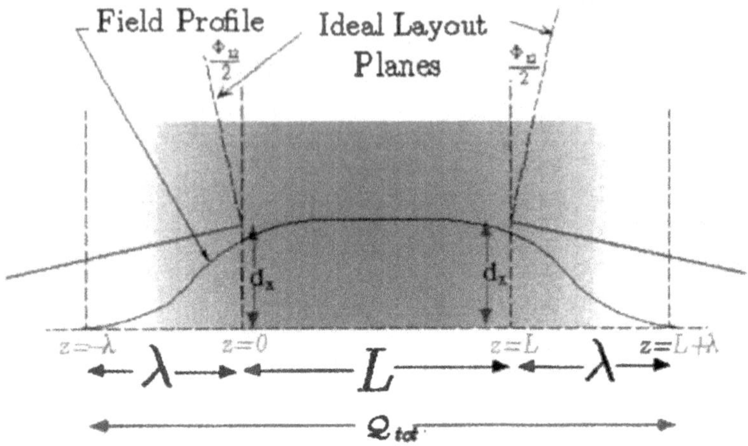

Figure 13.3 : Cartesian Element Fringe in a Curved Layout.

There is however one nontrivial[13.4] dynamical transformation available to us: it is the propagator itself. In expression (13.38), the true fringe effect $\mathcal{F}(\lambda)$, which has been integrated for a length λ along the z-axis, is imported by sandwiching it with an inverse drift and an inverse quadrupole: $\mathcal{T}_z(-\frac{\lambda}{2})\mathcal{F}_+(\lambda)\mathcal{Q}^{-1}(\frac{\lambda}{2}; b_2)$. From a purely geometric point of view, these two inverse maps are translations in their respective media (take another look at Figure 13.2).

Just to illustrate this in a more complex situation, we can imagine the following (realistic) situation. A bend with internal cylindrical symmetry is built, however the ends of the magnet, due to various construction constraints, has no particularly nice symmetry. It is then decided, to launch a probe to map the magnetic field (or run a fancy field code). These measurements (or calculations) are done in Cartesian coordinates. The probe is carefully inserted in planes parallel to the midplane.

In Figure 13.4, the Cartesian frame used in the field measurements is matched with the frame of the sector bend at an angle θ inside the sector. At that meeting point, the layout arc of the circle is tangent

[13.4] Transverse Euclidean transformations within a Cartesian element, as in the field-free case, are just the regular time-based Euclidean transformations: the x and y translations and the z-axis rotation.

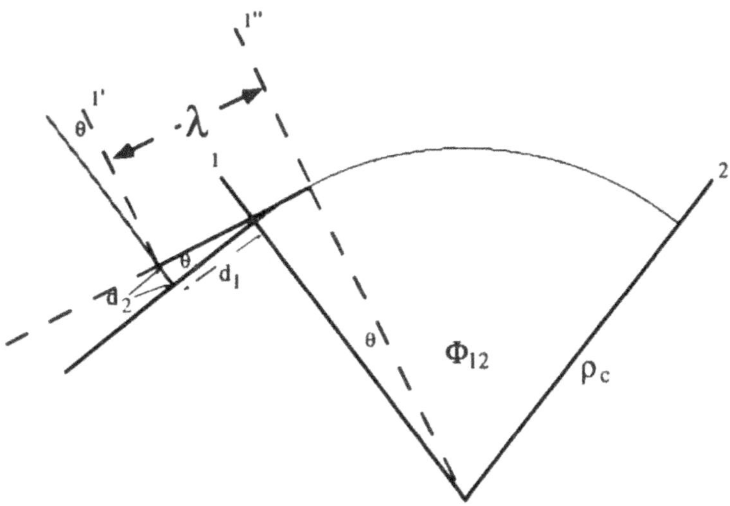

Figure 13.4 : Thick Fringe Field Map for a Sector Bend.

to the Cartesian frame of the fringe map. Denoting the map for the fringe $\mathcal{F}_+(\lambda)$, we can write the "thin" but exact entrance fringe map as

$$\mathcal{F}_+^1 = \mathcal{T}_z(-d_1)\mathcal{T}_x(d_2)\mathcal{Y}(\theta)\mathcal{F}_+(\lambda)\mathcal{S}(-\theta). \qquad (13.41)$$

The map $\mathcal{F}_+^1$ operates at the layout plane 1 which, in this example, is the nominal entrance plane of this sector bend. The map $\mathcal{S}(-\theta)$ is obtained by integrating the sector Hamiltonian of Equation (12.13) using one's favorite technique.

In conclusion, the techniques presented in this section take an exact fringe map representing a field that varies over a short[13.5] distance λ and imports it into the layout as a "thin map." The procedure is exact: no information is lost in the process. It is a convenient way to introduce fringe effects "on top" of the lattice already being simulated.

[13.5]It must be short compared to the ideal magnet. Otherwise one may as well import the full magnet.

CHAPTER 14
LARGE RING APPROXIMATIONS AND THE REST!

This chapter should really be a section of Chapter 12. However, historically and hysterically, the small angle approximation permeates the field of ring dynamics. It is such a good approximation that in the minds of many it has ceased to be an approximation. Quadrupoles and bends became linear in practice and in *theory*.

My decision to isolate this discussion on small angle approximations does not mean that I approve the permanent linearization of ideal elements—please read the previous chapters. I am acknowledging the fact that this approximation has been and continues to be a "damned" good approximation in the simulation of circular accelerators and linacs. In addition, by restricting nonlinearities to mostly chromatic sextupoles and sometimes wigglers, it permits a great simplification and speed increase of the simulation code.

Compromises will stop at the organizational level of this book: our description of the "small angle Hamiltonian" will start from the correct local treatment and will proceed with well-defined approximations. In Sect. 14.2 I will contrast this with approximations which are the result of a "low quality patching job." I have done it before, so I speak here of my own incompetence with competence!

Finally I will discuss briefly a few approximate elements for completeness.

14.1 THE CYLINDRICAL BEND

We start with the Hamiltonian of Equation (12.13) and we expand it in the momentum variables p_x and p_y while setting $b_1 = \frac{1}{\rho_c}$:

$$H = \frac{p_x^2 + p_y^2}{2\,(1+\delta)} + \frac{x^2}{2\rho_c^2} - \frac{x\delta}{\rho_c} + V(x, y; \rho_c) + \cdots . \tag{14.1}$$

The qualifier "small angles" also applies to the bending radii of the bends and thus it is customary to also neglect the term $\frac{x(p_x^2+p_y^2)}{2\rho_c}$. It should be noted that this Hamiltonian is exact in the energy deviation δ, this is a critical ingredient of the small angle approximation. Indeed, since the ideal machine is now linear in the transverse variable, one can solve for the transverse dynamics to *all orders* in δ. The angles may be small, but it is not clear that one can neglect higher powers of the energy deviation, and in fact, it is a bad idea to do so.

Exercise 14-1

1. *Assuming that one can write a global Hamiltonian for the ring which looks like Equation* (14.1), *show that the chromaticities are given by the formula*

$$\nu_i = -\frac{1}{4\pi}\oint \gamma_i ds \qquad i \equiv x, y. \tag{14.2}$$

2. *Using formula* (2.93) *and Equation* (14.2) *appropriately in each plane, show that the chromaticities are given by the alternate formula*

$$\nu_i = -\frac{1}{4\pi}\oint \beta_i k_i ds \tag{14.3}$$

$$k_x = k_{quad} + \frac{b_1}{\rho_c} \approx k_{quad}$$
$$k_y = -k_{quad}.$$

3. *So, what is going on here? Is chromaticity produced by "drifts" or produced by "quadrupoles." Answer: The question is meaningless if asked without a "design context." The chromaticity being a global quantity simply exists as a result of both the quadrupole and the drift parts of the Hamiltonian. However, in designing a ring, one can ask a question whose answer will be easier to find with one formula in mind versus the other. For example, a low beta insertion will "generate" a lot of chromaticity simply because γ must be large in that region. However, since we usually add quadrupoles to an existing ring*[14.1] *instead of chang-*

[14.1]This may be connected to the budgetary situation of the late 20th century in industrialized countries. Because money is in short supply, scientists are reusing existing rings. Hence the tunnel length is not a variable anymore!

ing its length, it may be more useful to say that quadrupoles generate chromaticity.

Formula (14.3) in Exercise 14-1 is more well-known than formula (14.2) because the standard kick codes, and thus the accompanying theory, use the scaled variables x' and y'. They are just

$$\begin{aligned} x' &= \frac{dx}{ds} = [x, H] = \frac{p_x}{1+\delta} \\ y' &= \frac{dy}{ds} = [y, H] = \frac{p_y}{1+\delta}. \end{aligned} \tag{14.4}$$

In the case absence of a cavity (coasting beam), the energy is a constant of the motion and thus we can treat it as a parameter of the reduced four-dimensional transverse phase space. It is easy to check that the Hamiltonian

$$H_\perp = \frac{x'^2 + y'^2}{2} + \frac{x^2}{2(1+\delta)\rho_c^2} - \frac{x\delta}{(1+\delta)\rho_c} + \frac{V(x,y;\rho_c)}{(1+\delta)} \tag{14.5}$$

generates the correct equation of motion for the transverse phase space if (x, x') and (y, y') are viewed as canonical pairs. This Hamiltonian does not generate the longitudinal motion and thus should not be used in tracking codes. However, in design and perturbative calculations, it may be a very useful object. First one notices that formula (14.3) emerges naturally from this object, thus it emphasizes the "chromaticity-is-produced-by-quads" bias. Also, since one abandons the canonical framework in the longitudinal phase space, one is now free to redefine δ. Following Irwin, it is more convenient in analytical perturbation theory to consider the variable $\overline{\delta}$ defined by

$$\frac{1}{1+\delta} = 1 - \overline{\delta}. \tag{14.6}$$

In many calculations, first order in $\overline{\delta}$ will be equivalent to higher order in δ. Of course, this simplifies the look of an analytical (and certain "back-of-the-envelope") calculation. It does not in any way help a TPSA/AD based calculation. Indeed, the computer code can use any convenient internal variable and then change the final map into the $\overline{\delta}$ variable. The resulting map is the same.

The transverse part of the motion must of course be integrated. Needless to say, all the splits used on the exact Hamiltonian also work here. However, historically the so-called "kick codes" have taken advantage of another possible split. Let us consider a combined function bend (bend+quadrupole) which is the result of a design decision (instead of a construction error). Ideally, we would like the "exact" solution for this ideal element. In the case of the correct Hamiltonian it does appear possible to solve the equations of motion in terms of elementary trigonometric functions. However, it is possible for the small angle Hamiltonian. Thus the split used in standard tracking codes is given by

$$
\begin{aligned}
H &= H_1 + H_2 \\
H_1 &= \frac{p_x^2 + p_y^2}{2\,(1+\delta)} + \frac{x}{\rho_c} + \left\{ k_2 + \frac{1}{\rho_c^2} \right\} \frac{x^2}{2} - k_2 \frac{y^2}{2} \\
H_2 &= V_{error}(x, y; \rho_c).
\end{aligned} \tag{14.7}
$$

The function V_{error} will contain the construction errors and the design nonlinear multipolar content. The Hamiltonian H_1 for each magnet can be integrated exactly in terms of trigonometric functions. Actually, the reader can convince himself of this fact using normal form theory. Depending on the sign of the focusing parameter ($k_2 + \frac{1}{\rho_c^2}$ in the horizontal plane and $-k_2$ in the vertical plane), the map produces circles or hyperbolas. For the special case of vanishing focusing strength it becomes a drift. The solutions will involve sines, cosines and their hyperbolic equivalents.

In the longitudinal plane, it is easy to check from the equations of motion that the solution for the change in path length will be in terms of the integral of a quadratic polynomial involving the transverse solutions. Clearly, all these integrals can be carried out. Traditionally, the longitudinal solution was ignored in kick codes. Recently these integrals of the path length were computed and incorporated in a code called "SIXTRACK."[71, 72]

Although during most of my career I was able to use "large machine codes" I have not recommended to my colleagues the traditional split presented here. I would like to explain my reasons again. The advantage of the traditional split is germane to the ideal machine. Indeed it gives us (neglecting fringe fields) the exact value for the linear system: Twiss parameters, tunes and phase advances are unambiguously computed and every one with an equivalent "deck" gets the same linear lattice. In the light of Chapter 11, and in particular, the reinterpretation of the integrator as the model, this advantage of the traditional split is less obvious. Certainly the omnipresence of this split in accelerator codes makes any other split difficult to introduce. However the advantages of the drift-kick split are numerous. First of all, the derivation of the path length effects are trivial. This is important to the extent that correct simulations of the synchrotron oscillations are a necessity. In addition, because the introduction of errors are important in any simulation, one may be forced to split the magnet anyway. Finally, we have mentioned that modern tracking codes should be equipped with the ability to take derivatives (through automatic TPSA) with respect to any reasonable set of design parameters. The various cases (circle, drift and hyperbola) appearing in the solution of the transverse motion of H_1 imply the existence of branching within the tracking code. At the junctions, the tracking code is *not differentiable* in the parameter k_2 even though it should be.

For these reasons, I always thought that it was simpler to convince my immediate colleagues to learn or accept the Talman reinterpretation than to equip a "SIXTRACK"-like[72, 71] program with the full TPSA-DA-LIE apparatus in addition to the radiative effects essential in electron machines.

14.2 FAKING EUCLIDEAN TRANSFORMATIONS?

In traditional codes, including those used by the author, the presence of a curved layout always implies the presence of a bend which ideally curves the ideal trajectory by the layout angle. When first presented with the task of introducing an error tilt in the $x-y$ plane of a bend, I automatically concluded that the field of the layout bend (which I assumed must exist), was also tilted. Thus the effect of an $x-y$ tilt on a bend could be represented by the following combination of maps to leading order in the layout angle and the tilt angle:

$$\mathcal{Z}(\theta)\;[S] = \mathcal{Z}(\theta)\exp(:-\frac{\theta L y}{2\rho_c}:)S\exp(:-\frac{\theta L y}{2\rho_c}:)\mathcal{Z}(-\theta). \qquad (14.8)$$

The effect of the map $\exp(:-\frac{\theta L}{2\rho_c}y:)$ is to kick the beam in the vertical direction

$$\Delta p_y = -\frac{\theta L}{2\rho_c}. \qquad (14.9)$$

There are many ways to "derive" this effect using the assumption that the map S is mostly a layout bend. If that is the case, then a rotation of the magnet by an angle θ will induce an x-component in the negative x-direction in the fixed laboratory frame (layout). Thus the reduced field $b_y = \frac{B_y}{Brho} = \frac{1}{\rho_c}$ will generate a skew component or horizontal field of strength $\sin(\theta)\frac{1}{\rho_c} \approx \frac{\theta}{\rho_c}$. The integrated effect of this field is then put symmetrically into Equation (14.8).

The reader will have no problem believing that, in the limit of small angles, the approximation of Equation (14.8) must be correct. But like any other hand waving approximation based on pseudo-mathematics and intellectual laziness, it presents a potential danger: the code structure and worse, our thoughts, are no longer compatible with a better approximate treatment. Historically, when pressed by my colleague Johan Bengtsson to "clean up the theory" to avoid programming ill-designed codes, I was immediately confronted with this problem. Of course the solution was the development of the dynamical Euclidean group.

Let us derive Equation (14.8) from the results of the dynamical Euclidean group. In the case of small layout angles, we may use the thin representation of Equation (10.31) or the locally exact representation of Equation (10.54). Let us use the locally exact representation. First, examining Equation (10.55.d), we need the matrix $Y^{1/2}$, which is the spatial $x - z$ rotation of half the layout angle:

$$Y^{1/2} = \begin{pmatrix} \cos\left(\frac{\Phi_{12}}{2}\right) & 0 & \sin\left(\frac{\Phi_{12}}{2}\right) \\ 0 & 1 & 0 \\ -\sin\left(\frac{\Phi_{12}}{2}\right) & 0 & \cos\left(\frac{\Phi_{12}}{2}\right) \end{pmatrix} \approx \begin{pmatrix} 1 & 0 & \frac{\Phi_{12}}{2} \\ 0 & 1 & 0 \\ -\frac{\Phi_{12}}{2} & 0 & 1 \end{pmatrix}. \tag{14.10}$$

The vector $\vec{D} = (1 - R^{-1}_{\theta\vec{\beta}})(0, 0, O_{12}/2)$ in Equations (10.55.a) and (10.55.c) vanishes because the rotation of angle θ in our example is in the $x - y$ plane. Thus we are left with Equations (10.55.b) and (10.55.d) to worry about—the vectors $\vec{\beta}^{in}$ and $\vec{\beta}^{out}$ are given by

$$\vec{\beta}^{out} = Y^{1/2}\vec{\beta} \overset{\text{Small Layout Angle}}{\approx} \begin{pmatrix} 1 & 0 & \frac{\Phi_{12}}{2} \\ 0 & 1 & 0 \\ -\frac{\Phi_{12}}{2} & 0 & 1 \end{pmatrix} \begin{pmatrix} 0 \\ 0 \\ 1 \end{pmatrix} = \begin{pmatrix} \frac{\Phi_{12}}{2} \\ 0 \\ 1 \end{pmatrix}$$

$$\vec{\beta}^{in} = Y^{-1/2}\vec{\beta} \approx \begin{pmatrix} -\frac{\Phi_{12}}{2} \\ 0 \\ 1 \end{pmatrix}. \tag{14.11}$$

Therefore, in accordance with the local representation of the Euclidean group in Equation (10.54), the rotated map $\mathcal{Z}(\theta)\,[S]$ is given by

$$\mathcal{Z}(\theta)\,[S] = \mathcal{R}_{\theta\vec{\beta}^{in}} S \mathcal{R}_{\theta\vec{\beta}^{out}}$$

$$\approx \mathcal{Z}(\theta)\mathcal{X}(-\frac{\theta\Phi_{12}}{2})\mathcal{S}\mathcal{X}(-\frac{\theta\Phi_{12}}{2})\mathcal{Z}(-\theta). \qquad (14.12)$$

But for small angles angles the map $\mathcal{X}(-\frac{\theta\Phi_{12}}{2})$ can be approximated:

$$\mathcal{X}(-\frac{\theta\Phi_{12}}{2}) = \exp\left(-\frac{\theta\Phi_{12}}{2} : y\sqrt{(1+\delta)^2 - p_x^2 - p_y^2} :\right)$$
$$\approx \exp\left(-\frac{\theta\Phi_{12}}{2} : y :\right). \qquad (14.13)$$

The layout angle $\frac{\Phi_{12}}{2}$ is just $\frac{L}{2\rho_c}$ and thus Equations (14.12) and (14.8) are identical.

Philosophically the point of view taken in deriving these two equations is radically different. In one case, a lot of hand waving and some unwarranted assumptions about the map had to be introduced. This last point is quite interesting; the correct derivation using the Euclidean group representation has a very surprising corollary: the presence of the corrective term $\exp(-\frac{\theta\Phi_{12}}{2} : y :)$ did not originate from any magnetic field! It is there for all possible maps including drift spaces! So, here are two weird exercises:

Exercise 14-2
If a bend in the layout is turned off completely, then it is a drift which must be left invariant under any displacement or rotation. Show that it is indeed the case! Hint: you must write the drift in polar coordinates or equivalently match it to the layout.

Exercise 14-3
The thin map representation of the dynamical Euclidean group is certainly very good in the small layout angle approximation, so to sharpen your skill or maybe to prove Exercise 14-2 *more directly, use Equation* (10.31) *to derive the result of this section.*

In conclusion, we have seen that a systematic expansion of all the dynamical Euclidean transformations leads to the physically intuitive results and more, of course. One possible large machine (small angle) representation of the Euclidean group in a simulation code is obtained by the following replacements:

$$\mathcal{T}_z(d_z) = \exp\left(d_z : \delta - \frac{p_x^2 + p_y^2}{(1+\delta)} :\right) \qquad (14.14.a)$$
$$\mathcal{X}(\theta) = \exp(\theta : y :) \qquad (14.14.b)$$
$$\mathcal{Y}(\theta) = \exp(\theta : x :). \qquad (14.14.c)$$

The expressions for the transverse translations $\mathcal{T}_x$ and $\mathcal{T}_y$ are left as is because they are already simple. The $x-y$ rotation $Z(\theta)$ should also stay the same for two reasons: first and foremost, a small angle approximation is not symplectic, and second, the full map is needed whenever torsion is required. And finally, because $Z(\theta)$ is a linear map, little is gained by trying to simplify it.

The transformations of the $x-z$ and $y-z$ rotations into translations of the momentum have a clear "physical" interpretation. They simply indicate that for small layout angles and transverse divergences, a rotation of the coordinate frame is tantamount to a projection of the longitudinal momentum (namely $p_z \approx 1$) along the new x or y direction. All other effects, which give rise to the complex nonlinear nature of the maps, are ignored.

14.3 AVOIDING THAT FAKE!

In Sect. 13.1.4 we introduced the formula $\mathcal{L}_{12} = \mathcal{Q}_1 \mathcal{W}_L \mathcal{Q}_2$ for the linear part of an ideal bend of entrance and exit angles ε_1 and ε_2 respectively. $\mathcal{W}_L$ is linear part of the sector bend map and these thin quadrupoles have a focusing strength of $k_i = -b_0 \tan(\varepsilon_i)$.

This being a "trick" one should replace it by the linearization of the real maps entering in the theory. However, I would be lying to the reader if I pretended not to have used this trick in a simulation code. In the small angle limit, this representation of the general ideal bend, is as good numerically as the linearization of the correct expression given by Equation (12.46). The only disadvantage of the quadrupole trick is to make the inclusion of the fringe field effect in an approximate code more tedious since the code is not properly structured. Hence I would recommend to anyone who writes a large machine code to structure his code in a way consistent with the inclusion of large angle effects (small machine). The author is preaching something that he has not done himself, but has paid the price for this mistake. In the design of the SLAC B-factory, I was asked what happens if various small ring effects are included in our simulation codes, which unfortunately used the quadrupole trick.

Exercise 14-4

The map $\mathcal{Q}_1$ of the quadrupole trick is the product of three maps: $\mathcal{Y}(\varepsilon_1)$, $\mathcal{F}_1$, and $\mathcal{U}(-\varepsilon_1)$. (See Equations (10.26), (13.7), and (12.41) respectively. Notice that the linearization of Equation (13.7) is discussed in Sect. 13.1.2.) A particle traveling on the ideal closed orbit $(0,0,0,0,0,0)$ will have the following changes as it passes through *these maps:*

$$\vec{0} \overset{\mathcal{Y}(\varepsilon_1)}{\longrightarrow} \begin{pmatrix} 0 \\ \sin(\varepsilon_1) \\ 0 \\ 0 \\ 0 \\ 0 \end{pmatrix} \overset{\mathcal{F}_1}{\longrightarrow} \begin{pmatrix} 0 \\ \sin(\varepsilon_1) \\ 0 \\ 0 \\ 0 \\ 0 \end{pmatrix} \overset{\mathcal{U}(-\varepsilon_1)}{\longrightarrow} \vec{0}. \tag{14.15}$$

Expanding around this trajectory, derive the linear maps corresponding to $\mathcal{Y}(\varepsilon_1)$, $\mathcal{F}_1$, and $\mathcal{U}(-\varepsilon_1)$. In other words derive the quadrupole trick.

14.4 SOLENOIDS AND CAVITIES

14.4.1 The Small Angle Solenoid

The ideal solenoid is represented by the Hamiltonian of Equation (12.21) which is exactly solvable. The small angle approximation is given by the Hamiltonian

$$H = \frac{\left(p_x + \frac{b_z}{2}y\right)^2 + \left(p_y - \frac{b_z}{2}x\right)^2}{2(1+\delta)} - \underbrace{\delta}_{\text{If you want the total path length}}. \tag{14.16}$$

This Hamiltonian can also be integrated exactly (no big surprise!) and the solution is gotten by replacing H in Equation (12.54) by $1 + \delta$, while the path length formula is just

$$\ell(z) = \ell + z\frac{\left(p_x + \frac{b_z}{2}y\right)^2 + \left(p_y - \frac{b_z}{2}x\right)^2}{2(1+\delta)^2}. \tag{14.17}$$

The reader will also notice that several splits of Equation (14.16) are exactly solvable. It might be more convenient in a code to isolate the drift part and integrate this solenoid using a multi-map integrator. It should be noticed that even if one uses the exact solution, we are still condemned to the two-map integrator due to the occasional presence of multipoles within a solenoid detector.

14.4.2 The Cavity

We discuss miscellaneous issues related to electric elements and, in particular, to the cavity.

14.4.2.1 From path length to time

In this section, time (or path length) is not measured relative to a given design clock, but is the real absolute quantity. For example, the absolute path length increments approximately by the circumference C at each turn. Discussion of the connection between relative and absolute longitudinal variables is found in Sect. 14.4.2.3.

The simplest model for a cavity and generally the only one used in rings dynamics, consists of a kick of the total energy as a function of time. The form of kick is given by the equation:

$$E^{fin} = E + Cav(T). \tag{14.18}$$

The function $Cav(T)$ must be a periodic function of time with smallest period $\frac{T_0}{h}$:

$$Cav(T + \frac{T_0}{h}) = Cav(T), \tag{14.19}$$

where grossly speaking the time T_0 is the revolution time of the design orbit. The integer h is called the "harmonic number" and represents the maximum number of "buckets" which could be present in the ring at once. In other words, in a ring with harmonic number h, the beam consists, at most, of a train of h beamlets approximately equally separated in longitudinal distance.

Even though energy and time form a natural canonical pair $[E, T] = 1$, we have not used it in this book because it is more convenient in ultrarelativistic machines to confuse path length with time. Now, we will remedy this situation. First, we need to switch from (E, T) to a set of scaled variables (e_τ, τ):

$$\tau = cT \tag{14.20.a}$$

$$e_\tau = \frac{E - E_0}{p_0 c}. \tag{14.20.b}$$

Using these new scaled variables, the sector bend Hamiltonian is

$$H = -(1 + \frac{x}{\rho_c})\sqrt{1 + \frac{2}{\beta_0}e_\tau + e_\tau^2 - p_x^2 - p_y^2}$$
$$+b_1(x + \frac{x^2}{2\rho_c}) + V(x, y; \rho_c), \tag{14.21}$$

where $\beta_0 = \frac{E_0}{p_0 c}$.

The relationship between the relative momentum deviation δ and the relative energy deviation is

$$(1+\delta)^2 = 1 + \frac{2}{\beta_0} e_\tau + e_\tau^2. \tag{14.22}$$

And, as expected, for an ultrarelativistic ring, all stable particles will have energies in the neighborhood of the design energy E_0. Such particles satisfy the relation $\beta_0 \approx 1$, and thus we conclude that $\delta \approx e_\tau$.

In contrast, for non-relativistic particles (proton machines), it is imperative to use time. While it is true that path length and momentum form a canonical pair in magnetic simulation of nonrelativistic particles (i.e., their changes are derivable from a Hamiltonian), this ceases to be true in electric elements. Even if only one thin lens cavity is present in the ring, the energy kick depends on the time of arrival, not on the path length. In Sect. 14.4.2.3, we address this issue from the point of view of an ultrarelativistic simulation code computing exclusively the path length: can it be patched up correctly at the electric elements?

14.4.2.2 Absolute or relative measures of time?

Consider a ring with a single cavity of design frequency ω_0. This design frequency must be related to the design revolution time T_0 through the relation

$$\omega_0 T_0 = 2\pi h. \tag{14.23}$$

To the extent that all the particles in the bucket will execute stable oscillations keeping their transit time in the neighborhood of T_0, we can introduce a relative time defined as follows:

$$\Delta\tau = c(T - T_0). \tag{14.24}$$

In evaluating the energy kick at the cavity, one can alternatively use the real time or the relative time. For example, let us assume that the function $Cav(T)$ is a simple sine function:

$$E^{fin} = E + \Delta E \ \sin(\omega_0 T) \tag{14.25.a}$$

$$\Leftrightarrow$$

$$e_\tau^{fin} = e_\tau + \frac{\Delta E \ \sin(\frac{\omega_0}{c}\Delta\tau)}{p_0 c}. \tag{14.25.b}$$

This exercise shows that one can use a relative time instead of true absolute time. Moreover, since the curved layout blocks usually correspond to places where the design orbit bends, it is usually correct to subtract locally the time of flight by assuming that the arc of the circle joining the origin of the entrance and exit plane represents the ideal trajectory of a particle. In other words, the ideal time of flight through a layout block of design curvature ρ_c^{-1} and angle Φ_{12} is just $L = \frac{\Phi_{12}\rho_c}{v_0}$.

For the sector bend of Equation (14.21), one can write a Hamiltonian which automatically subtracts this ideal time of flight or path length. It suffices to add $\frac{e_\tau}{\beta_0}$ to the Hamiltonian. This formalism has one virtue: round-off errors due to the manipulation of large numbers do not come into play. Thus many codes compute only the relative time and nothing else.

Now comes a crucial issue: should we have access to the full time or is the relative time alone sufficient? To answer this question, we must go back to some of the code design principles used in this book. The most important is the ability of a code to mimic reality—to respond to changes of the system in a way compatible with the real system. Can a code, that uses only the relative time, simulate all physical changes of parameters? The answer to this is no and here is the proof!

Consider again an ideal ring with a single sine cavity of frequency ω_0. Let us assume that the origin in six-dimensional phase space is sent into itself. Let us change the frequency of the cavity by $\Delta\omega_0$. It follows that the origin of phase space is still a fixed point in Equation (14.25.b). This is clearly nonsensical. Going back to the real time (Equation (14.25.a)), we see that if the frequency is suddenly changed, the "synchronous" particle will receive a kick of strength $\Delta e_\tau = \frac{\Delta E \ \sin(\Delta\omega_0 T/c)}{p_0 c}$ and will thus cease, in general, to be the synchronous particle.

So, simulation codes, in the purest mode, should subtract the total time of flight of the design orbit at one location around the ring. In doing so, the cavities become "realistic" elements whose parameter dependence can be studied by simulation or by producing a parameter dependent map with TPSA.

14.4.2.3 Restoring time out of path length

There are two types of effects found commonly in rings which modify the energy of a particle: electric elements and radiation. In the case of electric elements such as cavities and electric septa, which are localized, the time flight between them can be reconstructed exactly from a knowledge of the total path length and the velocity (energy) of the particle traversing the magnetic elements. This reconstruction exploits energy conservation in magnetic elements. The presence of radiation will mess that up. However, it is true that radiation is largely an electron effect which shows its ugly tail only at ultrarelativistic energies at which the substitution of path length in lieu of time is very adequate. Nevertheless it is conceivable to transform path length into time at every integration step and thus correctly modify a nonrelativistic electron code without rewriting the entire code.

Given the above caveats, it seems worthwhile to equip a path length code with a canonical transformation that restores time. This transformation is given by the map

$$\begin{aligned} \ell &= \beta\tau \\ \delta &= \sqrt{1+\frac{2}{\beta_0}e_\tau + e_\tau^2} - 1. \end{aligned} \tag{14.26}$$

The time τ and path length ℓ are the total values. The reader should remember that time is the physical variable, and therefore, in applying Equation (14.26) one tries to keep track of time and thus loses information about the total path length as the energy changes.[14.2] Hence, if we denote the transformation of Equation (14.26) by $\vec{\theta}$, then the map $\vec{\zeta}_{i\to i+1}$ through the ith element becomes

$$\vec{\zeta}^{new}_{i\to i+1} = \vec{\theta} \circ \vec{\zeta}_{i\to i+1} \circ \vec{\theta}^{-1}. \tag{14.27}$$

Exercise 14-5

Remembering that the relativistic β in Equation (14.26) is a function of the relative energy e_τ, show that $\vec{\theta}$ is a canonical transformation. In addition, how would you modify Equations (14.26) and (14.27) if your tracking code only computes the relative path length and uses only relative time at the cavities. You must assume that the design path length between location i and $i+1$, denoted L_i, is available to your code. Answer:

$$\begin{aligned} \Delta\tau = \vec{\theta}_L^{-1}(\delta,\ell) &= \frac{\ell+L}{\beta} - \frac{L}{\beta_0} \\ \vec{\zeta}^{new}_{i\to i+1} &= \vec{\theta}_{L=L_i} \circ \vec{\zeta}_{i\to i+1} \circ \vec{\theta}^{-1}_{L=0}. \end{aligned} \tag{14.28}$$

[14.2]The total path length can be computed by the code just like any function of the phase space variables $(x, p_x, y, p_y, e_\tau, \tau)$. It is no longer a canonical variable as it was in the ultrarelativistic régime.

CHAPTER 15
INCLUSION OF RADIATION

With the advent of TPSA, there is no excuse whatsoever to justify the avoidance of classical radiation in a tracking code. And, since the beam sizes depend in lowest approximation on stochastic fluctuations balancing the damping decrement due to classical radiation, I do not see any reason why the "stochastic " map that propagates moments should not be included. The resulting theory, the so-called beam envelope theory, is more accurate than Sands integral formalism and even more accurate than Chao's method. However it can be argued that this improved accuracy, especially over the Chao formalism, is academic. Our main goal is not to improve the physical results of these theories, but to show a method consistent with the modern tools presented in this book, and thus implementable in the same spirit. With TPSA/AD methods, the beam envelope approach is reliable and general. New complex elements are included and the programmer can be certain that all the "synchrotron integral" results (and more) will be delivered by the code correctly and self-consistently.

The Sands integrals are formulas whose usefulness in design is certainly unquestionable, but it is not the proper way to compute the equilibrium beam sizes in a tracking code for a given design.

In this chapter, we will continue a discussion which started in two Interludes: VII and IX. In Interlude VII we discussed the possibility of tracking quadratic moments beyond the case of a deterministic situation. The resulting beam envelope map was a new local concept defined completely at the local level. We will come back to this map in more detail in this chapter.

In Interlude IX we exploited the special nature of the quadratic moment map in order to use existing tools of normal forms to find the equilibrium beam sizes. We also showed the connection between the exact beam envelope treatment and Chao's treatment. The necessity of small damping and absence of linear resonance emerged clearly as a necessary condition for the existence of three nearly conserved equilibrium emittances. In this chapter we will use the formulas for the "equilibrium emittances" obtained from the beam envelope in Interlude IX to make the leap from Chao to Sands. Equilibrium emittances and damping decrements will be derived from the general formalism.

Our discussion will be based on new "tensorial" invariants of the linear normal form which will turn out to be related to the adiabatic dispersions defined in Sect. 7.3.3.1.

15.1 ULTRARELATIVISTIC CLASSICAL RADIATION

The addition of classical radiation to a local integrator is simple. According to Sands[37], the change in the variable δ is obtainable from the formula:

$$\frac{d\delta}{d\tau} = K_c(1+\delta)^2 \frac{\vec{B}_\perp \cdot \vec{B}_\perp}{Brho^2} \; ; \; K_c = 1.40789357\ 10^{-5}\ E_0^3. \quad (15.1)$$

The reference energy of the electron E_0 is measured in Gev. The field $\vec{B}_\perp$ is the component of the local magnetic field perpendicular to the direction of propagation. Needless to say that all these quantities are available to the tracking code.

Formula (15.1) is a derivative with respect to time which is a dependent variable in our layout description. Hence we rewrite it in terms of the integration variable s:

$$\begin{aligned} \frac{d\delta}{ds} &= \frac{d\delta}{d\tau}\frac{d\tau}{ds} \\ &= \frac{d\delta}{d\tau}\left[\tau, H\right] = -\frac{d\delta}{d\tau}\frac{\partial H}{\partial \delta}. \end{aligned} \quad (15.2)$$

During the radiation process,[15.1] an ultrarelativistic particle will emit a photon in the forward direction only. Thus, the usual directions $\frac{dx}{ds}$ and $\frac{dy}{ds}$ are left unchanged while the transverse momenta actually change. Therefore, to the symplectic integrator of step size Δs, we must add the radiative terms:

$$\delta^f = \delta + K_c(1+\delta)^2 \frac{\vec{B}_\perp \cdot \vec{B}_\perp}{Brho^2}\frac{\partial H}{\partial \delta}\Delta s \quad (15.3.a)$$

$$p_x^f = (p_x - a_x)\frac{1+\delta^f}{1+\delta} + a_x \quad (15.3.b)$$

$$p_y^f = (p_y - a_y)\frac{1+\delta^f}{1+\delta} + a_y. \quad (15.3.c)$$

[15.1] Path length and time are the same in the ultrarelativistic limit. One can substitute ℓ for τ in the entire chapter.

Equation (15.3) is a first order solution of Equation (15.2). Although it can be part of a multi-step Yoshida integrator, terms proportional to it will not converge with the rate predicted by the theory.

It goes without saying that the power series map can be computed with TPSA/AD and the damping rates extracted through a nonsymplectic normal form.

Exercise 15-1 (Radiation: Doing a bit better!)
Can one really add a correct radiation map and use it in a multi-map integrator? Let us see.

1. *Although Equation* (15.3) *applies to any bizarre orbit, argue on the basis of the ideal machine, that* (15.3) *should be merged with the "kick" part of the integrator for better convergence (Assume ideal constant field bends).*

2. *Write the Lie operator in non-canonical coordinates. Answer:*

$$\vec{F}^{rad} \cdot \vec{\nabla} = K_c(1+\delta)^2 \frac{\vec{B}_\perp \cdot \vec{B}_\perp}{Brho^2} \frac{\partial H}{\partial \delta} \partial_5. \qquad (15.4)$$

3. *Actually the above answer is slightly incomplete; the author should have read Sands after all! At the cavities, the variables (x', y') must be modified. This is a purely symplectic effect when using noncanonical variables. Answer:*

$$\vec{F}^{rad} \cdot \vec{\nabla} = K_c(1+\delta)^2 \frac{\vec{B}_\perp \cdot \vec{B}_\perp}{Brho^2} \frac{\partial H}{\partial \delta} \partial_5 + \frac{\partial^2 H}{\partial p_x \partial \delta} \frac{\partial H}{\partial \tau} \frac{\partial}{\partial x'} + \frac{\partial^2 H}{\partial p_y \partial \delta} \frac{\partial H}{\partial \tau} \frac{\partial}{\partial y'}. \qquad (15.5)$$

   The term $\frac{\partial H}{\partial \tau}$ is the cavity kick. Of course, when using thin cavities, it is much better to make sure that canonical variables are used. However for analytical purposes we will find this formula useful.

4. *Using the curvilinear coordinates of a bend, integrate the transfer map associated to the Lie operator of Equation* (15.4) *exactly. Express everything in the noncanonical variables. You may use the fact that $-\frac{\partial H}{\partial \delta}$ is the change in path length or use Hamiltonian theory. This map can be used in a Yoshida integrator without spoiling the rate of convergence of the integrator.*

15.2 INCLUSION OF STOCHASTIC EFFECTS

In this section we introduce the local stochastic theory of beam envelopes.

15.2.1 The Frobenius-Perron Operator: the Adjoint of our Compositional Map

In Interlude VII we introduced for the first time a map acting on the moments. Let us consider a monomial $P_{\vec{m}}$ as defined in Equation (4.10). The moment $\langle P_{\vec{m}} \rangle_{s=0}$ is a functional defined as

$$\langle P_{\vec{m}} \rangle_{s=0} = \int_{\mathbb{R}^{2N}} \rho(\vec{z}; s=0) P_{\vec{m}}(\vec{z}) d\vec{z}. \tag{15.6}$$

The function $\rho(\vec{z}; s=0)$ is a distribution function at $s = 0$. The moments at an arbitrary position s are defined in terms of the initial distribution at $s = 0$. Thus we can write

$$\langle P_{\vec{m}} \rangle_s = \int_{\mathbb{R}^{2N}} \rho(\vec{z}; s=0) P_{\vec{m}} \left(\vec{\zeta}_{0\to s}(\vec{z}) \right) d\vec{z}. \tag{15.7}$$

Equation (15.7) is a mathematical rendition of a simple statement: the average value of a function at s is given by evaluating this function in terms of the variables at $s = 0$ and averaging it with the known distribution at $s = 0$. It follows from Equation (15.7) that the distribution at s is given by

$$\langle P_{\vec{m}} \rangle_s = \int_{\mathbb{R}^{2N}} \frac{\rho \left(\vec{\zeta}^{-1}_{0\to s}(\vec{z}); s=0 \right)}{det \left| d\vec{\zeta}_{0\to s} \right|_{\vec{z}}} P_{\vec{m}} (\vec{z})\, d\vec{z}$$

$$\Downarrow$$

$$\forall \vec{z} \;\; \rho(\vec{z}; s) = \frac{\rho \left(\vec{\zeta}^{-1}_{0\to s}(\vec{z}); s=0 \right)}{det \left| d\vec{\zeta}_{0\to s} \right|_{\vec{z}}}$$

or

$$\rho_s = \frac{\mathcal{M}^{-1}_{0\to s} \rho_0}{det \left| d\vec{\zeta}_{0\to s} \right|_{\vec{z}}}. \tag{15.8}$$

In fact we have defined a new operator out of a physical process. In the literature on non-equilibrium statistical mechanics[73] it is sometimes called the Frobenius-Perron operator. This operator acts on a function by the rule:

$$U_{\vec{\zeta}} f = \frac{f \circ \vec{\zeta}^{-1}}{det \left| d\vec{\zeta}_{0\to s} \right|_{\vec{z}}}. \tag{15.9}$$

The Frobenius-Perron operator enters statistical mechanics in a natural way; one asks a physical question about averages over distributions. Its adjoint operator, sometimes called the Koopman operator in nonlinear dynamics and statistical mechanics, is none other than our compositional map $\mathcal{M}$:

$$U^{\dagger}_{\vec{\zeta}} f = \mathcal{M}_{\vec{\zeta}} f = f \circ \vec{\zeta}. \tag{15.10}$$

Going back to moments, we get the relation:

$$\begin{aligned} \langle P_{\vec{m}} \rangle_s &= \int_{\mathbb{R}^{2N}} \rho_0(\vec{z}) \, (\mathcal{M} P_{\vec{m}}) \, (\vec{z}) d\vec{z} \\ &= \int_{\mathbb{R}^{2N}} \rho_0(\vec{z}) \sum_{\vec{n}} \mathbf{M}_{\vec{m}\vec{n}} P_{\vec{n}} d\vec{z} \\ &= \sum_{\vec{n}} \mathbf{M}_{\vec{m}\vec{n}} \langle P_{\vec{n}} \rangle_0 . \end{aligned} \tag{15.11}$$

The moments are propagated forward in time with the matrix $\mathbf{M}$ which is precisely the transpose of the matrix $\mathbf{M}^T$ which represents the compositional map $\mathcal{M}$ in the monomial basis.

INTERLUDE XV
A CONTROVERSY BEYOND OUR CONTROL!

The reader must know by now that I have my bizarre views on several topics. But, as I said in the preface, the global/local approach is a mathematical exploitation of a physical fact which is not controversial. Magnets are usually dynamically independent so the theory should exploit this as much as possible. Of course, if they are not independent, such as a channel of quadrupoles with overlapping fields, then one has to exercise care in defining the magnet/object of the preface. In parallel to this, our introduction of the compositional map was completely natural. It is simply a way to rewrite a functional equation for invariants. The differential equation for the compositional map, first derived in Interlude I, is valid for a map whose domain is restricted to differentiable functions.

Dirac delta functions and other bizarre distributions are excluded. This means that the map equation

$$\frac{d}{ds}\mathcal{M} = \mathcal{M} : -H : \tag{15.12}$$

cannot apply to Dirac delta functions and thus cannot describe the propagation of a single particle distribution. This seems paradoxical because we use this map in "single particle dynamics." However, the reader will remember that we always act on the identity function with the differential representation; we never act on a delta function distribution with the Lie operators but on its argument $\vec{z} - \vec{z}_0$ which we view as a function of $\vec{z}_0$. In other words, we let the Lie map act only on the argument of the delta function distribution, not on the distribution itself.

In conclusion, Equation (15.12) alone does not permit the existence of point particles and, of course, neither can its inverse[15.2]

$$\frac{d}{ds}\mathcal{M}^{-1} = : H : \mathcal{M}^{-1}. \tag{15.13}$$

The reader may wonder about what kind of controversy looms under these trivialities. Well here it is: suppose we decide that Equation (15.13), which looks very much like a quantum mechanical equation, is not derived from the "plug in" map, but is far more fundamental. We have said that Equation (15.13) cannot be applied to delta function distributions, this means that the distribution $\delta\left(\vec{z} - \vec{\zeta}_{0\to s}(\vec{z}_0)\right)$, i.e., the trajectory of a single particle, is not allowed. Such trajectories do not exist: they are not permitted distributions of the operator $\mathcal{M}^{-1}$ as defined by Equation (15.13).

Going a bit further one may ask if it is possible to find eigenfunctions for the operator $\mathcal{M}^{-1}$ and study the evolution of a distribution by projecting it onto the eigenfunctions. It turns out that it is not possible: our compositional operators are not diagonizable in a regular Hilbert space, but possess a generalized spectral decomposition in terms of bizarre non-normalizable bras and kets!

Finally here is the punch line: the generalized eigenvectors can have eigenvalues smaller than unity, thus introducing irreversible effects in the propagations of distributions (time flows in one direction). What I have

[15.2] As pointed out by Dragt, the similarity between the equation for the inverse compositional map and the quantum operator equation is almost total.

just summarized is the "Brussels-Austin" view of non-equilibrium statistical mechanics of Prigogine and Petrosky. Needless to say that their views are not universally accepted.

There is one way in which this controversy leaks into our field—it is in the so-called emittance growth. According to Petrosky, emittance growth would happen naturally in any complex dynamical system. They claim that in complex Poincaré systems the wave function continuously collapses as if the system makes a measurement on itself. Irreversibility for the Brussels school is produced internally.

The standard explanation of irreversibility is based on a lack of precise measurements. The same can be said about emittance growth. If one looks at a volume in phase space, then it will stay the same forever as the system evolves under the map $\mathcal{M}$. However, it is well known that a distribution can filament as time goes on so that it seems to occupy a larger part of phase space. In such a case, the human eye, with its finite resolution, loses any knowledge of filamentation. The distribution has grown for all practical (i.e., experimental) purposes. There is another way to look at this issue. If one were to look at a Fourier mode decomposition of this distribution expressed around its center, it is probable that each Fourier mode would tend to a limit, but the convergence would not be absolute. In other words, given a time T less than infinity, there exists a mode m such that the modes smaller than m have settled down while those above m wildly oscillate. Thus, the lack of convergence is swept under the "rug of higher modes." The measuring apparatus cannot pick those high frequency oscillations and thus we say that the emittance has grown.[15.3]

END OF INTERLUDE

15.2.2 A Stochastic Map for the Moments

Instead of following the distribution, we may want to follow the moments themselves. Of course, as long as the system is deterministic everything is given by the Frobenius-Perron map $\mathbf{M}$, thus we will now add an extraneous stochastic effect which, in our case, will turn out to represent the quantum nature of the photon.

Let us assume that between positions s and $s + ds$, a particle changes its coordinates by an amount $\Delta\vec{z}$ governed by a probability[15.4]

[15.3] I am grateful to Alex Dragt and Kwangje Kim[74] for bringing these issues to my attention.

[15.4] It is far from obvious that this probability is local. In fact it is not local

$\Phi_s(\Delta\vec{z})$. We now substitute this in Equation (15.11):

$$
\begin{aligned}
\vec{\Sigma}_{s+ds} &\equiv \left\langle \vec{P} \right\rangle_{s+ds} \\
&= \int_{\mathbb{R}^{4N}} \Phi_{\vec{z};s}(\Delta\vec{z})\rho_s(\vec{z}) \left(\mathbf{M}_{s\to s+ds}\vec{P}\right)(\vec{z}+\Delta\vec{z})\, d\Delta\vec{z} d\vec{z} \\
&= \mathbf{M}_{s\to s+ds} \int_{\mathbb{R}^{4N}} \Phi_{\vec{z};s}(\Delta\vec{z})\rho_s(\vec{z}) \left(\vec{P}\right)(\vec{z}+\Delta\vec{z})\, d\Delta\vec{z} d\vec{z}.
\end{aligned} \tag{15.14}
$$

For ease of notation, we introduce the vector function of monomials $\vec{P}$ and the envelope vector $\vec{\Sigma} = \left\langle \vec{P} \right\rangle$ which is the vector of moments. In theory, Equation (15.14) can be integrated around the ring provided we know all the moments of the distribution. In practice, we will make some drastic approximations. I will list them:

$$
\int_{\mathbb{R}^{2N}} d\Delta\vec{z}\; P_{\vec{m}}(\Delta\vec{z})\; \Phi_{\vec{z}_c;s}(\Delta\vec{z}) = 0 \qquad \text{if } |\vec{m}| \text{ is odd} \tag{15.15.a}
$$

$$
\mathbf{M}_{s\to s+ds} \approx \mathbf{L}_{s\to s+ds} \qquad \text{if the map is linear.} \tag{15.15.b}
$$

Condition (15.15.a) states that the stochastic distribution has no odd moments around the deterministic orbit $\vec{z}_c$. Condition (15.15.b) states that we will only consider the action of the linearized map acting on quadratic moments around this orbit.

It is easy to check that this reduced Frobenius-Perron equation on the moment is self-consistent. The orbit will not change because $\langle \Delta\vec{z} \rangle = 0$ and the quadratic moments will be changed into themselves by the map $\mathbf{L}$.

Unlike the compositional map $\mathbf{M}^T$ which is "block-wise lower triangular," the adjoint map $\mathbf{M}$ is upper "triangular." Thus the higher moments and nonlinearities, if kept, would influence the lower moments. This situation is reminiscent of the quantum anharmonic oscillator where a quartic perturbation affects the ground state (see Equation (9.35)). Thus the truncation scheme is valid only if the entire distribution occupies a linear region around the orbit $\vec{z}_c$. If it does not we can expect the low order moments to be incorrect.

Accepting these important restrictions, we can proceed further with the moment map as follows:

$$
\vec{\Sigma}_{s+ds} = \mathbf{L}_{s\to s+ds} \int_{\mathbb{R}^{4N}} \Phi_{\vec{z}_c}(\Delta\vec{z})\rho_s(\vec{z})\vec{P}(\vec{z}+\Delta\vec{z}) d\Delta\vec{z} d\vec{z}
$$

---

and we will see spurious divergences in highly varying fields. The actual probability is a convolution in s. We will come back to this later.

$$= \mathbf{L}_{s\to s+ds}\left\{\vec{\Sigma}_s + \underbrace{\left\langle \Delta z_i \partial_i \vec{P} \right\rangle}_{=0} + \left\langle \vec{P}\,(\Delta\vec{z}\,) \right\rangle_s \right\}$$

$$\Downarrow$$

$$\frac{d\vec{\Sigma}_s}{ds} = \mathbf{F}_s \vec{\Sigma}_s + \left\langle \vec{P}\,(\Delta\vec{z}\,) \right\rangle_s . \tag{15.16}$$

In Equation (15.16) the middle term vanishes because of the assumption of vanishing odd moments. The first term belongs to deterministic dynamics and the second is a translation of the envelope $\vec{\Sigma}_s$ that is due to the stochastic nature of the (radiative) process represented by $\Phi_{\vec{z}_c}$. Thus, by assumption, Equation (15.16) applies self-consistently to quadratic moments only. The operator $\mathbf{F}_s$ is the adjoint of our usual compositional operator $\vec{F}_s \cdot \vec{\nabla}$, and by assumption generates the deterministic map $\mathbf{L}_{s_0\to s}$.

The construction of the one-turn map is now trivially obtained by integrating (15.16) around the ring:

$$\vec{\Sigma}_{s+C} = \mathbf{L}_{s\to s+C}\vec{\Sigma}_s + \underbrace{\oint_s^{s+C} \mathbf{L}_{\sigma\to s+C} \left\langle \vec{P}\,(\Delta\vec{z}\,) \right\rangle_\sigma d\sigma}_{D\Phi \text{ of Interlude VII}} . \tag{15.17}$$

This solution was first introduced in Interlude VII, Exercise 4-7. One notices that the parameter C is actually arbitrary and thus this map can represent a fraction of the ring.

15.2.3 Computation of the Stochastic Term

Now we are back in the world of local dynamics. We must compute the stochastic term $\left\langle \vec{P}\,(\Delta\vec{z}\,) \right\rangle_\sigma$ associated with photon emission. As before, let us assume that the map is expressed in noncanonical variables, something very easy to do with TPSA. We know that the positions and the directions stay constant during the emission and thus the only fluctuation possible happens in the energy direction. Thus if we revert to the matrix notation for an instant (instead of the column vector used thus far), we get

$$\left\langle \vec{P}\,(\Delta\vec{z}\,) \right\rangle_s = d\Phi_{ij} = 0 \ \text{ unless } i = j = 5, \tag{15.18}$$

in which case it is given by

$$
\begin{aligned}
d\Phi_{55} &= \int_{\text{spectrum}} \Phi^{\text{rad}}_{\bar{E}_c}(\delta)\ \delta^2\ d\delta \\
&= -K_c'(1+\delta)^4 \left\{ \frac{\vec{B}_\perp \cdot \vec{B}_\perp}{Brho^2} \right\}^{3/2} \frac{\partial H}{\partial \delta} ds \\
\text{where } K_c' &= 4.13209717\ 10^{-11}\ E_0^5,
\end{aligned} \tag{15.19}
$$

and as before the quantity E_0 is in GeV.

In matrix notation, Equation (15.16) takes the simplified form

$$
\frac{d\Sigma_{ij}}{ds} = F_{ia}\Sigma_{ab}F_{jb} + d\Phi_{ij}
$$

$$
\text{or} \tag{15.20}
$$

$$
\Sigma_{s+C} = L_{s\to s+C}\Sigma_s L^{\mathrm{T}}_{s\to s+C} + \oint_s^{s+C} L_{\sigma\to s+C}\, d\Phi_\sigma L^{\mathrm{T}}_{\sigma\to s+C}\, d\sigma.
$$

The matrix L is just the regular linear transfer map around the closed orbit (a six by six matrix) while F is its local derivative (the linear vector field).

Exercise 15-2 (Canonical Variables or Not)
Explain in detail how one would proceed if the underlying code is canonical. Your discussion should not involve complex algebra but should be an algorithm based on TPSA explaining what operations are necessary to connect the simple diffusion matrix $d\Phi_\sigma$ in noncanonical variables to the full one-turn map in canonical variables (*see [75] section* 8.2).

We must express some caveats concerning formula (15.19). The first nasty thing about this formula concerns non-differentiability in elements where the fields are near zero. This is a technical difficulty (due to the 3/2 power) which prevents us from studying the parameter dependence of $d\Phi$ in the usual TPSA/AD manner. One could restrict the parameter dependence to design bends only. This brings us to the next point, why is there a divergence? Clearly one must go back to the derivation[15.5] of formula (15.19): the probability of photon emission depends on the convolution of an expression containing

[15.5] I thank Kwangje Kim, Pisin Chen and Kaoru Yokoya for answering my dumb questions on this topic.

$B(s)B(s+\Delta s)$. It is assumed in the derivation of (15.19) that the field is nearly a constant throughout space and thus $B(s)B(s+\Delta s)$ is replaced by $B(s)B(s)$. This assumption is only strictly valid in the body of ideal bends which are larger than the so-called "radiation coherence length." It is not too discouraging to find out that formula (15.19) is differentiable where valid. One should go back to the exact theory in order to handle $B(s)B(s+\Delta s)$ correctly. This is beyond the scope of this book, but it is nevertheless a useful problem to untangle.

Finally the reader should be aware that the classical orbit does not exist strictly speaking. It is a mathematical construction which allows us to solve the problem in two parts. In fact, it is often the case that few photons are emitted in a single turn!

15.3 REVISITING THE LATTICE FUNCTIONS

The techniques of the beam envelope provide a powerful computational tool for the computation of equilibrium beam sizes: one can use the tools of automatic differentiation to extract the linear map and propagate the envelope Σ.

However in designing machines one needs approximate analytical formulas as a kind of rough guide. The so-called synchrotron integrals of Sands provide analytical approximations of the equilibrium beam sizes under the assumption of small damping and vanishing synchrobetatron coupling.

The Chao formalism is closest to the beam envelope formalism to the extent that it only assumes small damping. However analytical formulas based on this formalism are generally not used. Historically Chao wrote a computer code which multiplied matrices that had been manually computed for each type of element (pre-TPSA calculation). Sands and Chao formalisms are characterized by the "existence" of adiabatic emittances. The purpose of these methods is to compute the equilibrium emittances from the terms in $d\Phi_{55}(s)$ which stay invariant under the unperturbed symplectic map. The abstract connection between the beam envelope and the three emittances was spelled out in Interlude IX.

15.3.1 Derivation of the Damping Formulas

The problem is defined as follows: the unperturbed system is symplectic; to this symplectic system a tiny radiative term is added. The goal of the calculation is to compute the damping rates due to this term in each plane. We can apply our simple rules of analytical computation.

As we said before it is much simpler to use non-canonical variables. So I will assume that the unperturbed propagators are known in terms of the usual (x', y') or an equivalent set.

Outside the cavities, the Lie operator for the perturbation is given by the results of Exercise 15-1

$$\vec{F}^{rad} \cdot \vec{\nabla} \overset{\text{Non-Canonical}}{=} F_5^{rad} \partial_5. \tag{15.21}$$

The next step consists in normalizing the transfer matrix L'. I purposely distinguish this matrix from the canonical matrix L. Actually suppose we have diagonalized the symplectic matrix L

$$R = A^{-1} L A, \tag{15.22}$$

then the matrix L' is normalized using

$$R = A^{-1} B^{-1} L' B A. \tag{15.23}$$

Here B can be obtained with TPSA by expressing the noncanonical coordinates in terms of the canonical ones. This is a standard TPSA operation. Thus let us introduce the normalization matrix

$$A' = BA. \tag{15.24}$$

Following the rules of perturbation theory, we first expand the function F_5^{rad} around the closed orbit $\vec{z}_c$:

$$F_5^{rad} = F_5^{rad}(\vec{z}_c) + dk_i(\vec{z}_c) z_i + O(|\vec{z}|^2). \tag{15.25}$$

We then transform the linear Lie operator by A' using formula (4.29):

$$dk_i z_i \rightarrow dF_i^{\dagger} = dk_a A'_{ab} z_b A'^{-1}_{i5}. \tag{15.26}$$

We recall that in Floquet variables we are analyzing the following ring:

$$\left\{ \mathcal{R}_{0 \rightarrow s}, \exp\left(d\vec{F}_s^{\dagger} \cdot \vec{\nabla} \right), \mathcal{R}_{s \rightarrow C} \right\}. \tag{15.27}$$

Our analytical rules require us to extract the contribution of $d\vec{F}_s^{\dagger} \cdot \vec{\nabla}$ which produces tune shifts and damping. As usual we go into the complex resonance basis using the technique of Sect. 6.4.1, the result is

$$dF^{\dagger}_{2i-1} = \frac{dk_a}{2} A'^{-1}_{2i-1\ 5} \left(A'_{a\ 2b-1}(z_{2b-1}+z_{2b}) - iA'_{a\ 2b}(z_{2b-1}-z_{2b})\right) \\ +\frac{dk_a}{2} iA'^{-1}_{2i\ 5} \left(A'_{a\ 2b-1}(z_{2b-1}+z_{2b}) - iA'_{a\ 2b}(z_{2b-1}-z_{2b})\right).$$

The indices i and b denote the eigenplanes and thus take the values 1,2, and 3. Recalling the form of a diagonalized radiative map (Equation (4.62)), we conclude that the radiation contribution is given by:

$$d\alpha_i = \frac{dk_a}{2} \left\{A'^{-1}_{2i-1\ 5} A'_{a\ 2i-1} + A'^{-1}_{2i\ 5} A'_{a\ 2i}\right\}. \tag{15.28}$$

Next we look at the contribution of the cavities. The Lie operator in (15.5) for the cavities has the form $dC^2\partial_2 + dC^4\partial_4$ and depends on the transverse momenta, the energy and the time of arrival. Proceeding as before we expand the operator around the fixed point:

$$dC^{2k} = dC^{2k}(\vec{z}_c) + dC^{2k}_i(\vec{z}_c)z_i + O(|\vec{z}|^2). \tag{15.29}$$

Skipping identical intermediate steps, we get for the contribution to the damping coefficients:

$$d\alpha^{2k}_i = \frac{dC^{2k}_a}{2} \left\{A'^{-1}_{2i-1\ 2k} A'_{a\ 2i-1} + A'^{-1}_{2i\ 2k} A'_{a\ 2i}\right\}. \tag{15.30}$$

In first order perturbation theory the full value of the damping rates is given by integrating Equations (15.28) and (15.30) around the ring

$$\alpha_i = \oint_0^C \left(d\alpha_i + d\alpha^2_i + d\alpha^4_i\right). \tag{15.31}$$

Theorem 7 (Robinson's Condition) *Let us compute the sum $\alpha_1 + \alpha_2 + \alpha_3$. For the cavity contribution, it is given by*

$$\begin{aligned}\sum_{i=1,3} d\alpha^{2k}_i &= \sum_{i=1,3} \frac{dC^{2k}_a}{2} \left\{A'^{-1}_{2i-1\ 2k} A'_{a\ 2i-1} + A'^{-1}_{2i\ 2k} A'_{a\ 2i}\right\} \\ &= \sum_{i=1,6} \frac{dC^{2k}_a}{2} \left\{A'^{-1}_{i\ 2k} A'_{a\ i}\right\} \\ &= \frac{dC^{2k}_{2k}}{2}.\end{aligned} \tag{15.32}$$

Similarly for the other magnet contributions, one gets

$$\sum_{i=1,3} d\alpha_i = \frac{dk_5}{2}. \tag{15.33}$$

Hence, we obtain the general result

$$\alpha_1 + \alpha_2 + \alpha_3 = \frac{1}{2}\oint_0^C dk_5 + dC_2^2 + dC_4^4. \tag{15.34}$$

The final result comes about if we expand around the ideal orbit. To leading order, the cavity operator is given by:

$$\frac{\partial H}{\partial \tau}\left\{\frac{\partial^2 H}{\partial p_x \partial \delta}\frac{\partial}{\partial x'} + \frac{\partial^2 H}{\partial p_y \partial \delta}\frac{\partial}{\partial y'}\right\} = -\frac{\partial H}{\partial \tau}\left\{\frac{x'\frac{\partial}{\partial x'} + y'\frac{\partial}{\partial y'}}{1+\delta}\right\},$$

from which we deduce that

$$\oint dC_2^2 = \oint dC_4^4 = -\oint \frac{\partial H}{\partial \tau} ds = -\text{Energy Gain}. \tag{15.35}$$

If is easy to check from Equation (15.4) *that* dk_5*, the derivative with respect to* δ *of the energy change at a magnet, is twice the energy change. Since the energy gained at the cavity must balance the energy lost at the magnet, we conclude that*

$$\oint_0^C dC_2^2 = \oint_0^C dC_4^4 = \frac{1}{2}\oint_0^C dk_5. \tag{15.36}$$

Thus we have Robinson's condition:

$$\alpha_1 + \alpha_2 + \alpha_3 = \oint_0^C dk_5. \tag{15.37}$$

15.3.2 Freedom in the Normal Form: Gauge Invariance

The formulas for the damping rates contain some remarkable functions that are products of the canonical transformation with its inverse:

$$\eta^i_{j\ k} = A'^{-1}_{2i-1\ j} A'_{k\ 2i-1} + A'^{-1}_{2i\ j} A'_{k\ 2i}. \tag{15.38}$$

Before discussing these functions and their gauge invariance (i.e., they do not depend on the choice of A'), I will recall the symplectic results. In the symplectic case we chose the "physical route." We started with a ring concept—the Fourier decomposition of a quantity and computed it with a normal form. We found that normal form theory gives the Fourier peaks in terms of a phase and the invariants. It follows that these invariants cannot depend on the canonical transformation which was used to construct them. In the symplectic case the invariant actions (coupled Courant-Snyder invariants) are unique. Thus the component of these invariant functions in the monomial basis must be invariant under all possible choices of canonical transformations. In this manner we conclude that the linear lattice functions are true invariants of the theory and can be extended to the coupled and nonlinear case. The resulting lattice functions are universal; in Sect. 2.4.5 we quoted a paper which is remarkable for one figure in which the four coupled beta functions are mathematically defined by one single picture! These lattice functions differ from those of formalisms which are not canonically invariant; most notably the Teng-Edwards formalism. In such formalisms no attempt is made to ensure that the coupled lattice functions are unambiguous gauge invariant quantities. Instead the emphasis is on concocting something which vaguely resembles the uncoupled theory, allowing the user to take limits with ease. For a fully coupled case, these formalisms are certainly dangerous since the lattice functions they defined are totally arbitrary. Remarkably, the Teng-Edwards formalism can be extended to six by six matrices and used to derive perturbative results from the beam envelope formalism. This extension of the Teng-Edwards parametrization is well-suited for a derivation of Sands' integrals in terms of the beam envelope formalism. This was done by Ohmi, Hirata and Oide in reference [76]. Here we will take the gauge invariant route. Therefore we must study the functions of Equation (15.38) more carefully.

In the presence of damping, it is clear that we are lacking the physical equivalent of a time average since all the trajectory will eventually sink into the origin. Thus, sin of all sins, we will have to proceed on a purely mathematical level. I urge the reader to have some patience, the physical connection will be established later (partly). Our starting point is the freedom in the definition of A:

$$L = ARA^{-1}$$
$$\Downarrow$$
$$= ArRr^{-1}A^{-1}. \tag{15.39}$$

I have dropped the "prime" ignoring distinctions between the symplectic and general case. It follows from the above equation that one could have chosen a normalization matrix Ar instead of A. The matrix r is of the same nature as R:

$$r = \begin{pmatrix} \lambda_1 r_1 & 0 & 0 \\ 0 & \lambda_2 r_2 & 0 \\ 0 & 0 & \lambda_3 r_3 \end{pmatrix} \tag{15.40}$$
$$\text{where } r_i = \begin{pmatrix} \cos(\varphi_i) & \sin(\varphi_i) \\ -\sin(\varphi_i) & \cos(\varphi_i) \end{pmatrix}$$
$$\text{and } \lambda_i \in \mathbb{R}.$$

Now we proceed to build invariants of the normal form process. Let us think of $r^{-1}A^{-1}$ as three two by two rotations acting on eighteen minivectors of two components:

$$\begin{pmatrix} \lambda_1^{-1} r_1^{-1} & 0 & 0 \\ 0 & \lambda_2^{-1} r_2^{-1} & 0 \\ 0 & 0 & \lambda_3^{-1} r_3^{-1} \end{pmatrix} \begin{pmatrix} \begin{pmatrix} A^{-1}_{11} \\ A^{-1}_{21} \end{pmatrix} & \cdots & \begin{pmatrix} A^{-1}_{16} \\ A^{-1}_{26} \end{pmatrix} \\ \vdots & \vdots & \vdots \\ \begin{pmatrix} A^{-1}_{51} \\ A^{-1}_{61} \end{pmatrix} & \cdots & \begin{pmatrix} A^{-1}_{56} \\ A^{-1}_{66} \end{pmatrix} \end{pmatrix}.$$

If us denote the minivectors by

$$\vec{v}^{i,j} = \begin{pmatrix} A^{-1}_{2i-1\ j} \\ A^{-1}_{2i\ j} \end{pmatrix}, \tag{15.41}$$

we see that the product $r^{-1}A^{-1}$ will itself be made of minivectors:

$$\lambda_i^{-1} r_i^{-1} \begin{pmatrix} A^{-1}_{2i-1\ j} \\ A^{-1}_{2i\ j} \end{pmatrix}. \tag{15.42}$$

Finally, we do the same thing with the transpose of the matrix Ar

$$\lambda_i r_i^{-1} \underbrace{\begin{pmatrix} A_{j\ 2i-1} \\ A_{j\ 2i} \end{pmatrix}}_{=\vec{w}^{i\ j}}. \tag{15.43}$$

It follows that scalar products of the form $\vec{v}^{i\ j} \cdot \vec{w}^{i\ k}$ do not depend on the choice of canonical transformation. Explicitly they are

$$\vec{v}^{i\ j} \cdot \vec{w}^{i\ k} = A^{-1}_{2i-1\ j} A_{k\ 2i-1} + A^{-1}_{2i\ j} A_{k\ 2i}. \tag{15.44}$$

One notices that the quantities of Equation (15.38), which appeared in the formula for the damping, are just quadratic invariants (tensor products) of the normal form:

$$\eta^{i}_{j\ k} = \vec{v}^{i\ j} \cdot \vec{w}^{i\ k}. \tag{15.45}$$

We can keep going and ask if the wedge products

$$\beta^{i}_{j\ k} = \vec{v}^{i\ j} \wedge \vec{w}^{i\ k} = A^{-1}_{2i-1\ j} A_{k\ 2i} - A^{-1}_{2i\ j} A_{k\ 2i-1}, \tag{15.46}$$

also rotational invariants, are relevant to some analytical formulas. In fact, we will see that they are related to the quantities of the symplectic map measurable by turn averages.

15.3.2.1 What are the $\vec{v}^{i\ j} \wedge \vec{w}^{i\ k}$?

It is easy to check that a lot of these invariants are related: they are not a set of independent invariants. But, it is still interesting to discover what these wedges could be in the case of a symplectic map. Let us do a simple case and the reader will then get the idea.

$$\begin{aligned}\beta^{1}_{21} &= A^{-1}_{12} A_{12} - A^{-1}_{22} A_{11} \\ &= -A^{2}_{12} - A^{2}_{11} \\ &= -\beta_{11}.\end{aligned} \tag{15.47}$$

The function β_{11} is the usual beta function in one degree of freedom. In several degrees of freedom, it relates the average of x^2 to the first invariant $A^{-1}J_1$.

Out of the wedge products come all the symplectic lattice functions related to a pseudo-harmonic system with three distinct tunes. It is clear that if we compute the effect of a nonsymplectic force on the tunes (the imaginary part), the formula for the tune shifts will certainly involve the invariants $\beta^{i}_{j\ k}$.

15.3.2.2 What are the $\vec{v}^{\imath\ j} \cdot \vec{w}^{\imath\ k}$?

The scalar quantities $\eta^{i}_{j\ k}$ are more mysterious. Let us compute those involved in the formula for α_3. They are:

$$d\alpha_3 = \frac{dk_a}{2} \underbrace{\left\{A^{-1}_{55} A_{a5} + A^{-1}_{65} A_{a6}\right\}}_{\text{Equation (7.58)}} + \frac{dC^2_a}{2} \left\{A^{-1}_{52} A_{a5} + A^{-1}_{62} A_{a6}\right\} + \frac{dC^4_a}{2} \left\{A^{-1}_{54} A_{a5} + A^{-1}_{64} A_{a6}\right\} . \tag{15.48}$$

I should not leave the reader in the dark any longer. Please take a look at Equation (7.58)! The reader remembers that dispersion is, strictly speaking, a concept of the coasting beam normalization. Unless the energy δ is a constant, it is not possible to talk about the change in the closed orbit as a function of δ. However one can ask what happens in the adiabatic limit —if the tune of the third plane of oscillation is much smaller than the two other tunes. From the matrices A and A^{-1} one derives this slow change of the orbit as a function of δ: these four quantities must become the dispersions as the synchrotron tune approaches zero. It is somewhat of a surprise that a quantity which has only a physical sense in the adiabatic limit ends up being an invariant of the normal form algorithm for a general six by six matrix.

However the reader notices that we have other invariants multiplying the vector field of the cavities. These quantities are also well-defined invariants which are normally irrelevant because of the absence of strong synchro-betatron coupling at the cavities. So let us repeat the derivation of Equation (7.58) following the steps in Sect. 7.3.3.1. I invite the reader to read this section and come back here. Coffee break!

Back with our generalized dispersions: let us assume that a particle starts with initial conditions $\vec{z}(1) = (x_1, p_{x;1}, y_1, p_{y;1}, \delta_1, \tau_1)$ in contrast with $\vec{z}(1) = (0,0,0,0,\delta,0)$ used in Sect. 7.3.3.1. As before we transform this vector in Floquet space and average it over the "transverse" tunes. Clearly the word transverse refers to the two planes with the largest tunes whatever they may be:

$$\left\langle A^{-1}\vec{z}(1)\right\rangle_{\perp} = \left(0,0,0,0, A^{-1}_{5k} z_k(1), A^{-1}_{6k} z_k(1)\right) . \tag{15.49}$$

Then we transfer back into real space to get the generalization of Equation (7.58) and we denote the ath component of this "dispersion" as η_a^G:

$$\eta_a^G = \underbrace{\left\{A_{5k}^{-1} A_{a5} + A_{6k}^{-1} A_{a6}\right\}}_{\eta_k^3\ a} z_k(1). \tag{15.50}$$

We can rewrite the expression for damping rates in terms of these generalized dispersions:

$$d\alpha_3 = \frac{dk_a}{2}\eta_a' + \frac{dC_a^2}{2}\eta_a^G{}_2 + \frac{dC_a^4}{2}\eta_a^G{}_4. \tag{15.51}$$

In a canonical treatment, similar invariants would show up not from the cavity but from the transverse vector potential. This is quite apparent in [76] where the authors used an extension of the Teng-Edwards parametrization to obtain complex formulas for the damping coefficients. It should be noted that they also introduced new generalized dispersions which they denoted by the Greek letter ζ.

The other damping coefficients α_1 and α_2 depend on terms which would be adiabatic averages over tunes one and two should these have been small. They do not have an easy physical interpretation. It should be noted that if all three tunes are similar in magnitude then the numerical values of the invariants are far in magnitude from any coasting beam dispersion.

15.3.3 Reduction to Uncoupled Sands Integrals

It is interesting to take the limit used by Sands in his famous treatment. We will assume that the adiabatic averaging holds and thus the dispersion η_a' of Equation (15.51) reduces to the true coasting beam dispersions. We will also assume uncoupled transverse planes and mild coupling with the longitudinal planes. At first we will look at damping coefficients which we have been studying all along in this chapter. Then in Sect. 15.3.3.2 we will use the results of Interlude IX to derive the equilibrium emittances under the usual condition of Sands.

15.3.3.1 The damping rates

First we rewrite the formula for α_3 as

$$d\alpha_3 = \frac{dk_5}{2} + \frac{dk_1}{2}\eta_1' + \frac{dk_3}{2}\eta_3'. \tag{15.52}$$

Here we assume that, around the closed orbit, dk_2 and dk_4 are very small since the radiation force depends quadratically on the directions (x', y'). It should be noted that it is possible to have some vertical dispersion if vertical bends are present even if the system has no coupling. In the presence of weak coupling between the longitudinal and the transverse planes, η_5' must be equal to one. Actually it can be interpreted as the local two by two determinant or as the coefficient multiplying the average of δ under transverse oscillations: by the adiabatic assumption δ hardly moves and thus this coefficient is unity.

Next, let us look at α_1. We first examine the terms proportional to dk.

$$\begin{aligned}
\frac{d\alpha_1}{dk_a} &= \frac{1}{2}\left\{A_{15}^{-1}A_{a1} + A_{25}^{-1}A_{a2}\right\} \\
&= \frac{1}{2}\left\{\delta_{a5} - A_{35}^{-1}A_{a3} - A_{45}^{-1}A_{a4} - A_{55}^{-1}A_{a5} - A_{65}^{-1}A_{a6}\right\} \\
&= \begin{cases} 0 & \text{for } a = 3, 4 \text{ or } 5. \\ -\frac{1}{2}\eta_a' & \text{for } a = 1 \text{ or } 2. \end{cases}
\end{aligned} \tag{15.53}$$

Similarly for the terms proportional to dC:

$$\begin{aligned}
\frac{d\alpha_1^{2k}}{dC_a^{2k}} &= \frac{1}{2}\left\{A_{1\ 2k}^{-1}A_{a1} + A_{2\ 2k}^{-1}A_{a2}\right\} \\
&= \frac{1}{2}\left\{\delta_{a\ 2k} - A_{3\ 2k}^{-1}A_{a3} - A_{4\ 2k}^{-1}A_{a4} - A_{5\ 2k}^{-1}A_{a5} - A_{5\ 2k}^{-1}A_{a5}\right\} \\
&= \begin{cases} \frac{1}{2} & \text{for } a = 2 \text{ and } k = 1 \\ 0 & \text{for the rest} \end{cases}
\end{aligned} \tag{15.54}$$

Now we lump everything under one formula:

$$\begin{aligned}
\alpha_1 &= \frac{1}{2}\oint dC_2^2 - dk_1\eta_1' - dk_2\eta_2' \\
&= \frac{1}{2}\oint \frac{1}{2}dk_5 - dk_1\eta_1'.
\end{aligned} \tag{15.55}$$

We have again assumed that $dk_2 = 0$ and that the coefficient dC_2^2 integrated across all cavities gives the same value as the integral of $\frac{1}{2}dk_5$; the energy lost in the magnets equals the energy gained in the cavities. This was used in theorem 7. An identical formula holds for α_2:

$$\alpha_2 = \frac{1}{2}\oint \frac{1}{2}dk_5 - dk_1\eta_3'. \tag{15.56}$$

15.3.3.2 The equilibrium emittances

In Interlude IX we started with the existence of the stochastic map for the moments and derived an expression for the equilibrium moments. We then showed that away from linear resonances and for small damping rates, the equilibrium beam sizes depend very accurately on the lattice functions (i.e., the map A) and three quantities which are invariant under the linear map. These quantities are thus the so-called equilibrium emittances. They exist only in the limit of small damping.

Using the formalism of Interlude IX, we project Equation (15.19) along the invariant directions. Thus the equilibrium emittances are given by

$$\varepsilon_i = \frac{1}{-4\alpha_i}\oint_0^C \left[\left\{A^{-1}_{2i-1\ 5}\right\}^2 + \left\{A^{-1}_{2i\ 5}\right\}^2\right] d\Phi_{55}. \tag{15.57}$$

In contrast the results of Sands are given by the formulas:

$$\varepsilon_x = \frac{1}{-4\alpha_x}\oint_0^C \frac{\eta_x^2 + \left\{\beta_x\eta_x' + \alpha_x\eta_x\right\}^2}{\beta_x} d\Phi_{55} \tag{15.58.a}$$

$$\varepsilon_y = \frac{1}{-4\alpha_y}\oint_0^C \frac{\eta_y^2 + \left\{\beta_x\eta_y' + \alpha_y\eta_y\right\}^2}{\beta_y} d\Phi_{55} \tag{15.58.b}$$

$$\varepsilon_\delta = \frac{1}{-4\alpha_\delta}\oint_0^C \gamma_\delta d\Phi_{55}. \tag{15.58.c}$$

Here $(\eta_x, \eta_x', \eta_y, \eta_y')$ stands for $(\eta_1, \eta_2, \eta_3, \eta_4)$. Notice that the longitudinal gamma function appears because we selected the energy as the fifth variable. If time is selected this would be the longitudinal beta function: this is just a convention.

The first reaction I had while staring at Equations (15.57) and (15.58) was: how in the world can these two things be related? Certainly they are since they give similar results when put in tracking codes. Chao's formalism of Equations (15.57) and Sands' integral agree to a high degree when all the necessary conditions are met. So I will let the reader follow my train of thought. First, since things seems always easier in the longitudinal plane let us start there. Going back to the definition of the Courant-Snyder invariant in the longitudinal plane, we conclude that $\{A^{-1}_{5\,5}\}^2 + \{A^{-1}_{6\,5}\}^2$ is the coefficient of δ^2 and thus it is the longitudinal gamma function. Well, that was easy!

Now let us look at Equation (15.58.a) for the x-emittance. The first step consists in rewriting Equation (15.58.a) in terms of our six dimensional transformation A:

$$\frac{\eta_x^2 + \{\beta_x \eta_x' + \alpha_x \eta_x\}^2}{\beta_x} = \{A^{-1}_{11}\eta_x + A^{-1}_{12}\eta_x'\}^2 + \{A^{-1}_{21}\eta_x + A^{-1}_{22}\eta_x'\}^2. \quad (15.59)$$

We concentrate on the first term $A^{-1}_{11}\eta_x + A^{-1}_{12}\eta_x'$ and substitute the adiabatic expression for the dispersions:

$$\begin{aligned} A^{-1}_{11}\eta_x + A^{-1}_{12}\eta_x' &= A^{-1}_{11}\left(A_{15}A^{-1}_{55} + A_{16}A^{-1}_{65}\right) + A^{-1}_{12}\left(A_{25}A^{-1}_{55} + A_{26}A^{-1}_{65}\right) \\ &= A^{-1}_{55}\underbrace{\left(A^{-1}_{11}A_{15} + A^{-1}_{12}A_{25}\right)}_{=-A^{-1}_{15}A_{55} - A^{-1}_{16}A_{65}} + A^{-1}_{65}\underbrace{\left(A^{-1}_{11}A_{16} + A^{-1}_{12}A_{26}\right)}_{=-A^{-1}_{15}A_{56} - A^{-1}_{16}A_{66}} \\ &= -A^{-1}_{15}\underbrace{\left(A_{55}A^{-1}_{55} + A_{56}A^{-1}_{65}\right)}_{\approx 1} - A^{-1}_{16}\underbrace{\left(A_{65}A^{-1}_{55} + A_{66}A^{-1}_{65}\right)}_{\approx 0} \\ &= -A^{-1}_{15}. \end{aligned} \quad (15.60)$$

We used the weak synchro-betatron coupling and the symplectic condition in replacing $A_{55}A^{-1}_{55} + A_{56}A^{-1}_{65}$ by unity and $A_{65}A^{-1}_{55} + A_{66}A^{-1}_{65}$ by zero. Of course the same manipulation will work for the second term of Equation (15.59): $A^{-1}_{21}\eta_x + A^{-1}_{22}\eta_x' = -A^{-1}_{25}$. Thus in the limit of Sands, ε_1 obtained from Equation (15.57) agrees with ε_x of (15.58.a). The same result will apply (exchange the rôles of x and y) to ε_2 and ε_y.

Thus the equilibrium emittances and Sands formulas are directly obtainable from the moment or so-called beam envelope formalism.

CHAPTER 16
CONCLUSION

16.1 A MAN HAS GOT TO KNOW HIS LIMITATIONS!

There is a disastrous misconception about the entire "mappers" versus "trackers" controversy. At the center of this controversy there is the belief that those who introduce finite s maps into accelerator theory do it for a single purpose: to replace integrators (i.e., kick codes) by approximate powers series representation of the entire machine. Thus the word map refers to an attempt to speed up tracking studies by introducing transverse approximations of the dynamics. Then it follows logically that "mappers" fell in love with the "DA-Package" (i.e., automatic truncated power series algebra) because it extended their ability to produce high order power series for tracking. I remember in my days at SSC being told by my boss Alex Chao that it was important to reproduce 400 turns of tracking with "DA" to convince the authorities that money was not wasted on our salaries!

In this book I unequivocally stood on the side on the "trackers" when it came to tracking. By this I do not necessarily imply that Taylor series maps (symplectified or not) are useless and should never be used for tracking, but I am simply saying that they should be bench-marked against element by element tracking. This is why this book emphasized the topic of symplectic integration as *the preferred method* of ring modeling.

The need for a new paradigm, a new theoretical framework based on the common metaphors of the field, is not derived from the success of approximate maps as tracking engines. It is however derived from tracking, that is to say from our relative ability to produce reliable phase plots using "kick codes" and our relative inability to analyze precisely the equations of motion which produced these plots. Hence the tracking code becomes our most reliable map, but at the same time it is our most analytically distasteful representation of the one-turn map. The global/local paradigm does not try to fight this fact; instead it exploits it to a maximum.

"A man has got to know his limitations" says Clint Eastwood in his dirty Harry movies. The theory presented in this book is a precise acceptance of this piece of Hollywood wisdom. Our limited ability to normalize the complex Hamiltonian of a ring (not to mention radiative effects) is fully accepted. Thus we are free to formulate a local theory whose sole purpose is to increase our ability to write powerful simulation codes. We do not have to worry about action-angle variables, beta functions, canonical transformations or eta functions at the cavities: the electron does not need to normalize the Hamiltonian before entering the ring! So our codes can be real simulation codes. For example, the phase of the cavity in electron storage rings is no longer an input but actually an output.

At the other extreme, when looking at the tracking data, we undoubtly see things which are very much in tune with the pseudo-harmonic "solution" of the equations of motion. Thus we feel the need to re-incorporate in our description of this data beta functions, nonlinear distortions and other parameters of a properly normalized Hamiltonian. But, quite aware of the complexity of the underlying integrator, we do it at the level of the map, that is to say, we try to parameterize the tracking data shown to us by the tracking code—no more no less.

Automatic differentiation implemented in a power series algebra package is the glue that holds the local and global pieces together. In a first step it allows us to produce the dreadful approximate maps out of the integrator. Second we can write a sub-library of a TPSA package to analyze these maps and return our friendly beta functions and more. At no point in this loop is it important or necessary that the underlying Taylor map be good for dynamic aperture determination or even 400 turns of tracking. I would like to point out that my putting of MARYLIE into TEAPOT or RACETRACK[5, 6, 7] while at the SSC was an attempt to extract Taylor maps out of these codes using the concatenator of the code MARYLIE *for the purpose of analysis only*. In fact these two "kick" codes were my tracking engines. The arrival of TPSA changed the picture only because it automated an otherwise painful procedure. In addition TPSA provided routines which ultimately permitted the creation of the physics library LIELIB.

So, the framework of the theory is complete. We have optimized our integration tools and at the same time we have increased our ability to extract the usual perturbative parameters of the system. In this unified theory linear, nonlinear, coupled, decoupled, radiative, symplectic are all handled at the same time thanks to a greatly simplified theoretical edifice.

What we have lost is our ability to produce an s-dependent analytical solution for our perturbative concept. But actually we never lost it because we never had it for highly localized forces: "A man has got to know his limitations!" This book contains plenty of s-dependent results for simple systems which go beyond the standard treatments. Most notably we carried the one-resonance calculation correctly to second order for sextupoles thus beyond the so-called isolated resonance model. On real systems, numerical maps bring us light years ahead of the usual theory which hides its impotence behind a flood of Greek letters and Fourier modes in s.

It is my sincerely strongly held view that the main import of "mappers" to the field of ring dynamics and quasi-periodic systems is the development of the global/local theory exposed in this book. Therefore I never hide my wishes to see the standard Courant-Snyder theory (whatever that means), the Guignard theory, and all the variants derived or inspired by these works permanently discarded or at least properly modified. As I said in the preface, I do not say that Hamiltonian perturbation theory is useless; but I say that in *accelerator dynamics* it is an inferior way to look at our system because our system, our true Hamiltonian, is just too gross to be handle correctly by this approach.

16.2 DRAGT'S WORK ON APPROXIMATE MAPS

In this section I want to give to Caesar what belongs to Caesar. I have no doubts in my mind that Alex Dragt played an essential role in encouraging and developing (maybe indirectly) the ideas presented here. However, his emphasis has been mostly in the domain of approximate powers series maps and their Lie representation. I believe personally, as a ring physicist, that it is a mistake to emphasize how to compute Lie operators efficiently (fast, accurately, etc...) before it is made clear to the community at large why one should compute them in the first place. But having said that, I recognize that my emphasis on rings is paramount while Alex Dragt and his group have had more diverse interests.

So putting aside pedagogical issues and issues of relative importance there remains one fact: to my knowledge Dragt is the first person to see very clearly that Floquet's theorem applies not only to the one degree of freedom matrix of the Courant-Snyder theory, not only to the three degrees of freedom matrix of the full linear map, but to the full nonlinear compositional map whose matrix representation is an $\frac{(No+Nv)!}{No!Nv!}-1$ by $\frac{(No+Nv)!}{No!Nv!}-1$ matrix. This corresponds to the total number of monomials to degree No in Nv variables. This remark of Dragt, which he presented in a seminar circa 1983, and then published in an obscure Snowmass report, is for me the central justification which permits the total replacement of Hamiltonian theory and Courant-Snyder theory by a map-based theory. This book is based on this theoretical fact: the full map can be put in a Floquet form.

In parallel, the development of MARYLIE[5] gave us a testing ground for the computation and manipulation of approximate maps. Routines were installed to normalize fifth-order Taylor series maps and thus obtain invariants, beta functions, etc...The ability to write transfer maps and canonical transformations ($\mathcal{A}$) in resonance basis was known and programmed in MARYLIE as early as 1984.

In recent years, Dragt has continued to explore how one could improve the computation of approximate Taylor maps emphasizing longitudinal accuracy instead of transverse accuracy. I have explained at length why I tend to be in the camp of the "trackers" when it comes to this issue. However this work should not be ignored especially in single pass applications; one should simply recognize that whether or not these maps are useful for tracking is an issue disconnected from the validity of the global/local paradigm.

In the same vein, Dragt was a pioneer in the issue of "symplectification." This procedure consists in taking a Taylor series representation (i.e., a k-jet) and adding to it higher order terms to make it symplectic to machine accuracy. The code MARYLIE was first equipped with a symplectification procedure which used a characteristic function to represent the k-jet. Thus contrary to popular belief, MARYLIE was not a "Lie Algebraic Tracking Code." This brings me to the Irwin factorization, the first truly Lie Algebraic method of symplectification.

This method, actually first contemplated by Dragt, was formalized by Irwin and first implemented with Berz's DA-Package by myself. It consists in writing the original map in terms of a product of Lie maps which are nil-potent (i.e., kicks). This extremely clever trick of Irwin turns out to lack sophistication in the etymological sense of the word: lack of wisdom. Bob Warnock[16.1] of SLAC once pointed out to me that we had no idea, in a measure sense, of the accuracy or convergence of the Irwin factorization. In practice it is very difficult to get a correct selection of Irwin kicks which produce the correct k-jet and do not deteriorate the accuracy of the map. In other words if

[16.1] Warnock has himself developed fitted approximate maps. See [77] for example.

the original Taylor Series was "good" within an aperture A, it is easy to produce an Irwin representation which is terrible well within the outer edge of A. It is not easy to get a good Irwin representation. In the past three years, Dragt and his student Dan Abell have attacked this problem using very sophisticated tools of group theory. I can say publicly that I was skeptical and that I ended up being very wrong in the end. They started to uncover the reasons for this apparent failure of the original Irwin factorization. They have suggested new selections of Irwin kicks (or jolts as Dragt calls them) which are far better. In a sense, they have started to answer the criticisms of Warnock. This work is exciting, and according to Dragt himself, the most challenging problem of his scientific carrier. But, again I want to say that this is a welcome bonus, not a necessary condition to embrace the global/local theory.

16.3 THE BOLOGNA SCHOOL: TURCHETTI'S WORK

If one asks if Dragt was the only person to think in terms of maps, the answer is no. Let me quote the following passage[2, 25] from Turchetti et al.:

> In the present work we first show how the map **language** is the most appropriate to describe the transverse particle motion in a particle accelerator. Indeed, since the potential of the magnetic fields is **almost discontinuous** at the edges of the magnets, a Hamiltonian treatment is not only **innatural**, but due to a slow decaying queue in the Fourier spectrum (with respect to arc length), it raises very serious convergence problems, if the canonical perturbation theory is to be used.

I have put in bold the key words. In a sense the work of Turchetti is closest in spirit to the topic of this book. It seeks to describe a particle accelerator using an *alternate* language. Not because the authors are on an "ego-trip" or because it is fun to do so, not because they have time to waste on useless projects, but because the Hamiltonian approach is, in the last analysis, unnatural: it does not take into account the almost perfect dynamical existence of the magnet as a separate entity.

Thus it is fair to say that the global/local approach has been put forward independently in Europe by Turchetti's group and in America by Dragt's disciples. And, of course, being more involved in rings than either Turchetti or Dragt, I will take the blame for being the person who has tried the most to expand this approach beyond the transverse symplectic case. This book is one more attempt at popularizing these ideas.

It is also fair to say that our work was always closer to computer implementations than that of the Italians. And thus our wish to implement the necessary tools in actual computer codes, under more general conditions, has always dominated our thoughts. Recently Bazzani, Giovannozzi and Todesco [78] have developed a program more along the lines of our LIELIB. So, one might say that things are converging after all! This brings me to the so-called "DA-Package."

16.4 BERZ'S CONTRIBUTION

It is unfortunate that so few people know the reasons why I "put" MARYLIE in TEAPOT or RACETRACK[5, 6, 7]. I even remember clearly some old physicists declaring that one cannot trust the tracking in these modified codes because they must be using the same "stuff" as in MARYLIE.

The real reason to put MARYLIE in the integrators was connected to perturbation theory: the computation of nonlinear lattice functions. This was done in 1985-1986. MARYLIE contains special purpose algorithms which allowed me to extract and analyze fifth-order maps coming from kick codes. None of this was easy because I had to compute the "feed-down" around the actual closed orbit manually. The bottom line is this: I put MARYLIE in the kick codes to do "DA."

At this point came Martin Berz to the scene with his TPSA package. This package was far than perfect at first,[16.2] but it was optimized for a relatively small number of variables and a relatively high order. In any event, later versions of this package permitted two important things. First the extraction of maps using automatic differentiation (no manual computation of feed-down), and secondly, because of its generality, the construction of a complete sub-library of physics routines for the analysis and normalization of maps.

As I mentioned previously it provides the glue which holds the local and global parts together. The local part is based on kick codes (integrators) and the global part is based on normal forms: approximate maps must be coming from the tracking code if they are to be

[16.2] For example the polynomials were indexed according to their name in the version brought initially to SSC!

trusted. Computing feed-down by hand worked for awhile, but it is evidently not a final solution to such a problem. It was only feasible because of the simplicity of the "large machine Hamiltonians."

I would like to point out that Berz is the first person to write a "partial inverter" of a Taylor series map. One takes the series and re-expresses it in terms of mixed variables: old positions and new momenta as a function of the new positions and old momenta. The mixed generating function is then a simple integral of this map which can be done trivially by any user of "DA."

Since then Berz's work has been mostly related to the production of Taylor series maps from general Hamiltonians and in the creation of an interpreted environment (COSY-INFINITY) for these types of calculations. This is something of general interest to ring physicists, but somewhat at odds with our need for self-consistent transverse simulations. In single pass systems, the accuracy of the Taylor series and thus a powerful interpreted environment to extract them, is more in tune with the traditional way of thinking.

In any case, this book owes a great deal to the existence of a "TPSA" package which is flexible and bug-free. And Berz's contribution to this is central.

16.5 IRWIN'S WORK

Finally I would like to acknowledge Irwin's contribution to map methods. John Irwin is the first person to sit down and to try to figure out what these map "methods of Étienne" are all about. Anyway, Irwin not only learned a lot of things but became in my view a leading expert on map methods and the usage of Lie operators in accelerators.

As I said earlier, he developed the first "Lie Algebraic" symplectification scheme. The so-called Irwin kick factorization remains the only representation of a map which can be put on an integer grid and tracked without truncation errors. To get an idea of what this is all about, I suggest an article by Tremaine and Earn[79]. This work is still in an active stage of research in Dragt's group.

More important Irwin has applied map/Lie methods to the design and understanding of final focusing systems. More than any other piece of work, Irwin's use of Lie methods epitomizes the metaphorical nature of the theory. In Irwin's work, the Lie operator becomes an almost vivid representation of the magnet. Irwin's language and theoretical formalism tries as much as possible to remain intuitive. Irwin's work is an example of the local and global aspects being merged by a lucid understanding of the actual system under study. As I write these last lines, Irwin and his collaborators are trying to code the ideas of this book into an object-oriented environment. It is my hope that they succeed.

16.6 EPILOGUE

I would like to quote from the book[11] of my friend Leo Michelotti, whose eloquence in his epilogue I have failed to improve upon:

> Those who have the privilege and responsibility of guiding students should consider a different approach to teaching classical mechanics, one that does not assume it stopped developing shortly after the Napoleonic Wars, that makes connections with advanced subjects, and that emphasizes real applications.

The reader will appreciate that my book was supposed to be on the theoretical framework related to a particular "real application." Since it is quite natural to include more advanced and modern topics in a more academic book, such as Michelotti's, I will suggest that a Napoleonic treatment of classical mechanics leaves accelerator dynamics somewhere around the end of Louis XVIth reign. Maybe a revolution with a few heads rolling is overdue!

I would like to end at this point hoping that readers will continue in the footsteps of Irwin and others in applying the language of this book in ways beyond my wildest dreams.

BIBLIOGRAPHY

[1] A. J. Dragt, in *Proceedings of the 1984 Study on the Design and Utilization of the Superconducting Super Collider*, edited by R. Donaldson and J. Morfin (Am. Phys. Soc., New York, Snowmass, 1985).

[2] A. Bazzanni, P. Mazzanti, G. Servizi, and G. Turchetti, Nuovo Cimento **B**, 51 (1988).

[3] G. Lakoff and M. Johnson, *Metaphors we live by* (University of Chicago Press, Chicago, 1980).

[4] D. A. Edwards and M. J. Syphers, *An Introduction to the Physics of High Energy Accelerators, Wiley Series in Beam Physics and Accelerator Technology* (Wiley, New York, 1993).

[5] This is a code that parametrizes Taylor series maps around the design orbit using Lie operators in the Dragt-Finn factorization. Presently the most complete version of this code has a library of third order maps.

[6] L. Schachinger and R. Talman, Part. Accel. **22**, 35 (1987).

[7] RACETRACK: A kick code developed by A. Wrulich which uses the matrix-kick split on the small angle Hamiltonian.

[8] J. Irwin, invited talk at the 1994 Charged Particle Optics conference held in Tsukuba, Japan.

[9] K. Oide, Technical Report No. KEK Proc. 93-6, National Laboratory For High Energy Physics (KEK), (unpublished), in Japanese.

[10] F. Willeke and G. Ripken, in *Proc. Linear Accelerator Conf.* (American Institute of Physics, New York, 1989), p. 758.

[11] L. Michelotti, *Intermediate Classical Dynamics with Applications to Beam Physics, Wiley Series in Beam Physics and Accelerator Technology* (Wiley, New York, 1995).

[12] V. I. Arnold, *Mathematical Methods of Classical Mechanics* (Springer-Verlag, New York, 1978).

[13] E. Forest and K. Hirata, Technical Report No. 92-12, KEK (unpublished).

[14] H. S. Dumas and J. Laskar, Phys. Rev. Lett. **70**, 2975 (1993).

[15] G. Hori, Publ. Astr. Soc. Japan **18**, 287 (1966).

[16] G. Hori, Publ. Astr. Soc. Japan **23**, 567 (1971).

[17] K. L. Brown, D. C. Carey, C. Iselin, and F. Rothacker, Technical Report No. CERN 80-04, CERN (unpublished).

[18] S. Chattopadhyay, Technical Report No. CERN 84-11, CERN (unpublished).

[19] K. Hirata and F. Ruggiero, Technical Report No. LEP Note 611, CERN (unpublished).

[20] L. Perko, *Differential Equations and Dynamical Systems* (Springer-Verlag, New York, 1991).

[21] M. W. Hirsch and S. Smale, *Differential equations, dynamical systems, and linear algebra* (Academic Press, New York, 1974).

[22] A. W. Chao, in *XVth International Conference on High Energy Accelerators*, edited by J. Rossbach (World Scientific, Singapore, 1993), p. 981.

[23] L. Emery, in *XVth International Conference on High Energy Accelerators*, edited by J. Rossbach (World Scientific, Singapore, 1993), p. 1172.

[24] M. Hamermesh, *Group theory and its application to physical problems* (Addison-Wesley, Reading, Mass., 1962).

[25] A. Bazzani, E. Todesco, G. Turchetti, and G. Servizi, Technical Report No. CERN 94-02, CERN (unpublished).

[26] R. D. Ruth, in *Nonlinear Dynamics Aspects of Particle Accelerators*, edited by J. M. Jowett and M. Month (Springer-Verlag, Berlin, 1986), p. 37.

[27] E. Wilson, Technical Report No. CERN 84-19, CERN (unpublished).

[28] A. J. Dragt, in *Lectures on Nonlinear Orbit Dynamics* (American Institute of Physics, New York, 1981), Vol. 87, p. 147.

[29] P.-V. Koseleff, Technical Report No. LBID-2030 Rev., LBL, (unpublished), translated by D. Boucher and E. Forest.

[30] A. J. Dragt, Phys. Rev. Let. **75**, 1946 (1995).

[31] G. Guignard, Technical Report No. CERN 78-11, CERN (unpublished).

[32] A. Ando, Part. Accel. **15**, 177 (1984).

[33] J. Bengtsson, Technical Report No. CERN 88-05, CERN (unpublished).

[34] E. D. Courant, R. D. Ruth, and W. T. Weng, in *Physics of High Energy Particle Accelerators* (American Institute of Physics, New York, 1985), Vol. 127, p. 294.

[35] LIELIB is a post-processor of the original DA-Package of Berz for the analysis of nonlinear maps and quadratic moments. It is an integral component of the "ideal code" described in this book.

[36] A. W. Chao, J. Appl. Phy. **50**, 595 (1979).

[37] M. Sands, Technical Report No. SLAC-121, Stanford Linear Accelerator (unpublished).

[38] P. B. Kahn and Y. Zarmi, Physica D **54**, 65 (1991).

[39] C. L. Siegel and J. K. Möser, *Lectures on Celestial Mechanics* (Springer, Berlin, 1971).

[40] E. Forest and D. Murray, Physica D **74**, 181 (1994).

[41] E. Forest, Technical Report No. SSC-138, SSC-CDG (unpublished).

[42] D. A. Edwards and L. C. Teng, IEEE Trans. Nucl. Sci. **NS-20**, 885 (1983).

[43] THOR is an unpublished attempt (but apparently successful) by J. Bengtsson at producing a polymorphic environment and a class structure compatible with the requirements of this book using the C++ language.

[44] An effort, dubbed CLASSIC, is under way at SLAC, Fermilab and CERN to produce an object oriented particle dynamics code for accelerators.

[45] The original work of Michelotti falls in this category: the application of Lie methods to standard accelerator theory.

[46] D. Robin, E. Forest, C. Pellegrini, and A. Amiry, Phys. Rev. E. **48**, 2149 (1993).

[47] S. Gasiorowicz, *Quantum Physcis* (Wiley, New York, 1974), chapter 7 contains a very readable account of operator methods in quantum mechanics.

[48] S. Machida, in *Proceedings of the Computational Accelerator Physics Conference* (American Institute of Physics, New York, 1993).

[49] A. J. Dragt, Part. Accel. **12**, 205 (1982).

[50] What follows is a private communication from Chris Iselin of CERN. " The approach to rotation and displacement described in the MAD User's guide is also used to define the global machine geometry (in fact to tell where the different lego pieces sit in space). I found the idea in the old TRANSPORT program, which has an option to compute "floor coordinates". It uses a 3-vector and a 3-by-3-matrix to represent the Euclidean group, and it applies these both to the global layout and for computing the effect of misaligned elements. I just ported this mechanism to MAD.

[51] C. Iselin, The MAD User's guide, the most up-to-date version can be found at CERN and is available on the internet.

[52] L. M. Healy, Ph.D. thesis, University of Maryland, 1986.

[53] L. M. Healy, A. J. Dragt, and I. M. Gjaja, J. Math. Phys. **33**, 1948 (1992).

[54] S. G. Peggs and R. M. Talman, Ann. Rev. Nucl. Part. Sci. **36**, 287 (1986).

[55] H. Yoshida, Phys. Lett. A **150**, 262 (1990).

[56] E. Forest, J. Comp. Phys. **99**, 209 (1992).

[57] R. Ruth, IEEE Trans. Nucl. Sci. **Ns-30**, 2669 (1983).

[58] E. Forest and R. D. Ruth, Physica **D 43**, 105 (1990).

[59] J. M. Sanz-Serna, Phys. Rev. Lett. **70**, 2975 (1993).

[60] E. Forest, J. Bengtsson, and M. Reusch, Phys. Lett. A **5**, 99 (1991).

[61] M. Glasner, D. Yevick, and B. Hermansson, Math. and Comp. Mod. **16**, 177 (1992).

[62] D. Neuffer and E. Forest, Phys. Lett. A. **135**, 197 (1989).

[63] D. Ritson, Technical report, (unpublished), work performed at the now defunct SSC laboratory.

[64] K. Oide, mostly unpublished material concerning the SAD tracking code at KEK.

[65] A. J. D. et al., Technical report, University of Maryland Physics Department (unpublished).

[66] R. D. Ryne, Ph.D. thesis, University of Maryland, 1986.

[67] E. Forest and J. Milutinovic, Nucl. Instr. Meth. **A269**, 474 (1988).

[68] P. Krejcik, in *Proceedings of the IEEE PAC Conference* (IEEE, New York, 1987), p. 1278.

[69] K. Halbach, Nucl. Instr. and Meth. **A187**, 109 (1981).

[70] R. L. Gluckstern, in *Proc. Linear Accelerator Conf.* (LANL, Los Alamos, 1966), p. 250.

[71] G. Ripken, Technical Report No. 85–084, DESY, (unpublished), this is the oldest reference on SIXTRACK.

[72] F. Schmidt, Technical Report No. CERN SL/94–56 (AP), CERN, (unpublished), this is the latest reference on SIXTRACK.

[73] I. Antoniou and S. Tasaki, J. Phys. Math. Gen. **26**, 73 (1993).

[74] K.-J. Kim and R. Littlejohn, in *Particle Accelerator Conference* (IEEE, New York, 1995), also technical note LBL-36480 (CBP Note-111).

[75] E. Forest, M. F. Reusch, D. Bruhwiler, and A. Amiry, Part. Accel. **45**, 66 (1994).

[76] K. Ohmi, K. Hirata, and K. Oide, Phys. Rev. E. **49**, 751 (1994).

[77] J. S. Berg, R. L. Warnock, R. D. Ruth, and E. Forest, Phys. Rev. E **49**, 722 (1994), also see SLAC-PUB-6037.

[78] A. Bazzani, M. Giovannozzi, and E. Todesco, Comp. Phys. Comm. **86**, 199 (1995).

[79] D. J. D. Earn and S. Tremaine, Physica D **56**, 1 (1992).

INDEX

For Product Safety Concerns and Information please contact our EU representative GPSR@taylorandfrancis.com
Taylor & Francis Verlag GmbH, Kaufingerstraße 24, 80331 München, Germany

www.ingramcontent.com/pod-product-compliance
Lightning Source LLC
LaVergne TN
LVHW011103300926
849160LV00008B/261

* 9 7 8 9 0 5 7 0 2 5 7 4 7 *